Y0-AAA-605

Defects of Properties in Mathematics

Quantitative Characterizations

SERIES ON CONCRETE AND APPLICABLE MATHEMATICS

Published

Vol. 1 Long Time Behaviour of Classical and Quantum Systems
edited by S. Graffi & A. Martinez

Vol. 2 Problems in Probability
by T. M. Mills

Vol. 4 Stochastic Models with Applications to Genetics, Cancers, Aids and Other Biomedical Systems
by Tan Wai-Yuan

Forthcoming:

Vol. 3 Introduction to Matrix Theory
by F. Szidarovszky & S. Molnar

Series on Concrete and Applicable Mathematics Vol. 5

Defects of Properties in Mathematics

Quantitative Characterizations

Adrian I Ban & Sorin G Gal

University of Oradea, Romania

Published by

World Scientific Publishing Co. Pte. Ltd.

P O Box 128, Farrer Road, Singapore 912805

USA office: Suite 1B, 1060 Main Street, River Edge, NJ 07661

UK office: 57 Shelton Street, Covent Garden, London WC2H 9HE

British Library Cataloguing-in-Publication Data
A catalogue record for this book is available from the British Library.

DEFECTS OF PROPERTIES IN MATHEMATICS
QUANTITATIVE CHARACTERIZATIONS

ISBN 981-02-4924-1

Printed in Singapore by Uto-Print

To our lovely wives Olimpia Ban and Rodica Gal

Preface

The main aim is to reveal a method of research in mathematics, called by us quantitative study of the defect of property, which can be used in various fields of mathematics. Our viewpoint over the mathematical entities is that of an analyst, even if they belong to algebra, geometry, topology or logic. We examine in this monograph in a systematic way, the quantitative characterizations of the "deviation from a (given) property", by our terminology called "defect of property", in: Set Theory, Topology, Measure Theory, Real Function Theory, Complex Analysis, Functional Analysis, Algebra, Geometry, Number Theory and Fuzzy Mathematics. We also present a great variety of applications and open problems. To our knowledge, it is the first time in literature that a book has systematically studied this direction of research. This book is interdisciplinary in mathematics and contains the research of both authors over the past six years in these subjects. It also references most of the works of other main researchers in these areas. Each chapter can be read independently. The introduced concepts are simple and the proving methods are rather elementary, that is, making this material accessible to undergraduate and graduate students and researchers. The large spectrum covered by the topic, makes impossible to have a complete bibliography, which is only introductory and depends on the authors' preferences. The book is of wide audience and it is good for researchers, undergraduate and graduate students, courses and seminars.

Oradea, August 1, 2001 *The authors*

Preface

Contents

Chapter 1

Introduction

To convey some of the essence of this monograph to the reader, in this chapter we briefly present some important motivation for writing it and its main results. For the convenience of the reader, the results are numbered as they are in their respective chapters.

1.1 General Description of the Topic

It is well-known that the definition of a mathematical concept consists in the statement of one or several properties (axioms) that must be verified by some mathematical objects. More exactly, we can describe this by the followings.

Let U be a given abstract set of elements and let us denote by P a specific property of some elements in U. Obviously, P divides U into two disjoint sets:

$$U_P = \{x \in U; x \text{ has the property } P\}$$

and

$$U_{\overline{P}} = \{x \in U; x \text{ does not satisfy the property } P\}.$$

A powerful tool of study of U_P and $U_{\overline{P}}$ might be the introduction (not necessarily in an unique way) of a quantity $E(x)$, defined for all $x \in U$ with values in a normed space $(Y, \|\cdot\|)$ (which in general is $\mathbf{R}$ or $\mathbf{C}$), having the property

$$x \in U_P \text{ if and only if } E(x) = 0_Y.$$

In this case, for $x \in U_{\overline{P}}$ we necessarily have $E(x) \neq 0_Y$ and consequently the quantity $||E(x)||$ can be considered that it measures the "deviation" of x from the property P.

We will call $||E(x)||$ as defect of property P for the element $x \in U$.

As a consequence, the following kind of application holds: given a family of operators $A_\lambda : U \to Y, \forall \lambda \in \Gamma$, which satisfy $A_\lambda(x) = 0_Y, \forall x \in U_P, \lambda \in \Gamma$, another family of operators $B_\lambda : Y \to Y, \lambda \in \Gamma$, can be constructed such that $B_\lambda(0_Y) = 0_Y, \forall \lambda \in \Gamma$ and $A_\lambda(x) + B_\lambda(E(x)) = 0_Y, \forall x \in U, \lambda \in \Gamma$ (or more general, $||A_\lambda(x)|| \leq C||B_\lambda(E(x))||, \forall x \in U, \lambda \in \Gamma$, with C a positive constant). In other words, a formula valid for all $x \in U_P$, can be transformed by a "disturbance" factor (which depends on $E(x)$), into a more general formula, valid for all $x \in U$.

Many well-known concepts in various fields of mathematics can be described by this scheme: the measures of noncompactness (defects of compactness, in our terminology) that measure the deviation of a classical set in a topology from the property of compactness, the measure of nonconvexity (defect of convexity, in our terminology) that measures the deviation of a set in normed linear spaces from the property of convexity, the moduli of oscillation and of continuity (defects of continuity, in our terminology) that measure the deviation of a function from the property of continuity, the minimal displacement of points under mappings (defect of fixed point, in our terminology) that measures the deviation of a mapping from the property of fixed point, the areolar derivative (defect of holomorphy, in our terminology) that measures the deviation of a complex function from the property of holomorphy, and so on.

Beside these, in this monograph we introduce and study many other defects of property, as follows: measures of noncompactness for fuzzy sets, fuzzy and intuitionistic entropies, defect of additivity, subadditivity, superadditivity, complementarity, monotonicity for set functions, defect of monotonicity, convexity (concavity), differentiability, integrability for real functions, defect of equality for inequalities, defect of balancing, absorption, orthogonality for sets in normed linear spaces, defect of sublinearity for functionals, defect of symmetry, permutability, derivation for linear operators, defect of commutativity, associativity, identity element, invertibility, idempotency for binary operations, defect of orthogonality and parallelness in Euclidean and non-Euclidean Geometries, defect of curvature and of torsion in Geometry, defects of properties in Number Theory, defects of properties in Fuzzy Mathematics.

Many applications and open problems also are presented.

Because of the great variety, it would be impossible to make a very deep study of all the above concepts by this book. This task is left in many cases to the reader, the main goal of the book being only to put in evidence a method of research called quantitative study of the defect of property, which gives the opportunity to examine from the same viewpoint, basic concepts in various fields of mathematics.

1.2 On Chapter 2: Defect of Property in Set Theory

In this chapter we consider the measures of fuzziness as measuring the "deviation" of a fuzzy set from the concept of classical set, that is as measuring the defect of crisp (classical) set and the intuitionistic entropies as measuring the "deviation" of an intuitionistic fuzzy set from the concept of fuzzy set, that is as measuring the defect of fuzzy set. Then, some applications to the determination of degree of interference (mainly in the geography of population), to description of systems performance and to digital image processing are given.

We denote by $FS(X) = \{A \,|A : X \to [0,1]\}$ the class of all fuzzy sets on X.

A general definition of the measure of fuzziness is the following (see Rudas-Kaynak [181]):

Definition 2.1 A measure of fuzziness is a positive real function d_c defined on $\mathcal{F}(X) \subseteq FS(X)$, that satisfies the following requirements:

(*i*) If $A \in \mathcal{F}(X), A(x) \in \{0,1\}, \forall x \in X$ then $d_c(A) = 0$.

(*ii*) If $A \prec B$ then $d_c(A) \leq d_c(B)$, where $A \prec B$ means that A is sharper than B.

(*iii*) If A is maximally fuzzy then $d_c(A)$ assumes its maximum value.

Let us consider $(X, \mathcal{A}, \mu)$ a measure space and let us denote

$$\mathcal{F}_{\mathcal{A}}(X) = \{A \in FS(X); A \text{ is } \mathcal{A}\text{-measurable}\}.$$

Definition 2.4 Let $(X, \mathcal{A})$ be a measurable space and $t \in (0,1)$. A t-measure of fuzziness is a function $d_c^t : \mathcal{F}_{\mathcal{A}}(X) \to \mathbf{R}$ that satisfies the following conditions:

(*i*) If $A(x) \in \{0,1\}, \forall x \in X$ then $d_c^t(A) = 0$;

(*ii*) If $A \prec_t B$ then $d_c^t(A) \leq d_c^t(B)$, where $A \prec_t B$ if and only if $A(x) \leq B(x)$ for $B(x) \leq t$ and $A(x) \geq B(x)$ for $B(x) \geq t$;

(*iii*) If $A(x) = t, \forall x \in X$ then $d_c^t(A)$ is the maximum value of d_c^t.

Definition 2.5 Let $(X, \mathcal{A}, \mu)$ be a measure space and $t \in (0,1)$. A function $s_c^t : \mathcal{F}_{\mathcal{A}}(X) \to \mathbf{R}$ that satisfies the following conditions:

(*i*) $s_c^t(A) = 0$ if and only if $A(x) \in \{0,1\}, \mu$-a.e $x \in X$;

(*ii*) If $A \prec_t B$ then $s_c^t(A) \leq s_c^t(B)$;

(*iii*) $s_c^t(A)$ is the maximum value of s_c^t if and only if $A(x) = t, \mu$-a.e. $x \in X$,

is called strict t-measure of fuzziness with respect to μ.

Definition 2.6 Let N be a fuzzy complement. A t-measure of fuzziness is called symmetrical with respect to N (or N-symmetrical) if $d_c^t(A) = d_c^t(\overline{A}_N), \forall A \in \mathcal{F}_{\mathcal{A}}(X)$.

A family of t-measures of fuzziness is given by

Theorem 2.3 *Let X be a finite set, $t \in (0,1)$ and $h : \mathbf{R}_+ \to \mathbf{R}_+$ increasing, such that $h(0) = 0$, $(g_x)_{x \in X}, g_x : [0,1] \to \mathbf{R}_+$ increasing on $[0,t]$ and decreasing on $[t,1]$, such that $g_x(0) = g_x(1) = 0, \forall x \in X$ and $g_x(t)$ is the maximum value of all functions $g_x, x \in X$. The function $d_c^t : FS(X) \to \mathbf{R}$ defined by*

$$d_c^t(A) = h\left(\sum_{x \in X} g_x(A(x))\right)$$

is a t-measure of fuzziness. If, in addition,

$$g_x(a) = g_x(N(a)), \forall a \in [0,1], \forall x \in X,$$

where N is a fuzzy complement, then d_c^t is a N-symmetrical t-measure of fuzziness.

Under additional conditions, we obtain a family of strict t-measures of fuzziness.

Theorem 2.4 *If the functions h and $(g_x)_{x \in X}$ satisfy the hypothesis in Theorem 2.3 and, in addition, h is strictly increasing, $(g_x)_{x \in X}$ are strictly increasing on $[0,t]$ and strictly decreasing on $[t,1]$, then the function s_c^t :*

$FS(X) \to \mathbf{R}$ *defined by*

$$s_c^t(A) = h\left(\sum_{x \in X} g_x(A(x))\right)$$

is a strict t-measure of fuzziness which is N-symmetrical if $g_x(a) = g_x(N(a)), \forall a \in [0,1], \forall x \in X$, *where N is a fuzzy complement.*

Other families of t-measures of fuzziness can be obtained by using the concept of t-norm function in the sense of Vivona [217].

Definition 2.7 Let $t \in (0,1)$ and $(X, \mathcal{A}, \mu)$ be a measure space. An $\mathcal{A}$-measurable function with respect to the first variable, $\varphi_t : X \times [0,1] \to [0,1]$ that satisfies the properties

(*i*) $\varphi_t(x,0) = \varphi_t(x,1) = 0, \forall x \in X$;

(*ii*) $\varphi_t(x,\cdot)$ is increasing on $[0,t]$ and decreasing on $[t,1]$;

(*iii*) $\varphi_t(x,t) = 1, \forall x \in X$,

is called t-norm function.

Theorem 2.5 *Let* $t \in (0,1), (X, \mathcal{A}, \mu)$ *be a measure space and* $\varphi_t : X \times [0,1] \to [0,1]$ *a t-norm function. The function* $d_c^{\varphi_t} : \mathcal{F}_{\mathcal{A}}(X) \to [0,1]$ *defined by*

$$d_c^{\varphi_t}(A) = \int_X \varphi_t(x, A(x)) \, d\mu$$

is a t-measure of fuzziness. If $N : [0,1] \to [0,1]$ *is a fuzzy complement and the t-norm function* φ_t *verifies*

$$\varphi_t(x, N(a)) = \varphi_t(x, a), \forall x \in X, \forall a \in [0,1],$$

then the t-measure of fuzziness $d_c^{\varphi_t}(A)$, *defined as above, is N-symmetrical.*

An intuitionistic fuzzy set (see *e.g.* Atanassov [5]) A on X, is an object having the form

$$A = \{\langle x, \mu_A(x), \nu_A(x)\rangle : x \in X\},$$

where the functions $\mu_A, \nu_A : X \to [0,1]$ define the degree of membership and the degree of non-membership of the element $x \in X$ to the set $A \subseteq X$, respectively, and for every $x \in X$,

$$\mu_A(x) + \nu_A(x) \le 1.$$

We denote by $IFS(X)$ the family of intuitionistic fuzzy sets on X.

Let us consider $(X, \mathcal{A}, m)$ a finite measure space and let us denote by $\mathcal{I}_{\mathcal{A}}(X)$ the family of all $\mathcal{A}$-measurable intuitionistic fuzzy sets on X.

Definition 2.9 A real function $d_f : \mathcal{I}_{\mathcal{A}}(X) \to \mathbf{R}_+$ is called an intuitionistic entropy (on $\mathcal{I}_{\mathcal{A}}(X)$) if the following properties are satisfied:

(*i*) $d_f(A) = 0$ if and only if $\mu_A(x) + \nu_A(x) = 1$, m-a.e. $x \in X$;

(*ii*) $d_f(A)$ is maximum if and only if $\mu_A(x) = \nu_A(x) = 0$, m-a.e. $x \in X$;

(*iii*) $d_f(A) = d_f(\overline{A}), \forall A \in \mathcal{I}_{\mathcal{A}}(X)$;

(*iv*) If $A, B \in \mathcal{I}_{\mathcal{A}}(X)$ and $A \preceq B$ then $d_f(A) \geq d_f(B)$.

Definition 2.10 Let $(X, \mathcal{A}, m)$ be a measure space and let us denote $D = \{(\mu, \nu) \in [0,1] \times [0,1] : \mu + \nu \leq 1\}$. An intuitionistic norm function is an $\mathcal{A}$-measurable function with respect to the first variable, $\Phi : X \times D \to [0,1]$ with the following properties, for every element $x \in X$:

(*i*) $\Phi(x, \mu, \nu) = 0$ if and only if $\mu + \nu = 1$;

(*ii*) $\Phi(x, \mu, \nu) = 1$ if and only if $\mu = \nu = 0$;

(*iii*) $\Phi(x, \mu, \nu) = \Phi(x, \nu, \mu)$;

(*iv*) If $\mu \leq \mu'$ and $\nu \leq \nu'$ then $\Phi(x, \mu, \nu) \geq \Phi(x, \mu', \nu')$.

Theorem 2.6 *Let $(X, \mathcal{A}, m)$ be a finite measure space and Φ an intuitionistic norm function. Then, the function $d_f^{\Phi} : \mathcal{I}_{\mathcal{A}}(X) \to [0,1]$ defined by*

$$d_f^{\Phi}(A) = \frac{1}{m(X)} \int_X \Phi(x, \mu_A(x), \nu_A(x))\, dm$$

with $A = \{\langle x, \mu_A(x), \nu_A(x)\rangle : x \in X\}$, is an intuitionistic entropy.

The intuitionistic entropies can be characterized by using I_φ-functions.

Definition 2.11 Let $(X, \mathcal{A}, m)$ be a measure space and $\varphi : [0,1] \to [0,1]$ be a continuous function such that if $\alpha + \beta \leq 1$ then $\varphi(\alpha) + \varphi(\beta) \leq 1$. The function $I_\varphi : \mathcal{I}_{\mathcal{A}}(X) \to [0,1]$ defined by

$$I_\varphi(A) = \frac{1}{m(X)} \int_X (1 - \varphi(\mu_A(x)) - \varphi(\nu_A(x)))\, dm$$

with $A = \{\langle x, \mu_A(x), \nu_A(x)\rangle : x \in X\}$, is called I_φ-function.

Theorem 2.8 *Let $(X, \mathcal{A}, m)$ be a finite measure space, $\varphi : [0,1] \to [0,1]$ be a continuous function and $d_\varphi : \mathcal{I}_{\mathcal{A}}(X) \to [0,1]$. The function d_φ is an* *intuitionistic entropy and an I_φ-function if and only if*

$$d_\varphi(A) = \frac{1}{m(X)} \int_X (1 - \varphi(\mu_A(x)) - \varphi(\nu_A(x)))\, dm,$$

where φ satisfies the conditions:

(i) φ is increasing;

(ii) $\varphi(\alpha) = 0$ if and only if $\alpha = 0$;

(iii) $\varphi(\alpha) + \varphi(\beta) = 1$ if and only if $\alpha + \beta = 1$.

Between intuitionistic fuzzy measures and intuitionistic entropies we can establish the following connection.

Theorem 2.9 *Let $(X, \mathcal{A}, m)$ be a measure space. If $d_\varphi : \mathcal{I}_\mathcal{A}(X) \to [0,1]$ is an intuitionistic entropy and an I_φ-function then d_φ is an intuitionistic fuzzy T_M-measure on $\mathcal{I}_\mathcal{A}(X)$, where T_M is the triangular norm given by $T_M(x,y) = \min(x,y)$. Moreover, if φ is additive then d_φ is an intuitionistic fuzzy h-measure on $\mathcal{I}_\mathcal{A}(X)$, for every continuous triangular norm h which verifies $h(x,y) + h^c(x,y) = x + y, \forall x, y \in [0,1]$.*

At the end of this chapter we give some applications to the determination of degree of interference, applications to the geography of population and to description of the systems performance. Also, a method that permits to obtain a binary digital image from a fuzzy digital image is presented.

For example, let us consider V a country and let us denote by $X = \{S_1, ..., S_n\}$ a partition of V, that is $S_i, i \in \{1, ..., n\}$ represent all the districts of V. We can introduce a normal indicator $\Gamma_{U,X}$ (that is $0 \leq \Gamma_{U,X} \leq 1$) which estimates the degree of homogeneity of the territorial distribution of population on V, with respect to the organization corresponding to X. We notice that, the density of population, the usual indicator used in this situation, is a global indicator which does not take into account the pointwise aspects. For example, if there exists a great concentration of population in certain zones, this indicator is not significant.

Let us denote by Γ_{med}, the density of population in V, that is

$$\Gamma_{med} = \frac{\text{total population in } V}{\text{area of } V},$$

by Γ_{MAX}, the maximum possible density of population in a district of V, that is

$$\Gamma_{MAX} = \frac{\text{total population in } V}{\text{area of the smaller district } S_i}$$

and by Γ_i, the density of population in $S_i, i \in \{1, ..., n\}$, that is

$$\Gamma_i = \frac{\text{population in } S_i}{\text{area of } S_i}.$$

We define the fuzzy set $P : X \to [0,1]$ by

$$P(S_i) = \frac{\Gamma_i}{\Gamma_{MAX}},$$

and we denote $t = \frac{\Gamma_{med}}{\Gamma_{MAX}}$.

We put $\Gamma_{V,X} = s_c^t(P)$, where s_c^t is a strict t-measure of fuzziness in normalized form. Following the axioms of strict t-measures of fuzziness (see Definition 2.5), we get properties of the indicator $\Gamma_{V,X}$ which are in concordance with our intuition:

$(O1)$ $\Gamma_{V,X}$ has the minimum value (equal to 0) if and only if the entire population of V is completely situated in the district with the smallest area;

$(O2)$ If $\Gamma_j \leq \Gamma_i \leq t$ or $t \leq \Gamma_i \leq \Gamma_j$ then the contribution of the district S_i to $\Gamma_{V,X}$ is greater than the contribution of the district S_j;

$(O3)$ $\Gamma_{X,U}$ has the maximum value (equal to 1) if and only if in all districts of V the density of population is equal to Γ_{med}.

Choosing a suitable strict t-measure of fuzziness we can get interesting results concerning the territorial distribution of population.

1.3 On Chapter 3: Defect of Property in Topology

This chapter mainly discusses the concept of measure of noncompactness (defect of compactness, in our terminology) in classical setting, in random setting and in fuzzy setting.

Let (X, ρ) be a metric space and let us denote

$$\mathcal{P}_b(X) = \{Y \subset X; Y \neq \emptyset, Y \text{ is bounded}\}.$$

Definition 3.1 (i) (Kuratowski [129], [130]). The Kuratowski's measure of noncompactness for $Y \in \mathcal{P}_b(X)$ is given by

$$\alpha(Y) = \inf \left\{\varepsilon > 0; \exists n \in \mathbf{N}, A_i \in X, i = \overline{1,n} \text{ with } Y \subset \bigcup_{i=1}^n A_i \text{ and } diam(A_i) < \varepsilon\right\},$$

where $diam(A_i) = \sup\{\rho(x,y); x, y \in A_i\}, i = \overline{1,n}$.

(*ii*) (see *e.g.* Istrătescu [105]) The Hausdorff's measure of noncompactness of $Y \subset X$ is given by

$$h(Y) = \inf\left\{\varepsilon > 0; \exists n \in \mathbf{N}, y_i \in Y, i = \overline{1,n} \text{ with } Y \subset \bigcup_{i=1}^{n} B(y_i, \varepsilon)\right\}.$$

Theorem 3.1 (*i*) *(see e.g. Rus [182], p.85-87 or Banas-Goebel [32]) Let* $X, Y \in \mathcal{P}_b(X)$. *We have:*
$0 \le \alpha(A) \le diam(A)$;
$A \subseteq B$ *implies* $\alpha(A) \le \alpha(B)$;
$\alpha(\overline{A}) = \alpha(A)$ *and moreover* $\alpha(A) = 0$ *if and only if* $\overline{A}$ *is compact;*
$\alpha(V_\varepsilon(A)) \le \alpha(A) + 2\varepsilon$, *where* $V_\varepsilon(A) = \{x \in X; \rho(x, A) < \varepsilon\}$ *and* $\rho(x, A) = \inf\{\rho(x, y); y \in A\}$;
$\alpha(A \cup B) = \max\{\alpha(A), \alpha(B)\}$;
$\alpha(A \cap B) \le \min\{\alpha(A), \alpha(B)\}$;
Let $A_i \subset X, A_{i+1} \subset A_i$ *be with* A_i *closed and nonempty,* $i = 1, 2, ...$
If $\lim_{n\to\infty} \alpha(A_n) = 0$ *then* $\bigcap_{n=1}^{\infty} A_n$ *is nonempty and compact.*
If, in addition, X *is a Banach space, then* $\alpha(A + B) \le \alpha(A) + \alpha(B)$, $\alpha(cA) = |c|\,\alpha(A), c \in \mathbf{R}, \alpha(convA) = \alpha(A)$, *where* $convA$ *is the convex hull of* $A \in \mathcal{P}_b(X)$.

(*ii*) *(see e.g. Beer [36], Banas-Goebel [32])*
$h(A) = 0$ *if and only if* A *is totally bounded (that is,* $\forall \varepsilon > 0, \exists x_1, ..., x_p \in A$ *such that* $\forall x \in X, \exists x_i$ *with* $\rho(x, x_i) < \varepsilon$*);*
$A \subset B$ *implies* $h(A) \le 2h(B)$;
$h(\overline{A}) = h(A)$;
$h(A \cup B) \le \max\{h(A), h(B)\}$;
h *is continuous on* $CL(X) = \{Y \subset X; Y \text{ is closed}, Y \ne \emptyset\}$ *with respect to Hausdorff topology, i.e. if* $Y_n, Y \in CL(X), n = 1, 2, ...,$ *satisfy*
$D_H(Y_n, Y) \overset{n\to\infty}{\to} 0$ *(where* D_H *is the Hausdorff-Pompeiu distance), then* $\lim_{n\to\infty} h(Y_n) = h(Y)$. *Also,* h *is upper semicontinuous on* $CL(X)$ *with respect to the so-called Vietoris topology.*

In addition,

$$h(A) \le \alpha(A) \le 2h(A), \forall A \in \mathcal{P}_b(X)$$

and if moreover X *is Hilbert space, then*

$$\sqrt{2}h(A) \le \alpha(A) \le 2h(A), \forall A \in \mathcal{P}_b(X).$$

The properties in Theorem 3.1 suggest an axiomatic approach in Banach spaces $(X, \|\cdot\|)$, as follows. Firstly we need the notations:

$$\begin{aligned} RC(X) &= \{Y \subset X; Y \neq \emptyset, Y \text{ is relatively compact}\}, \\ CO(X) &= \{Y \subset X; Y \neq \emptyset, Y \text{ is compact}\}. \end{aligned}$$

Definition 3.2 (Banas-Goebel [32]). $\mathcal{K} \subset RC(X)$ is called kernel (of a measure of noncompactness) if it satisfies:

(i) $A \in \mathcal{K}$ implies $\overline{A} \in \mathcal{K}$;

(ii) $A \in \mathcal{K}, B \subset A, B \neq \emptyset$ implies $B \in \mathcal{K}$;

(iii) $A, B \in \mathcal{K}$ implies $\lambda A + (1-\lambda) B \in \mathcal{K}, \forall \lambda \in [0,1]$;

(iv) $A \in \mathcal{K}$ implies $convA \in \mathcal{K}$;

(v) The set $\mathcal{K}^c = \{A \in \mathcal{K}; A \text{ is compact}\}$ is closed in $CO(X)$ with respect to Hausdorff topology (i.e. the topology induced on $CO(X)$ by the Hausdorff-Pompeiu distance).

Definition 3.3 (Banas-Goebel [32]). The function $\mu : \mathcal{P}_b(X) \to [0, +\infty)$ is called measure of noncompactness (or defect of compactness, in our terminology) with the kernel $\mathcal{K}$ (denoting $\ker \mu = \mathcal{K}$) if satisfies:

(i) $\mu(A) = 0$ if and only if $A \in \mathcal{K}$;

(ii) $\mu(\overline{A}) = \mu(A)$;

(iii) $\mu(convA) = \mu(A)$;

(iv) $A \subset B$ implies $\mu(A) \leq \mu(B)$;

(v) $\mu(\lambda A + (1-\lambda) B) \leq \lambda \mu(A) + (1-\lambda) \mu(B), \forall \lambda \in [0,1]$;

(vi) If $A_n \in \mathcal{P}_b(X), \overline{A}_n = A_n$ and $A_{n+1} \subset A_n, n = 1, 2, \ldots$ and if $\lim_{n \to \infty} \mu(A_n) = 0$, then $\bigcap_{n=1}^{\infty} A_n \neq \emptyset$.

If $\mathcal{K} = RC(X)$ then μ will be called full (or complete) measure.

Many examples satisfying Definitions 3.2, 3.3 are given.

Let π be the set of all open coverings of a topological space X. Let us consider the family $\mathcal{C} = 2^{\pi}$ of all subsets of π ordered by the relation

$$x \leq y \text{ if and only if } x \supset y, \text{ for } x, y \in \mathcal{C}.$$

Then $(\mathcal{C}, \leq)$ is a complete lattice with the minimal element $\Theta = \pi$. Now we can define the $\mathcal{C}$-measure of noncompactness as follows:

Definition 3.4 (Kupka-Toma [128]). If $A \subset X$ then the measure of noncompactness of A is the element of $\mathcal{C}$ defined by

$$m(A) = \left\{ \mathcal{P} \in \pi \,\middle|\, \exists \text{ finite } \delta \subset \mathcal{P} : A \subset \bigcup \delta \right\}.$$

Theorem 3.2 *(Kupka-Toma [128]). The mapping $m : 2^X \to \mathcal{C}$ has the following properties:*

(i) If A is a compact subset of X then $m(A) = \Theta$. If A is closed then the converse is true;

(ii) If $A \subset B$ then $m(A) \leq m(B)$;

(iii) $m(A \cup B) = m(A) \cap m(B) = \sup\{m(A), m(B)\}$;

(iv) $m(A \cap B) \leq m(A) \cup m(B) = \inf\{m(A), m(B)\}$.

Theorem 3.3 *(Kupka-Toma [128]). Let X be a complete uniform space. Then we have the followings:*

(i) For every closed subset $A \subset X, m(A) = \Theta$ if and only if $\Phi(A) = 0$;

(ii) For every decreasing net $(A_\gamma : \gamma \in \Gamma)$ of closed nonempty subsets of X,

$$\inf\{\Phi(A_\gamma) : \gamma \in \Gamma\} = 0 \Rightarrow \lim_{\gamma \in \Gamma} m(A_\gamma) = \Theta.$$

(Here Φ denotes Lechicki's measure in Lechicki [132]).

Theorem 3.4 *(Kupka-Toma [128]). Let $(A_\gamma : \gamma \in \Gamma)$ be a decreasing net of closed nonempty subsets of a topological space X. Then the following implication is true:*

$$\lim_{\gamma \in \Gamma} m(A_\gamma) = \Theta \Rightarrow A = \bigcap_{\gamma \in \Gamma} A_\gamma \text{ is nonempty and compact.}$$

Let us denote by Δ the set of all nondecreasing and left continuous functions $f : \mathbf{R} \to [0,1]$, such that $f(0) = 0$ and $\lim_{x \to +\infty} f(x) = 1$.

Definition 3.5 (Menger [145]). A probabilistic metric space (PM-space, shortly) is an ordered pair (S, F), where S is an arbitrary set and $F : S \times S \to \Delta$ satisfies:

(*i*) $F_{p,q}(x) = 1, \forall x > 0$ if and only if $p = q$;

(*ii*) $F_{p,q}(x) = F_{q,p}(x), \forall p, q \in S$;

(*iii*) If $F_{p,q}(x) = 1$ and $F_{q,r}(y) = 0, \forall p, q, r \in S$ then $F_{p,q}(x + y) = 1$.

(Here $F_{p,q}(x) = F(p,q)(x)$).

Definition 3.6 (Egbert [69]). Let $A \subset S, A \neq \emptyset$. The function $D_A(\cdot)$ defined by

$$D_A(x) = \sup_{t<x} \inf_{p,q \in A} F_{p,q}(x), x \in \mathbf{R}$$

is called the probabilistic diameter of A. If $\sup\{D_A(x); x \in \mathbf{R}\} = 1$ then A is called bounded.

Definition 3.7 (Bocşan-Constantin [41], see also Istrătescu [105], Constantin- Istrătescu [58]). Let A be a bounded subset of S. The mapping

$$\alpha_A(x) = \sup \left\{\varepsilon \geq 0; \exists n \in \mathbf{N}, A_i, i = \overline{1,n} \text{ with } A = \bigcup_{i=1}^{n} A_i \text{ and } D_{A_i}(x) \geq \varepsilon\right\},$$

where $x \in \mathbf{R}$, is called the random Kuratowski's measure of noncompactness.

Theorem 3.6 (Bocşan-Constantin [41]). *We have:*

(*i*) $\alpha_A \in \Delta$;

(*ii*) $\alpha_A(x) \geq D_A(x), \forall x \in \mathbf{R}$;

(*iii*) *If* $\emptyset \neq A \subset B$ *then* $\alpha_A(x) \geq \alpha_B(x), \forall x \in \mathbf{R}$;

(*iv*) $\alpha_{A \cup B}(x) = \min\{\alpha_A(x), \alpha_B(x)\}, \forall x \in \mathbf{R}$;

(*v*) $\alpha_A(x) = \alpha_{\overline{A}}(x)$, *where* $\overline{A}$ *is the closure of* A *in the* (ε, λ)*-topology of* S *(where by* (ε, λ)*-neighborhood of* $p \in S$*, we understand the set* $V_p(\varepsilon, \lambda) = \{q \in S; F_{p,q} > 1 - \lambda\}, \varepsilon > 0, \lambda \in [0,1]$*).*

Theorem 3.8 *(Bocşan-Constantin [41]). Let* (S, ρ) *be an usual metric space and* (S, F) *the corresponding* PM*-space generated by* (S, ρ)*.*

(*i*) $A \subset (S, F)$ *is probabilistic precompact if and only if* $\alpha_A(x) = H(x), \forall x \in \mathbf{R}$;

(*ii*) $A \subset (S, \rho)$ *is precompact if and only if* A *is probabilistic precompact subset of* (S, F);

(*iii*) *For any bounded* $A \subset (S, \rho)$ *we have*

$$\alpha_A(x) = H(x - \alpha(A)), \forall x \in \mathbf{R},$$

where $\alpha(A)$ *is the usual Kuratowski's measure of noncompactness in* (S, ρ) *and* $\alpha_A(x)$ *is the random Kuratowski's measure of noncompactness in the generated* PM*-space* (S, F).

Definition 3.9 (Menger [145]). (S, F) is called PM-space of Menger-type with the t-norm T, if F satisfies the first two properties in Definition 3.5 and the third one is replaced by

$$F_{p,q}(x + y) \geq T(F_{p,r}(x), F_{r,q}(x)), \forall p, q, r \in S.$$

We denote it as the triplet (S, F, T).

Definition 3.10 (see *e.g.* Istrătescu [105]). If (S,F,T) is a PM-space of Menger-type and $A,B \subset S$, then the probabilistic (random) Hausdorff-Pompeiu distance between A and B is given by

$$D^H_{A,B}(x) = \sup_{t<x} T\left(\inf_{p\in A}\sup_{q\in B} F_{p,q}(t), \inf_{q\in B}\sup_{p\in A} F_{p,q}(t)\right), x \in \mathbf{R}.$$

Definition 3.11 (see *e.g.* Istrătescu [105]). Let (S,F,T) be a PM-space of Menger-type and $A \subset S$, bounded. The random Hausdorff's measure of noncompactness of A is given by

$$h_A(x) = \sup\left\{\varepsilon > 0; \exists \text{ finite } F_\varepsilon \subset A \text{ such that } D^H_{A,F_\varepsilon}(x) \geq \varepsilon\right\}.$$

Further, we extend the above concepts and results to fuzzy subsets of usual metric spaces.

Definition 3.12 The sets

$$G_0(\varphi_A) = \{(x,y); 0 < y = \varphi_A(x), x \in X\} = Graph(F)$$

and

$$HG_0(\varphi_A) = \{(x,y); 0 < y \leq \varphi_A(x), x \in X\} = hypo(F) \cap (A \times (0,1])$$

are called the support graph and the support hypograph of (A,φ_A) respectively, where $F: A \to (0,1], F(x) = \varphi_A(x), \forall x \in A$.
The diameter and the hypo-diameter of the fuzzy set (A,φ_A) are given by

$$D(\varphi_A) = \sup\{d^*(a,b); a,b \in G_0(\varphi_A)\}$$

and

$$hD(\varphi_A) = \sup\{d^*(a,b); a,b \in HG_0(\varphi_A)\},$$

respectively. If $D(\varphi_A) < +\infty$ ($hD(\varphi_A) < +\infty$) we say that (A,φ_A) is bounded (hypo-bounded). Here $d^*: X \times [0,1] \to \mathbf{R}_+$, where (X,d) is metric space, is given by $d^*((x_1,r_1),(x_2,r_2)) = \max\{d(x_1,x_2), |r_1 - r_2|\}$.

Definition 3.13 Let (A,φ_A) be bounded (or hypo-bounded, respectively). The Kuratowski's measure and the Kuratowski's hypo-measure of noncompactness of (A,φ_A) are given by

$$\begin{aligned} K(\varphi_A) &= \inf\left\{\varepsilon > 0; \exists n \in \mathbf{N}, (A_i,\varphi_{A_i}) \text{ with } D(\varphi_{A_i}) \leq \varepsilon, i = \overline{1,n}, \right. \\ &\quad \left. \text{such that } \varphi_A(x) \leq \sup\left\{\varphi_{A_i}(x); i = \overline{1,n}\right\}, \forall x \in X\right\} \end{aligned}$$

and by

$$hK(\varphi_A) = \inf\{\varepsilon > 0; \exists n \in \mathbf{N}, (A_i, \varphi_{A_i}) \text{ with } hD(\varphi_{A_i}) \leq \varepsilon, i = \overline{1,n}, \text{ such that } \varphi_A(x) \leq \sup\{\varphi_{A_i}(x); i = \overline{1,n}\}, \forall x \in X\},$$

respectively.

The Hausdorff's measure and the Hausdorff's hypo-measure of noncompactness of (A, φ_A) are given by

$$H(\varphi_A) = \inf\{\varepsilon > 0; \exists n \in \mathbf{N}, \exists P_i = (x_i, r_i) \in G_0(\varphi_A), i = \overline{1,n}, \text{ such that } \forall P = (x,r) \in G_0(\varphi_A), \exists P_k \text{ with } d^*(P, P_k) \leq \varepsilon\}$$

and by

$$hH(\varphi_A) = \inf\{\varepsilon > 0; \exists n \in \mathbf{N}, \exists P_i = (x_i, r_i) \in HG_0(\varphi_A), i = \overline{1,n}, \text{ such that } \forall P = (x,r) \in HG_0(\varphi_A), \exists P_k \text{ with } d^*(P, P_k) \leq \varepsilon\}$$

respectively.

Theorem 3.9 *(i)* $K(\varphi_A) \leq hK(\varphi_A)$.

(ii) $K(\varphi_A) \leq K^*(G_0(\varphi_A))$, $H(\varphi_A) = H^*(G_0(\varphi_A))$, $hH(\varphi_A) = H^*(HG_0(\varphi_A))$, *where* K^* *and* H^* *represent the usual Kuratowski's and Hausdorff's measures of noncompactness (respectively) of the subsets in the metric space* $(X \times [0,1], d^*)$. *(Here* (A, φ_A) *is considered hypo-bounded).*

Theorem 3.10 *Let* $(A, \varphi_A) \subset (B, \varphi_B)$ *be two 1-bounded fuzzy subsets of* (X, d). *We have:*

(i) $K(\varphi_A) \leq D(\varphi_A)$.

(ii) $(A, \varphi_A) \subset (B, \varphi_B)$ *implies* $K(\varphi_A) \leq K(\varphi_B)$, *where* $(A, \varphi_A) \subset (B, \varphi_B)$ *means* $\varphi_A(x) \leq \varphi_B(x), \forall x \in X$.

(iii) $K(\varphi_A \vee \varphi_B) = \max\{K(\varphi_A), K(\varphi_B)\}$, *where* $(\varphi_A \vee \varphi_B)(x) = \max\{(\varphi_A)(x), (\varphi_B)(x)\}, \forall x \in X$.

Theorem 3.11 *(i)* $hH(\varphi_A) = +\infty$ *if and only if* $HG_0(\varphi_A)$ *is unbounded.*

(ii) If $(A, \varphi_A) \subset (B, \varphi_B)$ *then* $hH(\varphi_A) \leq 2hH(\varphi_B)$.

(iii) $hH(\varphi_A) = 0$ *if and only if* $HG_0(\varphi_A)$ *is totally bounded.*

(iv) $hH(\varphi_A \vee \varphi_B) \leq \max\{hH(\varphi_A), hH(\varphi_B)\}$, *where* $(\varphi_A \vee \varphi_B)(x) = \max\{\varphi_A(x), \varphi_B(x)\}, \forall x \in X$.

By using the level sets method, we can introduce

Definition 3.14 The (α)-Kuratowski measure of noncompactness of (A, φ_A) is given by

$$\alpha K(\varphi_A) = \sup\{K_0(A_\lambda); \lambda \in (0,1]\},$$

where $A_\lambda = \{x \in X; \varphi_A(x) \geq \lambda\}$ and K_0 is the usual Kuratowski's measure of noncompactness of usual subsets in (X, d).
The (α)-Hausdorff measure of noncompactness of (A, φ_A) is given by

$$\alpha H(\varphi_A) = \sup\{H_0(A_\lambda); \lambda \in (0,1]\},$$

where H_0 represents the usual Hausdorff's measure of noncompactness of usual subsets in (X, d). Obviously, $\alpha H(\varphi_A)$ can take the $+\infty$ value.

Definition 3.15 (see *e.g.* Weiss [220]) If (X, d) is a metric space, then the induced fuzzy topology on (X, d) is the collection of all fuzzy sets of (X, d) with $\varphi_A : X \to [0,1]$ lower semicontinuous on X.
A fuzzy set (A, φ_A) is called (α)-bounded if for each $\lambda \in (0,1], A_\lambda$ is bounded in the metric space (X, d).
A fuzzy set (A, φ_A) is called (α)-compact if for each $\lambda \in (0,1], A_\lambda$ is compact in the metric space (X, d).

Theorem 3.13 *Let $(A, \varphi_A), (B, \varphi_B)$ be (α)-bounded.*

(i) $\alpha K(\varphi_A) = 0$ if and only if each $\overline{A_\lambda}, \lambda \in (0,1]$ is compact in (X, d), where $\overline{A_\lambda}$ denotes the closure of A_λ in (X, d). Also, if (A, φ_A) is (α)-compact then $\alpha K(\varphi_A) = 0$.

(ii) If $(A, \varphi_A) \subset (B, \varphi_B)$ (i.e. $\varphi_A(x) \leq \varphi_B(x), \forall x \in X$) then $\alpha K(\varphi_A) \leq \alpha K(\varphi_B)$.

(iii) $\alpha K(\varphi_A) \leq K_0(A)$, where K_0 represents the usual Kuratowski's measure for subsets in (X, d), and in general we have no equality.

(iv) $\alpha K(\varphi_A \vee \varphi_B) = \max\{\alpha K(\varphi_A), \alpha K(\varphi_B)\}$, where $(\varphi_A \vee \varphi_B)(x) = \max\{\varphi_A(x), \varphi_B(x)\}, \forall x \in X$.

(v) $\alpha H(\varphi_A) = 0$ if and only if each $A_\lambda, \lambda \in (0,1]$, is totally bounded.

(vi) If $(A, \varphi_A) \subset (B, \varphi_B)$ then $\alpha H(\varphi_A) \leq 2 \cdot \alpha H(\varphi_B)$.

(vii) $\alpha H(\varphi_A \vee \varphi_B) \leq \max\{\alpha H(\varphi_A), \alpha H(\varphi_B)\}$.

In what follows one consider measures of noncompactness of fuzzy subsets in (fuzzy) topological spaces.

Definition 3.21 Let (X, T) be a classical topological space and A a fuzzy subset of X (i.e. $A : X \to [0,1]$). Then, the (α)-measure of noncompactness of the fuzzy set A will be

$$\alpha M(A) = \bigcap_{\alpha \in (0,1]} m(A_\alpha),$$

where $A_\alpha = \{x \in X; A(x) \geq \alpha\}$ and $\alpha M : I^X \to \mathcal{C}$, I^X denoting the class of all fuzzy subsets of X.

Definition 3.22 Let $A_i \in T_F, i \in I$ and $A^1, ..., A^n \in T_F$. The family $\mathcal{K} = \left\{(A_i)_{i \in I}, A^1, ..., A^n\right\}$ of fuzzy sets is an open $\widetilde{s}$-cover of a fuzzy set A if

$$A \subset (\vee_{i \in I} A_i)\, \widetilde{s} A^1 \widetilde{s} ... \widetilde{s} A^n,$$

where $(\vee_{i \in I} A_i)(x) = \sup\{A_i(x); i \in I\}, x \in X$ and $A \subset B$ means $A(x) \leq B(x), \forall x \in X$.
We say that $\left\{(A_j)_{j \in J}, A^{k_1}, ..., A^{k_p}\right\}$ is a finite open $\widetilde{s}$-subcover in $\mathcal{K}$ of the fuzzy set A, if $A \subset (\vee_{j \in J} A_j)\, \widetilde{s} A^{k_1} \widetilde{s} ... \widetilde{s} A^{k_p}$, where $J \subset I$ is finite and $\{k_1, ..., k_p\} \subset \{1, ..., n\}$.

Definition 3.23 A fuzzy set $A \in I^X$ is $(C, \widetilde{s})$-compact if each open $\widetilde{s}$-cover of A has a finite $\widetilde{s}$-subcover of A. A fuzzy set A is $(L, \widetilde{s})$-compact if for all open $\widetilde{s}$-cover of A and for all $\varepsilon > 0$, there exists a finite $\widetilde{s}$-subcover of B, where B is the fuzzy set defined by $B(x) = \max(0, A(x) - \varepsilon), \forall x \in X$.

Definition 3.24 Let $A \in I^X, (X, T_F)$ be a quasi-fuzzy topological space and s a triangular conorm. The (C, s)-measure of noncompactness of A is the element of $\mathcal{C}^S$ defined by

$$m_C^S(A) = \left\{\mathcal{D} \in \pi^S; \exists \delta \subset \mathcal{D}, \text{ finite open } \widetilde{s}\text{-subcover of } A\right\}.$$

The (L, s)-measure of noncompactness of A is the element of $\mathcal{C}^S$ defined by

$$m_L^S(A) = \Big\{\mathcal{D}_* \in \pi^S; \forall \varepsilon > 0, \exists \delta_* \subset \mathcal{D}_*, \text{ finite, } \delta_* = \left\{(A_j)_{j \in J}, A^1, ..., A^p\right\}$$

$$\text{with } \max\{0, A(x) - \varepsilon\} \leq \left((\vee_{j \in J} A_j)\, \widetilde{s} A^1 \widetilde{s} ... \widetilde{s} A^p\right)(x), \forall x \in X\Big\}.$$

Theorem 3.15 *Let (X, T) be a topological space. The mapping $\alpha M : I^X \to \mathcal{C}$ has the following properties:*

(i) If $A \in I^X$ is compact in the sense of Weiss [220], Definition 3.5, then $\alpha M(A) = 0_{\mathcal{C}}$. If A is closed (in the sense of Proposition 3.3 in Weiss

[220]) and $\alpha M(A) = 0_{\mathcal{C}}$, then A is compact (in the sense of Definition 3.5 *in Weiss [220]).*

(ii) If $A \subset B$ then $\alpha M(A) \leq \alpha M(B)$.

(iii) $\alpha M(A \vee B) = \alpha M(A) \cap \alpha M(B)$.

(iv) $\alpha M(A \wedge B) \leq \alpha M(A) \cup \alpha M(B)$.

(Here $(A \vee B)(x) = \max\{A(x), B(x)\}, (A \wedge B)(x) = \min\{A(x), B(x)\}$, $\forall x \in X$).

Theorem 3.16 *Let (X, T_F) be a quasi-fuzzy topological space. The mapping $m_C^S : I^X \to \mathcal{C}^S$ has the following properties:*

(i) If $A \in I^X$ is $(C, \widetilde{s})$-compact then $m_C^S(A) = 0_{\mathcal{C}}^S$.

(ii) If $A \subset B$ then $m_C^S(A) \leq m_C^S(B)$.

(iii) $m_C^S(A \vee B) = m_C^S(A) \cap m_C^S(B)$.

(iv) $m_C^S(A \wedge B) \leq m_C^S(A) \cup m_C^S(B)$.

Theorem 3.17 *Let (X, T_F) be a quasi-fuzzy topological space. The mapping $m_L^S : I^X \to \mathcal{C}^S$ has the following properties:*

(i) If $A \in I^X$ is $(L, \widetilde{s})$-compact then $m_L^S(A) = 0_{\mathcal{C}}^S$.

(ii) If $A \subset B$ then $m_L^S(A) \leq m_L^S(B)$.

(iii) $m_L^S(A \vee B) = m_L^S(A) \cap m_L^S(B)$.

(iv) $m_L^S(A \wedge B) \leq m_L^S(A) \cup m_L^S(B)$.

Applications to upper semicontinuity of fuzzy multifunctions also are presented.

At the end of this chapter one present:

Definition 3.34 Let (X, ρ) be a metric space and $D(A, B)$ be a certain distance between the subsets $A, B \subset X$. For $Y \subset X$, we call:

(i) defect of opening of Y, the quantity $d_{OP}(D)(Y) = D(Y, intY)$;

(ii) defect of closure of Y, the quantity $d_{CL}(D)(Y) = D\left(Y, \overline{Y}\right)$.

Theorem 3.23 If *(X, ρ) is a metric space and $Y \in \mathcal{P}_b(X)$, then:*

(i) $d_{OP}(D^*)(Y) = 0$ if and only if Y is open;

(ii) $d_{CL}(D^*)(Y) = 0$ if and only if Y is closed.

In addition,

$$\begin{aligned} d_{CL}(D^*)(\lambda Y) &= \lambda d_{CL}(D^*)(Y), \forall \lambda > 0; \\ d_{OP}(D^*)(\lambda Y) &= \lambda d_{OP}(D^*)(Y), \forall \lambda > 0. \end{aligned}$$

(Here D^* is a special distance).

1.4 On Chapter 4: Defect of Property in Measure Theory

This chapter introduces and studies some defects of property for set function (especially, for fuzzy measures): defect of additivity, defect of complementarity, defect of monotonicity, defect of subadditivity and of superadditivity. Also, in Section 4.5, the defect of measurability for sets is discussed.

Let $\mathcal{A}$ be a σ-algebra of subsets of X.

Definition 4.1 (see *e.g.* Ralescu-Adams [168]) A set function $\mu : \mathcal{A} \to [0, \infty)$ is said to be a fuzzy measure if the followings hold:

(*i*) $\mu(\emptyset) = 0$;

(*ii*) $A, B \in \mathcal{A}$ and $A \subseteq B$ implies $\mu(A) \leq \mu(B)$.

Definition 4.2 The defect of additivity of order $n, n \in \mathbf{N}, n \geq 2$, for the measure μ is given by

$$a_n(\mu) = \sup \quad \{|\mu(\bigcup_{i=1}^{n} A_i) - \sum_{i=1}^{n} \mu(A_i)| : A_i \in \mathcal{A}, \forall i \in \mathbf{N}, \\ A_i \cap A_j = \emptyset, i \neq j\}$$

The defect of countable additivity for the measure μ is given by

$$a_\infty(\mu) = \sup \quad \{|\mu(\bigcup_{i=1}^{\infty} A_i) - \sum_{i=1}^{\infty} \mu(A_i)| : A_i \in \mathcal{A}, \forall i \in \mathbf{N}, \\ A_i \cap A_j = \emptyset, i \neq j\}$$

Theorem 4.1 (*i*) $a_n(\mu) \leq a_{n+1}(\mu), \forall n \geq 2$;

(*ii*) *If μ is a continuous from below fuzzy measure then* $a_\infty(\mu) = \lim_{n\to\infty} a_n(\mu)$;

(*iii*) $a_n(\mu) \leq a_{n-1}(\mu) + a_2(\mu), \forall n \geq 3$;

(*iv*) $a_n(\mu) \leq (n-1) a_2(\mu), \forall n \geq 3$;

(*v*) $a_n(\mu) \leq (n-1) \mu(X), \forall n \geq 2$;

(*vi*) *If μ is a superadditive fuzzy measure then* $a_n(\mu) \leq \mu(X), \forall n \geq 2$.

An important class of fuzzy measures are the λ-additive fuzzy measures.

Definition 4.3 (see *e.g.* Kruse [126]) Let $\lambda \in (-1, \infty), \lambda \neq 0$ and $\mathcal{A}$ be a σ-algebra on the set X. A continuous fuzzy measure μ with $\mu(X) = 1$ is called λ-additive if, whenever $A, B \in \mathcal{A}, A \cap B = \emptyset$,

$$\mu(A \cup B) = \mu(A) + \mu(B) + \lambda \mu(A) \mu(B).$$

Theorem 4.2 *(Wierzchon [222]) Let m be a classical finite measure, $m : \mathcal{A} \to [0, \infty)$. Then $\mu = t \circ m$ is a λ-additive fuzzy measure if and only* *if the function* $t : [0, \infty) \to [0, \infty)$ *has the form* $t(x) = \frac{c^x - 1}{\lambda}, c > 0, c \neq 1$, $\lambda \in (-1, 0) \cup (0, \infty)$.

Concerning the defect $a_{n,\lambda}(\mu)$ of the above measures, we can prove the following.

Theorem 4.3 *If* $\mu = t \circ m$ *(with t and m as above) then we have*

$$a_{n,\lambda}(\mu) \leq \frac{\lambda + n - n \sqrt[n]{\lambda + 1}}{|\lambda|}, \forall n \geq 2, \forall \lambda \in (-1, 0) \cup (0, \infty)$$

and

$$a_{\infty,\lambda}(\mu) \leq \left| 1 - \frac{\ln(\lambda + 1)}{\lambda} \right|, \forall \lambda \in (-1, 0) \cup (0, \infty).$$

An useful result is

Theorem 4.4 *(see Ghermănescu [89], p.260, Th. 4.9) The measurable solutions of the functional equation* $f(x + y) - f(x) - f(y) = \varphi(x, y)$ are given by

$$\begin{aligned} f(x) &= B(x) + ax \\ \varphi(x, y) &= B(x + y) - B(x) - B(y) \end{aligned}$$

where B *is an arbitrary measurable function and* a *is an arbitrary real constant.*

Theorem 4.4 is used to generate classes of fuzzy measures as follows.

Theorem 4.5 *Let* $B(x)$ *be such that* $B(0) = 0$ *and* $f(x) = B(x) + ax$ *is strictly increasing and* $f : [0, \infty) \to [0, \infty)$. *Then* $\mu(A) = (f \circ m)(A)$ *(where m is a classical finite measure) is a fuzzy measure which satisfies the functional equation*

$$\mu(A \cup B) = \mu(A) + \mu(B) + \varphi\left(f^{-1}(\mu(A)), f^{-1}(\mu(B))\right),$$

$\forall A, B \in \mathcal{A}, A \cap B = \emptyset$, *where* φ *and* f *are those in Theorem 4.4 and* f^{-1} *represents the inverse function of* f *(here we assume* $f(0) = 0$ *and* $\lim_{x \to +\infty} f(x) = +\infty$*).*

We estimate the defects of additivity for fuzzy measures obtained by Theorem 4.5 as follows.

Theorem 4.6 *If* $\mu : \mathcal{A} \to [0,\infty)$ *is the fuzzy measure defined by* $\mu(A) = \lambda m^2(A)$, *where* m *is a classical finite measure, then* $a_n(\mu) \leq \frac{n-1}{n}\mu(X)$, *for all* $n \geq 2$ *and* $\lambda > 0$.

Theorem 4.7 *If* $\mu : \mathcal{A} \to [0,\infty)$ *is the fuzzy measure defined by the relation*

$$\mu(A \cup B) = \frac{\ln\left((\lambda+1)^{\mu(A)} + (\lambda+1)^{\mu(B)} - 1\right)}{\ln(\lambda+1)},$$

for every $A, B \in \mathcal{A}, A \cap B = \emptyset$, *then*

$$a_{n,\lambda}(\mu) \leq n\mu(X) - \frac{\ln\left(n(\lambda+1)^{\mu(X)} - n + 1\right)}{\ln(\lambda+1)},$$

for every $n \geq 2, \lambda \in (0,\infty)$.

Next, one present three kinds of applications of the previous results: to approximative calculation of the fuzzy integrals, to best approximation of a fuzzy measure and to the definition of a metric on the family of fuzzy measures. Thus, for example, an estimation of the quantity of best approximation of a fuzzy measure $\mu : \mathcal{A} \to [0,1]$ by elements from

$$\mathcal{M} = \{m : \mathcal{A} \to [0,1]; m \text{ additive on the } \sigma\text{-algebra } \mathcal{A}\},$$

is given by

$$\frac{a_n(\mu)}{n+1} \leq E(\mu) \leq 1, \forall n \geq 2,$$

where

$$E(\mu) = \inf\{d(\mu, m); m \in \mathcal{M}\} \leq 1$$

and $d(\mu,\nu) = \sup\{|\mu(A) - \nu(A)|; A \in \mathcal{A}\}$ is a metric on

$$\mathcal{F} = \{\mu : \mathcal{A} \to [0,1]; \mu(\emptyset) = 0, \mu \text{ nondecreasing on the } \sigma\text{-algebra } \mathcal{A}\}.$$

Another introduced concept is given by

Definition 4.6 The value

$$c(\mu) = \frac{\sup\{|\mu(X) - \mu(A) - \mu(A^c)| : A \in \mathcal{A}\}}{\mu(X)}$$

where $A^c = X \setminus A$, is called defect of complementarity of the fuzzy measure μ.

Theorem 4.8 *Let $\mu, \mu' : \mathcal{A} \to [0, \infty)$ be fuzzy measures. We have:*
(i) $0 \leq c(\mu) \leq 1$;
(ii) $c(\mu) \leq \frac{a_2(\mu)}{\mu(X)}$;
(iii) $c(\mu) = c(\mu^c)$, *where μ^c is the dual of μ, that is $\mu^c(A) = \mu(X) - \mu(A^c)$, A^c being the complementary of A;*
(iv) $c(\alpha\mu) = c(\mu), \forall \alpha > 0$;
(v) $c(\mu + \mu') \leq \frac{\mu(X)}{(\mu+\mu')(X)} c(\mu) + \frac{\mu'(X)}{(\mu+\mu')(X)} c(\mu')$.

We estimate the defects of complementarity for fuzzy measures obtained by Theorem 4.5 as follows.

Theorem 4.11 *If μ is a λ-additive fuzzy measure then*

$$c_\lambda(\mu) \leq \frac{\left(1 - \sqrt{\lambda + 1}\right)^2}{|\lambda|}.$$

Theorem 4.12 *If $\mu : \mathcal{A} \to [0, \infty)$ is the fuzzy measure defined by $\mu(A) = \lambda m^2(A), \forall A \in \mathcal{A}$, where m is a finite measure, then $c(\mu) \leq \frac{1}{2}, \forall \lambda \in (0, \infty)$.*

Theorem 4.13 *If $\mu : \mathcal{A} \to [0, \infty)$ is the fuzzy measure defined by $\mu(A) = \frac{\ln(\lambda m(A)+1)}{\ln(\lambda+1)}, \forall A \in \mathcal{A}$, where m is a finite measure and $\lambda > 0$, then*

$$c(\mu) \leq 2\frac{\ln\left(\lambda \frac{m(X)}{2} + 1\right)}{\ln(\lambda m(X) + 1)} - 1.$$

Also, we prove that a cardinality of a convenient fuzzy set is a good indicator of complementarity in pointwise sense.

At the end of Section 4.2, we give interpretations, applications and numerical examples with respect to the proved results.

Concerning other properties of set functions, we consider

Definition 4.12 Let $\nu : \mathcal{P}(X) \to \mathbf{R}$ be a set function. The defect of monotonicity of ν is given by

$$d_{MON}(\nu) = \frac{1}{2} \sup \{|\nu(A_1) - \nu(A)| + |\nu(A_2) - \nu(A)| - |\nu(A_1) - \nu(A_2)|;$$

$$A_1, A, A_2 \in \mathcal{P}(X), A_1 \subseteq A \subseteq A_2\}.$$

Theorem 4.15 *The set function $\nu : \mathcal{P}(X) \to \mathbf{R}$ is monotone if and only if $d_{MON}(\nu) = 0$.*

Theorem 4.16 *Let $\nu : \mathcal{P}(X) \to \mathbf{R}$ be a set function. We have:*

(i) $0 \leq d_{MON}(\nu) \leq 2\sup\{|\nu(A)|; A \in \mathcal{P}(X)\}$. If ν is non-negative then $0 \leq d_{MON}(\nu) \leq \sup\{\nu(A); A \in \mathcal{P}(X)\}$.

(ii) $d_{MON}(\alpha\nu) = |\alpha| d_{MON}(\nu), \forall \alpha \in \mathbf{R}$, where $(\alpha\nu)(A) = \alpha\nu(A)$, $\forall A \in \mathcal{P}(X)$.

(iii) $d_{MON}(\nu_B) \leq d_{MON}(\nu)$, where $\nu_B(A) = \nu(A \cap B), \forall A \in \mathcal{P}(X)$, is the induced set function on $B \in \mathcal{P}(X)$.

(iv) If ν is an additive set function then $d_{MON}(\nu) \leq a_2(|\nu|)$, where $|\nu|(A) = |\nu(A)|, A \in \mathcal{P}(X)$.

(v) If ν is a signed fuzzy measure then $d_{MON}(\nu) = d_{MON}(\nu^c)$, where ν^c is the dual of ν, that is $\nu^c(A) = \nu(X) - \nu(A^c), \forall A \in \mathcal{P}(X)$.

Theorem 4.19 *For any $\nu \in \mathcal{S} = \{\nu : \mathcal{P}(X) \to \mathbf{R}; \nu \text{ is bounded}\}$, we have*

$$E_{MON}(\nu) \geq \frac{d_{MON}(\nu)}{6},$$

where

$$E_{MON}(\nu) = \inf\{d(\nu, m); m \in \mathcal{M}\},$$

$$\mathcal{M} = \{m : \mathcal{P}(X) \to \mathbf{R}; m \textit{ is monotone and bounded}\}$$

and

$$d(\nu_1, \nu_2) = \sup\{|\nu_1(A) - \nu_2(A)|; A \in \mathcal{P}(X)\}, \forall \nu_1, \nu_2 \in \mathcal{S}.$$

Definition 4.15 Let $\nu : \mathcal{A} \to \mathbf{R}, \nu(\emptyset) = 0$, where $\mathcal{A} \subseteq \mathcal{P}(X)$ is a σ-algebra. The defect of subadditivity of order $n, n \geq 2$, of ν is given by

$$\begin{aligned} d_{SUB}^{(n)}(\nu) &= \sup\left\{\nu\left(\bigcup_{i=1}^{n} A_i\right) - \sum_{i=1}^{n}\nu(A_i); A_i \in \mathcal{A},\right. \\ &\left. A_i \cap A_j = \emptyset, i \neq j, i, j = \overline{1,n}\right\} \end{aligned}$$

and the defect of superadditivity of order $n, n \geq 2$, of ν is given by

$$d_{SUP}^{(n)}(\nu) = \sup\left\{\sum_{i=1}^{n}\nu(A_i) - \nu\left(\bigcup_{i=1}^{n} A_i\right); A_i \in \mathcal{A},\right.$$

$$A_i \cap A_j = \emptyset, i \neq j, i, j = \overline{1, n}\}.$$

Similarly, the defect of countable subadditivity of ν is given by

$$d_{SUB}^{\infty}(\nu) = \sup\left\{\nu\left(\bigcup_{i=1}^{\infty} A_i\right) - \sum_{i=1}^{\infty} \nu(A_i); A_i \in \mathcal{A}, A_i \cap A_j = \emptyset, i \neq j, i, j \in \mathbf{N}\right\}$$

and the defect of countable superadditivity of ν is given by

$$d_{SUP}^{\infty}(\nu) = \sup\left\{\sum_{i=1}^{\infty} \nu(A_i) - \nu\left(\bigcup_{i=1}^{\infty} A_i\right); A_i \in \mathcal{A}, A_i \cap A_j = \emptyset, i \neq j, i, j \in \mathbf{N}\right\}.$$

The main properties of the above defined defects are the followings.

Theorem 4.20 *(i)* $0 \leq d_{SUB}^{(n)}(\nu) \leq a_n(\nu), 0 \leq d_{SUP}^{(n)}(\nu) \leq a_n(\nu)$, $\forall n \in \mathbf{N}, n \geq 2$, *where* $a_n(\nu)$ *is the defect of additivity of order* n *of* ν*;*

(ii) $0 \leq d_{SUB}^{\infty}(\nu) \leq a_{\infty}(\nu), 0 \leq d_{SUP}^{\infty}(\nu) \leq a_{\infty}(\nu)$, *where* $a_{\infty}(\nu)$ *is the countable defect of additivity of* ν*;*

(iii) If ν *is subadditive or countable subadditive then* $d_{SUP}^{(n)}(\nu) = a_n(\nu)$ *or* $d_{SUB}^{\infty}(\nu) = a_{\infty}(\nu)$*, respectively;*

(iv) If ν *is superadditive or countable superadditive then* $d_{SUB}^{(n)}(\nu) = a_n(\nu)$ *or* $d_{SUP}^{\infty}(\nu) = a_{\infty}(\nu)$*, respectively;*

(v) $d_{SUB}^{(n)}(\nu) \leq d_{SUB}^{(n+1)}(\nu), d_{SUP}^{(n)}(\nu) \leq d_{SUP}^{(n+1)}(\nu), \forall n \in \mathbf{N}, n \geq 2$*;*

(vi) $d_{SUB}^{(n)}(\nu) = 0$ *if and only if* ν *is subadditive;* $d_{SUP}^{(n)}(\nu) = 0$ *if and only if* ν *is superadditive;* $d_{SUB}^{\infty}(\nu) = 0$ *if and only if* ν *is countable subadditive;* $d_{SUP}^{\infty}(\nu) = 0$ *if and only if* ν *is countable superadditive;*

(vii) $d_{SUB}^{(n)}(\nu) \leq d_{SUB}^{(n-1)}(\nu) + d_{SUB}^{(2)}(\nu)$ *and* $d_{SUP}^{(n)}(\nu) \leq d_{SUP}^{(n-1)}(\nu) + d_{SUP}^{(2)}(\nu), \forall n \in \mathbf{N}, n \geq 3$.

(viii) $d_{SUB}^{(n)}(\nu) \leq (n-1) d_{SUB}^{(2)}(\nu)$ *and* $d_{SUP}^{(n)}(\nu) \leq (n-1) d_{SUP}^{(2)}(\nu)$, $\forall n \in \mathbf{N}, n \geq 3$.

Let (X, ρ) be a bounded metric space, $CL(X) = \{A \subset X; X \setminus A \in \mathcal{T}_{\rho}\}$, where $\mathcal{T}_{\rho}$ is the topology generated by the metric ρ and $\varphi : \mathcal{T}_{\rho} \to \mathbf{R}_+$. We define $\overline{\varphi} : CL(X) \to \mathbf{R}_+$ by

$$\overline{\varphi}(A) = diam(X) - \varphi(X \setminus A), \forall A \in CL(X),$$

and $\varphi^*, \varphi_* : \mathcal{P}(X) \to \mathbf{R}_+$ by

$$\varphi^*(A) = \sup\{\overline{\varphi}(F); F \in CL(X), F \subseteq A\},$$

$$\varphi_*(A) = \inf\{\varphi(G); G \in \mathcal{T}_\rho, A \subseteq G\}.$$

Definition 4.17 Keeping the above notations and assumptions, we say that $A \in \mathcal{P}(X)$ is φ-measurable if $\varphi^*(A) = \varphi_*(A)$. The quantity

$$d_{MAS}(A) = \varphi^*(A) - \varphi_*(A)$$

is called defect of φ-measurability of A.

Theorem 4.17 *(i)* $0 \le d_{MAS}(A) \le \varphi(X), \forall A \in \mathcal{P}(X)$ *and* $d_{MAS}(A) = 0$ *if and only if* A *is* φ*-measurable;*

(ii) $d_{MAS}(A) = d_{MAS}(A^c), \forall A \in \mathcal{P}(X)$;

(iii) $d_{MAS}(A \cup B) \le d_{MAS}(A) + d_{MAS}(B), \forall A, B \in \mathcal{P}(X), A \cap B = \emptyset$;

(iv) Let T *be the one to one transformation of the entire real line into itself, defined by* $T(x) = \alpha x + \beta$, *where* α *and* β *are real numbers and* $\alpha \neq 0$. *If for every bounded* $A \subset \mathbf{R}$ *we denote* $T(A) = \{\alpha x + \beta; x \in A\}$, *then*

$$d_{MAS}(T(A)) = |\alpha| d_{MAS}(A).$$

(Here ρ *is the usual metric induced on* A *or* $T(A)$*, respectively).*

1.5 On Chapter 5: Defect of Property in Real Function Theory

This chapter discusses various defects of properties of the real functions of real variable.

Definition 5.1 (see *e.g.* Sireţchi [199], p.151, p.165, [200], p.239). Let $f : E \to \mathbf{R}$. For $x_0 \in E$, the quantity defined by

$$\omega(x_0; f) = \inf\{\delta[f(V \cap E)]; V \in \mathcal{V}(x_0)\},$$

is called oscillation of f at x_0, where $\mathcal{V}(x_0)$ denotes the class of all neighborhoods of x_0 and $\delta[A] = \sup\{|a_1 - a_2|; a_1, a_2 \in A\}$ represents the diameter

of the set $A \subset \mathbf{R}$.
For x_0 limit point of E, the quantity defined by

$$\widehat{\omega}(x_0; f) = \inf \{\delta [f(V \cap E \setminus \{x_0\})]; V \in \mathcal{V}(x_0)\}$$

is called pointed oscillation of f at x_0.
For $\varepsilon > 0$, the quantity defined by

$$\omega(f; \varepsilon)_E = \sup \{|f(x_1) - f(x_2)|; |x_1 - x_2| \leq \varepsilon, x_1, x_2 \in E\}$$

is called modulus of continuity of f on E with step $\varepsilon > 0$.

Theorem 5.1 *(see e. g. Sireţchi [199], p.166, p.154, [200], p.211, p.239).*
(i) f is continuous at $x_0 \in E$ if and only if $\omega(x_0; f) = 0$;
(ii) f has finite limit at $x_0 \in E'$ if and only if $\widehat{\omega}(x_0; f) = 0$. Also

$$\overline{\lim_{x \to x_0}} f(x) - \underline{\lim}_{x \to x_0} f(x) \leq \widehat{\omega}(x_0; f)$$

and if, in addition, f is bounded on E, then

$$\widehat{\omega}(x_0; f) = \lim_{x \to x_0} f(x) - \underline{\lim}_{x \to x_0} f(x),$$

where

$$\underline{\lim}_{x \to x_0} f(x) = \sup \{\inf f(V \cap E \setminus \{x_0\}); V \in \mathcal{V}(x_0)\}$$

and

$$\overline{\lim_{x \to x_0}} f(x) = \inf \{\sup f(V \cap E \setminus \{x_0\}); V \in \mathcal{V}(x_0)\}.$$

(iii) f is uniformly continuous on E if and only if $\inf \{\omega(f; \varepsilon); \varepsilon > 0\} = 0$.

Definition 5.2 The quantities

$$d_C(f)(x_0) = \omega(x_0; f),$$

$$d_{\lim}(f)(x_0) = \widehat{\omega}(x_0; f)$$

and

$$d_{uc}(f)(E) = \inf \{\omega(f; \varepsilon); \varepsilon > 0\},$$

can be called defect of continuity of f on x_0, defect of limit of f on x_0 and defect of uniform continuity of f on E, respectively.

Definition 5.3 (see Burgin-Ŝostak [44], [45]). Let $f : X \to Y$ be with $X, Y \subset \mathbf{R}$ and $x_0 \in X$. The defect of continuity of f at x_0 can be defined by

$$\mu(x_0; f) = \sup\{|y - f(x_0)|; y \text{ is limit point of } f(x_n) \text{ when } x_n \to x_0\}.$$

The defect of continuity of f on X is defined by

$$\mu(f)_X = \sup\{\mu(x_0; f); x_0 \in X\}.$$

Theorem 5.2 *(see Burgin-Ŝostak [44], [45]).*
(i) f is continuous on x_0 if and only if $\mu(x_0; f) = 0$;
(ii) f is continuous on X if and only if $\mu(f)_X = 0$;
(iii)

$$\begin{aligned}
|\mu(x_0; f) - \mu(x_0; g)| &\leq \mu(x_0; f + g) \leq \mu(x_0; f) + \mu(x_0; g), \\
|\mu(x_0; f) - \mu(x_0; g)| &\leq \mu(x_0; f - g), \\
\mu(x_0; -f) &= \mu(x_0; f)
\end{aligned}$$

and similar inequalities for $\mu(f)_X$ hold.
(iv)

$$\mu(x_0; f \cdot g) \leq \mu(x_0; f) \cdot \|g\|_{x_0} + \mu(x_0; g) \cdot \|f\|_{x_0},$$

where $\|g\|_{x_0} = \sup\{|g(t)|; t \in V_{x_0}\}, \|f\|_{x_0} = \sup\{|f(t)|; t \in V_{x_0}\}, V_{x_0}$ is a neighborhood of x_0 and $f, g : X \to Y$, with $X, Y \subset \mathbf{R}$.
(v) If we define $\mu_H(x_0; f) = \frac{\mu(x_0;f)}{M}, \mu_H(f)_X = \frac{\mu(f)_X}{M}$, where $M = \sup_{x \in X} f(x) - \inf_{x \in X} f(x), f : X \to Y$, then $\mu_H(f)_X = \mu(t \circ f)_X$, where $t(x) = kx, k > 0$ fixed.

Definition 5.4 Let $f : [a, b] \to \mathbf{R}$ and $x_0 \in [a, b]$. The quantity

$$d_{dif}(f)(x_0) = \widehat{\omega}(x_0; F),$$

where $F : [a, b] \setminus \{x_0\} \to \mathbf{R}$ is given by $F(x) = \frac{f(x) - f(x_0)}{x - x_0}$, is called defect of differentiability of f on x_0.

Corollary 5.1 (i) f is differentiable on x_0 if and only if $d_{dif}(f)(x_0) = 0$.
(ii) If f is locally Lipschitz on x_0 (i.e. Lipschitz in a neighborhood of x_0) then

$$d_{dif}(f)(x_0) = \overline{\lim_{x \to x_0}} F(x) - \underline{\lim_{x \to x_0}} F(x).$$

Theorem 5.3 *Let $f : [a,b] \to \mathbf{R}$ and $x_0 \in (a,b)$. If f is locally Lipschitz on x_0, then*

$$\widehat{\omega}(x_0; f) \leq (b-a)\, d_{dif}(f)(x_0).$$

Definition 5.6 Let $f : [a,b] \to \mathbf{R}$ be bounded. The real number

$$d_{int}(f)([a,b]) = \overline{\int_a^b} f(x)\,\mathrm{d}x - \underline{\int_a^b} f(x)\,\mathrm{d}x$$

is called defect of integrability of f on $[a,b]$.

Let us denote

$$\begin{aligned} B[a,b] &= \{f : [a,b] \to \mathbf{R}; f \text{ bounded on } [a,b]\}, \\ C[a,b] &= \{g : [a,b] \to \mathbf{R}; g \text{ continuous on } [a,b]\} \end{aligned}$$

and for $f \in B[a,b]$ the quantity of best approximation

$$E_c(f) = \inf\{\|f-g\|; g \in C[a,b]\},$$

where $\|f\| = \sup\{|f(t)|; t \in [a,b]\}$.

Theorem 5.4 *Let $f \in B[a,b]$ be with*

$$C_f = \{x \in [a,b]; x \text{ is point of discontinuity of } f\}.$$

Then

$$E_c(f) \geq \frac{\sup\{\omega(x;f); x \in C_f\}}{2}.$$

Let us denote

$$\begin{aligned} L[a,b] &= \{f : [a,b] \to \mathbf{R}; \|f\|_L < +\infty\}, \\ C_1[a,b] &= \{g : [a,b] \to \mathbf{R}; g \text{ is differentiable on } [a,b]\}, \end{aligned}$$

where $\|f\|_L = \sup\left\{\left|\frac{f(x)-f(y)}{x-y}\right|; x,y \in [a,b], x \neq y\right\}$ is the so-called Lipschitz norm. If we denote $E_L(f) = \inf\{\|f-g\|_L; g \in C_1[a,b]\}$, we can present

Theorem 5.5 *Let $f \in L[a,b]$ be with*

$$D_f = \{x \in [a,b]; x \text{ is point of non-differentiability of } f\}.$$

Then

$$E_L(f) \geq \sup\{d_{dif}(f)(x); x \in D_f\}.$$

Theorem 5.6 *For each $f \in B[a,b]$ we have*

$$E_I(f) \geq d_{int}(f)([a,b]),$$

where $E_I(f) = \inf\{\|f-g\|; g$ *is Riemann integrable on* $[a,b]\}$.

Definition 5.7 Let $f : E \to \mathbf{R}$. The quantity

$$\begin{aligned} d_M(f)(E) &= \sup\{|f(x_1)-f(x)| + |f(x_2)-f(x)| \\ &\quad -|f(x_1)-f(x_2)|; x_1, x, x_2 \in E, x_1 \leq x \leq x_2\} \end{aligned}$$

is called defect of monotonicity of f on E.

For $f \in B[a,b]$, let us define

$$E_M(f)([a,b]) = \inf\{\|f-g\|; g \in M[a,b]\},$$

where $M[a,b] = \{g : [a,b] \to \mathbf{R}; g$ is monotone on $[a,b]\}$.

Theorem 5.8 *For any $f \in B[a,b]$ we have*

$$E_M(f)([a,b]) \geq \frac{d_M(f)([a,b])}{6}.$$

Definition 5.8 Let $f : [a,b] \to \mathbf{R}$ be bounded on $[a,b]$. The quantities

$$d_{CONC}(f)([a,b]) = \sup\{K(f)(x) - f(x); x \in [a,b]\}$$

and

$$d_{CONV}(f)([a,b]) = \sup\{f(x) - k(f)(x); x \in [a,b]\}$$

are called defect of concavity and of convexity of f on $[a,b]$, respectively, where $K(f) : [a,b] \to \mathbf{R}$ is the least concave majorant of f and $k(f) : [a,b] \to \mathbf{R}$ is the greatest convex minorant of f on $[a,b]$.

Theorem 5.9 *Let $f \in B[a,b]$. Then:*

(*i*) $d_{CONC}(f)([a,b]) = 0$ *if and only if f is concave on* $[a,b]$;

(*ii*) $d_{CONV}(f)([a,b]) = 0$ *if and only if f is convex on* $[a,b]$.

Definition 5.9 Let $p \in \mathbf{N}$ and $f : [a,b] \to \mathbf{R}$ be such that the derivative of order p, $f^{(p)}(x)$, exists, Lebesgue measurable and bounded on $[a,b]$ and

$\int_a^b \left|f^{(p)}(x)\right| dx \neq 0$. The quantities

$$K_+^{(p)}(f)([a,b]) = \frac{\int_a^b f_+^{(p)}(x)\,dx}{\int_a^b \left|f^{(p)}(x)\right| dx}$$

and

$$K_-^{(p)}(f)([a,b]) = \frac{\int_a^b f_-^{(p)}(x)\,dx}{\int_a^b \left|f^{(p)}(x)\right| dx}$$

are called degree of convexity (or defect of concavity) of order p of f on $[a,b]$, respectively. Here $f_+^{(p)}(x) = f^{(p)}(x)$ if $f^{(p)}(x) \geq 0, f_+^{(p)}(x) = 0$ if $f^{(p)}(x) < 0$ and $f_-^{(p)}(x) = -f^{(p)}(x)$ if $f^{(p)}(x) \leq 0, f_-^{(p)}(x) = 0$ if $f^{(p)}(x) > 0$.

Theorem 5.10 *Let $f : [a,b] \to \mathbf{R}$ be satisfying the conditions in Definition 5.9. We have:*

(i) $K_+^{(p)}(f)([a,b]), K_-^{(p)}(f)([a,b]) \geq 0$ and $K_+^{(p)}(f)([a,b]) + K_-^{(p)}(f)([a,b]) = 1$.

(ii) If $f^{(p)}(x) \geq 0$, a.e. $x \in [a,b]$ then $K_+^{(p)}(f)([a,b]) = 1$, if $f^{(p)}(x) \leq 0$, a.e. $x \in [a,b]$ then $K_-^{(p)}(f)([a,b]) = 1$.

(iii) Let us suppose, in addition, that $f^{(p)}$ is continuous on $[a,b]$. Then f is convex (concave) of order p on $[a,b]$ if and only if $K_+^{(p)}(f)([a,b]) = 1$ ($K_-^{(p)}(f)([a,b]) = 1$, respectively).

Further we study two concepts that measure the "quality" of an inequality: the absolute defect of equality and the averaged defect of equality.

Definition 5.10 Let L, R be two functions, $L : D_1 \to \mathbf{R}, R : D_2 \to \mathbf{R}$, where $D_1, D_2 \subseteq \mathbf{R}^n, n \in \mathbf{N}$ such that

$$L(x) \leq R(x), \forall x \in D \subseteq D_1 \cap D_2.$$

The absolute defect of equality for this inequality is defined by

$$d_D^{ab}(L,R) = \sup\{R(x) - L(x); x \in D\}.$$

If, in addition, D is Lebesgue measurable and $L(x), R(x)$ are continuous on D then the averaged defect of equality on $[-r,r]^n$ for the above inequality is defined by

$$d_D^{av}(L,R)(r) = \frac{\int\limits_{D\cap[-r,r]^n} (R(x) - L(x))\mathrm{d}\mu_n}{\mu_n(D\cap[-r,r]^n)},$$

where $[-r,r]^n = \underbrace{[-r,r]\times ... \times[-r,r]}_{n}$ and μ_n is the Lebesgue measure on $\mathbf{R}^n$.

Theorem 5.11 *(i) If $L(x)$ and $R(x)$ are given by Definition 5.10 then we have:*

$$\begin{aligned} d_D^{ab}(L+Q,R+Q) &= d_D^{ab}(L,R), \\ d_D^{av}(L+Q,R+Q)(r) &= d_D^{av}(L,R)(r), \forall r>0 \end{aligned}$$

for any $Q: D\to\mathbf{R}$ and

$$\begin{aligned} d_D^{ab}(kL,kR) &= kd_D^{ab}(L,R), \\ d_D^{av}(kL,kR)(r) &= kd_D^{av}(L,R)(r), \forall r>0 \end{aligned}$$

for any constant $k\geq 0$.

(ii) If $D\subseteq\mathbf{R}^n, 0\leq L_1(x)\leq R_1(x), \forall x\in D$ and $0\leq L_2(x)\leq R_2(x), \forall x\in D$ then

$$d_D^{ab}(L_1L_2,R_1R_2)\leq d_D^{ab}(L_1,R_1)\sup_{x\in D}L_2(x) + d_D^{ab}(L_2,R_2)\sup_{x\in D}R_1(x)$$

and

$$d_D^{av}(L_1L_2,R_1R_2)(r)\leq d_D^{av}(L_1,R_1)(r)\sup_{x\in D}L_2(x)+d_D^{av}(L_2,R_2)(r)\sup_{x\in D}R_1(x)$$

for every $r>0$.

(iii) $d_D^{av}(L,R)(r)\leq d_{D\cap[-r,r]^n}^{ab}(L,R)$.

(iv) $A(x)\leq B(x)\leq C(x), \forall x\in D$ implies $d_D^{ab}(A,C)\geq d_D^{ab}(B,C)$ and $d_D^{av}(A,C)(r)\geq d_D^{av}(B,C)(r), \forall r>0$.

Also, many concrete examples of inequalities are studied.

Open problem 5.2 A central problem in approximation theory is that of shape preserving approximation by operators. One of the most known result in this sense is that the Bernstein polynomials preserve the convexity of order p of f, for any $p \in \mathbf{N}$. If, for example, $f : [0,1] \to \mathbf{R}$ is monotone nondecreasing or convex, etc., then the Bernstein polynomials

$$B_n(f)(x) = \sum_{k=0}^{n} C_n^k x^k (1-x)^{n-k} f\left(\frac{k}{n}\right)$$

are monotone nondecreasing, or convex, etc., respectively.
But if f is not, for example, monotone on $[0,1]$, then it is natural to ask how much of its degree of monotonicity is preserved by the Bernstein polynomials $B_n(f)(x)$. More exactly, it is an open question if there exists a constant $r \geq 1$ (independent of at least n, but possibly independent of f too), such that

$$d_M(B_n(f))([0,1]) \leq r d_M(f)([0,1]), \forall n \in \mathbf{N},$$

where d_M is the defect of monotonicity in Definition 5.7.
Similarly, are open questions if there exist constants $r, s, t, u, v \geq 1$ (independent of at least n, but possibly independent of f too), such that

$$\begin{aligned} d_{CONV}(B_n(f))([0,1]) &\leq r d_{CONV}(f)([0,1]), \forall n \in \mathbf{N}, \\ K_-^{(0)}(B_n(f))([0,1]) &\leq s K_-^{(0)}(f)([0,1]), \forall n \in \mathbf{N}, \\ K_-^{(1)}(B_n(f))([0,1]) &\leq t K_-^{(1)}(f)([0,1]), \forall n \in \mathbf{N}, \\ d_{CONV}^*(B_n(f))([0,1]) &\leq u d_{CONV}^*(f)([0,1]), \forall n \in \mathbf{N}, \end{aligned}$$

and

$$d_{LIN}(B_n(f))([0,1]) \leq v d_{LIN}(f)([0,1]), \forall n \in \mathbf{N}.$$

1.6 On Chapter 6: Defect of Property in Functional Analysis

This chapter introduces and studies some defects of property in functional analysis: defect of orthogonality, defect of convexity, of linearity, of balancing for sets, defect of subadditivity (additivity), of convexity for functionals, defect of symmetry, of normality, of idempotency, of permutability for linear operators, defect of fixed point.

Let $(E, \|\cdot\|)$ be a real normed space.

Definition 6.1 Let $X, Y \subseteq E$. We say that X is orthogonal to Y in the:

(*i*) a-isosceles sense if

$$\|x - ay\| = \|x + ay\|, \forall x \in X, \forall y \in Y,$$

where $a \in \mathbf{R} \setminus \{0\}$ is fixed. We write $X \perp_I Y$ (see James [107]).

(*ii*) a-Pythagorean sense if

$$\|x - ay\|^2 = \|x\|^2 + a^2 \|y\|^2, \forall x \in X, \forall y \in Y,$$

where $a \in \mathbf{R} \setminus \{0\}$ is fixed. We write $X \perp_P Y$ (see James [107]).

(*iii*) Birkhoff sense if

$$\|x\| \leq \|x + \lambda y\|, \forall \lambda \in \mathbf{R}, \forall x \in X, \forall y \in Y.$$

We write $X \perp_B Y$ (see Birkhoff [39]).

(*iv*) Diminnie-Freese-Andalafte sense if

$$(1 + a^2) \|x + y\|^2 = \|ax + y\|^2 + \|x + ay\|^2, \forall x \in X, \forall y \in Y,$$

where $a \in \mathbf{R} \setminus \{1\}$ is fixed. We write $X \perp_{DFA} Y$ (see Diminnie-Freese-Andalafte [63]).

(*v*) Kapoor-Prasad sense if

$$\|ax + by\|^2 + \|x + y\|^2 = \|ax + y\|^2 + \|x + by\|^2, \forall x \in X, \forall y \in Y,$$

where $a, b \in (0, 1)$ are fixed. We write $X \perp_{KP} Y$ (see Kapoor-Prasad [113]).

(*vi*) Singer sense if $\|x\| \cdot \|y\| = 0$ or

$$\left\| \frac{x}{\|x\|} + \frac{y}{\|y\|} \right\| = \left\| \frac{x}{\|x\|} - \frac{y}{\|y\|} \right\|, \forall x \in X, \forall y \in Y.$$

We write $X \perp_S Y$ (see Singer [197]).

(*vii*) usual sense if

$$\langle x, y \rangle = 0, \forall x \in X, \forall y \in Y,$$

where, in addition, the norm $\|\cdot\|$ is generated by the inner product $\langle \cdot, \cdot \rangle$. We write $X \perp Y$.

Corresponding to the concepts of orthogonality in Definition 6.1, we introduce the following

Definition 6.2 Let $(E, \|\cdot\|)$ be a real normed space and $X, Y \subseteq E$. We call defect of orthogonality of X with respect to Y, of:

(i) a-isosceles kind, $a \in \mathbf{R} \setminus \{0\}$ fixed, the quantity

$$d_I^{\perp}(X, Y; a) = \sup_{x \in X, y \in Y} \left| \|x - ay\|^2 - \|x + ay\|^2 \right|.$$

(ii) a-Pythagorean kind, $a \in \mathbf{R} \setminus \{0\}$ fixed, the quantity

$$d_P^{\perp}(X, Y; a) = \sup_{x \in X, y \in Y} \left| \|x\|^2 + a^2 \|y\|^2 - \|x - ay\|^2 \right|.$$

(iii) Birkhoff kind, the quantity

$$d_B^{\perp}(X, Y) = \sup_{x \in X, y \in Y} \sup_{\lambda \in \mathbf{R}} \left\{ \|x\|^2 - \|x + \lambda y\|^2 \right\}.$$

(iv) Diminnie-Freese-Andalafte kind, the quantity

$$d_{DFA}^{\perp}(X, Y; a) = \sup_{x \in X, y \in Y} \left| (1 + a^2) \|x + y\|^2 - \left(\|ax + y\|^2 + \|x + ay\|^2 \right) \right|,$$

where $a \in \mathbf{R} \setminus \{1\}$ is fixed.

(v) Kapoor-Prasad kind, the quantity

$$\begin{aligned} d_{KP}^{\perp}(X, Y; a, b) = \sup_{x \in X, y \in Y} & \left| \|ax + by\|^2 + \|x + y\|^2 \right. \\ & - \left. \left(\|ax + y\|^2 + \|x + by\|^2 \right) \right|, \end{aligned}$$

where $a, b \in (0, 1)$ are fixed.

(vi) Singer kind, the quantity $d_S^{\perp}(X, Y)$ defined by $d_S^{\perp}(X, Y) = 0$ if $X = \{0\}$ or $Y = \{0\}$ and

$$d_S^{\perp}(X, Y) = \sup_{x \in X \setminus \{0\}, y \in Y \setminus \{0\}} \left| \left\| \frac{x}{\|x\|} + \frac{y}{\|y\|} \right\|^2 - \left\| \frac{x}{\|x\|} - \frac{y}{\|y\|} \right\|^2 \right|,$$

contrariwise.

(vii) usual kind

$$d^{\perp}(X, Y) = \sup_{x \in X, y \in Y} |\langle x, y \rangle|,$$

if, in addition, the norm $\|\cdot\|$ is generated by the inner product $\langle \cdot, \cdot \rangle$ on E.

Theorem 6.1 *$X \perp_* Y$ if and only if $d_*^{\perp}(X, Y) = 0$, where $*$ represents any kind of orthogonality in Definition 6.1.*

Theorem 6.2 *Let $(E, \langle \cdot, \cdot \rangle)$ be a real inner product space endowed with the norm $\|x\| = \sqrt{\langle x, x \rangle}$, $x \in E$. We have:*

(i) $d_I^{\perp}(X,Y;a) = 4|a| \cdot d^{\perp}(X,Y)$.

(ii) $d_P^{\perp}(X,Y;a) = 2|a| \cdot d^{\perp}(X,Y)$.

(iii) $d_B^{\perp}(X,Y) = 0$ *if* $Y = \{0\}$ *and* $d_B^{\perp}(X,Y) = \sup_{x\in X, y\in Y\setminus\{0\}} \frac{\langle x,y\rangle^2}{\langle y,y\rangle}$ $\leq \sup_{x\in X} \|x\|^2 \leq (\mathrm{diam}(X))^2$ *if* $Y \neq \{0\}$.

(iv) $d_{DFA}^{\perp}(X,Y;a) = 2(a-1)^2 \cdot d^{\perp}(X,Y)$.

(v) $d_{KP}^{\perp}(X,Y;a,b) = 2(1-a)(1-b) \cdot d^{\perp}(X,Y)$.

(vi) $d_S^{\perp}(X,Y) = 0$ *if* $X = \{0\}$ *or* $Y = \{0\}$ *and* $d_S^{\perp}(X,Y) = d^{\perp}(X',Y')$, *where* $X' = P(X), Y' = P(Y), P(x) = \frac{x}{\|x\|}$ *(i. e.* X' *and* Y' *are the projections of* X *and* Y *on the unit sphere* $\{u \in E; \|u\| = 1\}$*).*

Theorem 6.3 *Let* $(E, \|\cdot\|)$ *be a real normed space.* E *is an inner product space if and only if for all* $X, Y \subset E$ *we have*

$$d_I^{\perp}(X,Y;1) = 2d_P^{\perp}(X,Y;1)$$

and

$$d_I^{\perp}(X,Y;-1) = 2d_P^{\perp}(X,Y;-1).$$

Remark. From the proof of Theorem 6.3 it easily follows that $(E, \|\cdot\|)$ is an inner product space if and only if, for all $X, Y \subset E$, we have

$$2d_P^{\perp}(X,Y;1) \leq d_I^{\perp}(X,Y;1)$$

and

$$2d_P^{\perp}(X,Y;-1) \leq d_I^{\perp}(X,Y;-1).$$

Concerning the defect of orthogonality of Cartesian product we present

Theorem 6.4 *Let* $(E_1, \|.\|_1), (E_2, \|\cdot\|_2)$ *be two real normed linear spaces and let us introduce on* $E_1 \times E_2$ *the norm* $\|(u_1,u_2)\| = \sqrt{\|u_1\|_1^2 + \|u_2\|_2^2}$, $\forall (u_1,u_2) \in E_1 \times E_2$. *For all* $X_1, Y_1 \subseteq E_1, X_2, Y_2 \subseteq E_2$, *we have*

$$d_*^{\perp}(X_1 \times X_2, Y_1 \times Y_2) \leq d_*^{\perp}(X_1,Y_1) + d_*^{\perp}(X_2,Y_2),$$

for any kind " $*$ " of orthogonality in Definition 6.1, excepting Singer orthogonality.

Next we present some applications.

Definition 6.3 Let $(E, \langle\cdot,\cdot\rangle)$ be a real inner product space and $X \subseteq E$. We call defect of orthogonality of X, the quantity

$$d^{\perp}(X) = \sup\{|\langle x,y\rangle|; x, y \in X, x \neq y\}.$$

Remark. Obviously X is orthogonal system in E if and only if $d^{\perp}(X) = 0$.

Theorem 6.8 *Let* $(E, \langle\cdot,\cdot\rangle)$ *be real inner product space and* $X = \{x_1, ..., x_n\} \subset E$ *with* $\langle x_k, x_k\rangle = 1, \forall k \in \{1, ..., n\}$. *Then for any* $x \in E$ *the following estimate*

$$\left| \|x - s_n\|^2 - \left(\|x\|^2 - \sum_{i=1}^{n} |c_i|^2 \right) \right| \leq d^{\perp}(X) \cdot \sum_{i,j=1, i \neq j}^{n} |c_i| \cdot |c_j|$$

holds, where $s_n = \sum_{k=1}^{n} c_k x_k, c_k = \langle x, x_k\rangle$.

Corollary 6.1 *In the hypothesis of Theorem 6.8, for any* $x \in E$ *we have*

$$\left| \|x - s_n\|^2 - \left(\|x\|^2 - \sum_{i=1}^{n} |c_i|^2 \right) \right| \leq d^{\perp}(X) \cdot \|x\|^2 \cdot n(n-1).$$

Corollary 6.2 *If* $X = \{x_1, ..., x_n, ...\}$ *is countable, such that* $\langle x_i, x_i\rangle = \frac{1}{i^4}, i \in \mathbf{N}$, *then*

$$\left| \left\| x - \sum_{k=1}^{\infty} c_k x_k \right\|^2 - \left(\|x\|^2 - \sum_{i=1}^{\infty} |c_i|^2 \right) \right| \leq d^{\perp}(X) \cdot \|x\|^2 \cdot \frac{\pi^4}{36}.$$

Remark. If X is orthonormal then $d^{\perp}(X) = 0$ and the inequality in Theorem 6.8 becomes the well-known equality $\|x - s_n\|^2 = \|x\|^2 - \sum_{i=1}^{n} |c_i|^2$.

Definition 6.4 Let $(E, \|\cdot\|)$ be a real normed space, $G \subset E$ and $x \in E$ be fixed. We say that $g_0 \in G$ is element of best approximation of x by elements by G, with respect to the orthogonality $\perp_*$ if $x - g_0 \perp_* G$, where $\perp_*$ can be any orthogonality in Definition 6.1.

Theorem 6.9 *Let us suppose that* $G \subset E$ *is a linear subspace of* E *and* $x \in E \backslash G$. *If there exists* $g_0 \in G$, *such that for a given* $a \in \mathbf{R} \backslash \{0\}$, *we have* $x - g_0 \perp_P G$ *(i.e. if* g_0 *is element of best approximation of* x *with respect to* $\perp_P$*), then* g_0 *is element of best approximation in usual sense (that is,* $\|x - y_0\| = \inf\{\|x - g\|; g \in G\}$*),* $x - g_0 \perp_R g_0$ *(*$\perp_R$ *is orthogonality in Roberts sense (see e.g Singer [198], p.86)) and* $\|g_0\| \leq \|x\|, \|x - 2g_0\| = \|x\|$.

Theorem 6.10 *Let* $(E, \|\cdot\|)$ *be a real normed space,* $f \in E^*$, $H = \{y \in E; f(y) = 0\}$. *If there exists* $z \in E$ *satisfying* $|f(z)| = 1$ *and* $z \perp_P H$ *for* $a = 1$, *(that is, if* 0 *is best approximation of* z *by elements in* H, *with respect to* $\perp_P$ *for* $a = 1$*), then for all* $x \in E \setminus G$ *with* $|f(x)| < 1$, *there exists element of best approximation in* H*, in classical sense (i.e.* $\exists g_0 \in H$ *with* $\|x - g_0\| = \inf\{\|x - g\|; g \in H\}$*).*

Definition 6.5 The quantity $0 < E_*^{\perp}(x) = \inf\{d_*^{\perp}(x - g, G); g \in G\}$ will be called almost best approximation of x with respect to $\perp_*$ and an element $g_0 \in G$ with $E_*^{\perp}(x) = d_*^{\perp}(x - g_0, G)$ will be called element of almost best approximation of x (by elements in G), with respect to $\perp_*$. If, in addition, $E_*^{\perp}(x) = d_*^{\perp}(x - g_0, G) = 0$, then g_0 is element of best approximation defined by Definition 6.4.

Theorem 6.11 *Let* $(E, \|\cdot\|)$ *be a real normed space,* $G \subset E$ *be compact and* $x \in E$ *(*$x \in E \backslash G$ *for* $\perp_S$*). Then for each* $\perp_*$ *in Definition 6.1 (excepting* $\perp_B$*), there exists* $g^* \in G$ *such that* $E_*^{\perp}(x) = d_*^{\perp}(x - g^*, G)$*.*

Let $(X, \|\cdot\|)$ be a real normed space.

Definition 6.7 A subset $Y \subseteq X$ is called:

(i) convex, if $\lambda y_1 + (1 - \lambda) y_2 \in Y, \forall y_1, y_2 \in Y, \lambda \in [0, 1]$.

(ii) linear, if

$$\alpha y_1 + \beta y_2 \in Y, \forall y_1, y_2 \in Y, \alpha, \beta \in \mathbf{R}.$$

(iii) balanced, if

$$\alpha y \in Y, \forall y \in Y, |\alpha| \leq 1.$$

(iv) absorbent, if

$$\forall x \in X, \exists \lambda > 0 \text{ such that } x \in \lambda Y.$$

Remark. The following characterizations also are well-known:
Y is convex if and only if $Y = convY$, where

$$convY = \left\{ \sum_{i=1}^{n} \alpha_i y_i; n \in \mathbf{N}, y_i \in Y, \alpha_i \geq 0, i \in \{1, ..., n\}, \sum_{i=1}^{n} \alpha_i = 1 \right\}$$

represents the convex hull of Y;
Y is linear if and only if $Y = spanY$, where

$$spanY = \left\{ \sum_{i=1}^{n} \alpha_i y_i; n \in \mathbf{N}, y_i \in Y, \alpha_i \in \mathbf{R}, i \in \{1, ..., n\} \right\};$$

Y is balanced if and only if $Y = balY$, where

$$balY = \{\alpha y; y \in Y, |\alpha| \leq 1\}.$$

Definition 6.8 Let $(X, \|\cdot\|)$ be a real normed space and $D(A, B)$ be a certain distance between the subsets $A, B \subset X$ (D will be specified later). For $Y \subseteq X$, we call:

(i) defect of convexity of Y (with respect to D), the quantity

$$d_{CONV}(D)(Y) = D(Y, convY);$$

(ii) defect of linearity of Y, the quantity

$$d_{LIN}(D)(Y) = D(Y, spanY);$$

(iii) defect of balancing of Y, the quantity

$$d_{BAL}(D)(Y) = D(Y, balY);$$

(iv) defect of absorption of bounded Y, the quantity $d_{ABS}(D)(Y) = 0$, if Y is absorbent and

$$d_{ABS}(D)(Y) = D\left(Y, \overline{B}(0, R_Y)\right)$$

if Y is not absorbent, where

$$\begin{aligned} R_Y &= \sup\{\|y\|; y \in Y\}, \\ \overline{B}(0, R_Y) &= \{x \in X; \|x\| \leq R_Y\}. \end{aligned}$$

A natural candidate for D might be the Hausdorff-Pompeiu distance

$$D_H(Y_1, Y_2) = \max\{d(Y_1, Y_2), d(Y_2, Y_1)\},$$

where $d(Y_1, Y_2) = \sup\{d(y_1, Y_2); y_1 \in Y_1\}$,
$d(y_1, Y_2) = \inf\{\|y_1 - y_2\|; y_2 \in Y_2\}$. In this case we present

Theorem 6.13 *Let $(X, \|\cdot\|)$ be a real normed space and let us suppose that $Y \subseteq X, Y \neq \emptyset$ is closed. Then:*

(i) Y is convex if and only if $d_{CONV}(D_H)(Y) = 0$.

(ii) Y is linear if and only if $d_{LIN}(D_H)(Y) = 0$.

(iii) Y is balanced if and only if $d_{BAL}(D_H)(Y) = 0$.

(iv) bounded Y is absorbent if and only if $d_{ABS}(D_H)(Y) = 0$.

Theorem 6.14 *Let $(X, \|\cdot\|)$ be a real normed space. For all bounded sets $Y, Y_1, Y_2 \subset X$, with $Y_1, Y_2 \neq \emptyset$, we have*

(*i*)

$$\begin{aligned} d_{LIN}(D_H)(Y) &= d_{LIN}(D_H)(\overline{Y}); \\ d_{LIN}(D_H)(Y) &= 0 \textit{ if and only if } \overline{Y} \textit{ is linear;} \\ d_{LIN}(D_H)(\lambda Y) &= \lambda d_{LIN}(D_H)(Y), \forall \lambda > 0; \\ d_{LIN}(D_H)(Y_1 + Y_2) &\leq d_{LIN}(D_H)(Y_1) + d_{LIN}(D_H)(Y_2). \end{aligned}$$

(*ii*)

$$\begin{aligned} d_{BAL}(D_H)(Y) &= d_{BAL}(D_H)(\overline{Y}); \\ d_{BAL}(D_H)(Y) &= 0 \textit{ if and only if } \overline{Y} \textit{ is balanced;} \\ d_{BAL}(D_H)(\lambda Y) &= \lambda d_{BAL}(D_H)(Y), \forall \lambda > 0; \\ d_{BAL}(D_H)(Y_1 + Y_2) &\leq d_{BAL}(D_H)(Y_1) + d_{BAL}(D_H)(Y_2); \\ D_H(balY_1, balY_2) &\leq D_H(Y_1, Y_2); \end{aligned}$$

$$|d_{BAL}(D_H)(Y_1) - d_{BAL}(D_H)(Y_2)| \leq 2D_H(Y_1, Y_2).$$

(*iii*)

$$\begin{aligned} d_{ABS}(D_H)(Y) &= d_{ABS}(D_H)(\overline{Y}); \\ d_{ABS}(D_H)(\lambda Y) &= \lambda d_{ABS}(D_H)(Y), \forall \lambda > 0; \end{aligned}$$

$$\begin{aligned} |d_{ABS}(D_H)(Y_1) - d_{ABS}(D_H)(Y_2)| &\leq 2D_H(Y_1, Y_2) \\ &+ D_H\left(\overline{B}(0, R_{Y_1}), \overline{B}(0, R_{Y_2})\right). \end{aligned}$$

For $Y \in \mathcal{P}_b(Y) = \{Y \subset X; Y \text{ is bounded}\}$, let us denote

$$\begin{aligned} E_{CONV}(Y) &= \inf\{D_H(Y, A); A \in \mathcal{P}_b(X), A \text{ convex}\}, \\ E_{BAL}(Y) &= \inf\{D_H(Y, A); A \in \mathcal{P}_b(X), A \text{ balanced}\}, \end{aligned}$$

the best approximation of a bounded set by convex bounded and by balanced bounded sets, respectively.

Theorem 6.19 *Let* $(X, \|\cdot\|)$ *be a real normed space. Then for any* $Y \in \mathcal{P}_b(X)$ *we get*

$$\begin{aligned} E_{CONV}(Y) &\geq \frac{1}{2} d_{CONV}(D_H)(Y), \\ E_{BAL}(Y) &\geq \frac{1}{2} d_{BAL}(D_H)(Y). \end{aligned}$$

Definition 6.9 Let $(X, +, \cdot)$ be a real linear space.

(i) Let $Y \subset X$ be convex. Then $f : Y \to \mathbf{R}$ is called convex if

$$f(\alpha_1 y_1 + \alpha_2 y_2) \leq \alpha_1 f(y_1) + \alpha_2 f(y_2), \forall y_i \in Y, \alpha_i \geq 0, i = 1, 2, \alpha_1 + \alpha_2 = 1$$

(It is well-known that this is equivalent with $f\left(\sum_{i=1}^{n} \alpha_i y_i\right) \leq \sum_{i=1}^{n} \alpha_i f(y_i)$, $\forall n \in \mathbf{N}$, $\forall y_i \in Y, \alpha_i \geq 0, i \in \{1, ..., n\}, \sum_{i=1}^{n} \alpha_i = 1$);

(ii) $f : X \to \mathbf{R}$ is called subadditive if

$$f(x + y) \leq f(x) + f(y), \forall x, y \in X;$$

(iii) $f : X \to \mathbf{R}$ is called positive homogeneous if

$$f(\lambda x) = \lambda f(x), \forall \lambda \geq 0, \forall x \in X;$$

(iv) $f : X \to \mathbf{R}$ is called absolute homogeneous if

$$f(\lambda x) = |\lambda| f(x), \forall \lambda \in \mathbf{R}, \forall x \in X;$$

(v) $f : X \to \mathbf{R}$ is called sublinear if it is subadditive and positive homogeneous;

(vi) $f : X \to \mathbf{R}$ is called quasi-seminorm if it is sublinear and $f(x) \geq 0$, $\forall x \in X$;

(vii) $f : X \to \mathbf{R}$ is seminorm if it is subadditive and absolute homogeneous;

$(viii)$ $f : X \to \mathbf{R}$ is norm if it is seminorm and $f(x) = 0$ implies $x = 0$.

Concerning these properties we can introduce the following defects.

Definition 6.10 Let $(X, +, \cdot)$ be a real linear space.

(i) If $Y \subset X$ is convex and $f : Y \to \mathbf{R}$, then the n-defect of convexity of f on Y, $n \geq 2$, is defined by

$$d_{CONV}^{(n)}(f)(Y) = \sup\left\{ f\left(\sum_{i=1}^{n} \alpha_i y_i\right) - \sum_{i=1}^{n} \alpha_i f(y_i), y_i \in Y, \alpha_i \geq 0, i \in \{1, ..., n\}, \sum_{i=1}^{n} \alpha_i = 1 \right\}.$$

(ii) Let $Y \subset X$ be a linear subspace of X and $f : Y \to \mathbf{R}$. The defect of subadditivity of f on Y is defined by

$$d_{SADD}(f)(Y) = \sup\{f(y_1 + y_2) - (f(y_1) + f(y_2)); y_1, y_2 \in Y\}.$$

The defect of absolute homogeneity of f on Y is defined by

$$d_{AH}(f)(Y) = \sup\{|f(\lambda y) - |\lambda| f(y)|; \lambda \in \mathbf{R}, y \in Y\}.$$

Theorem 6.20 *Let $(X, +, \cdot)$ be a real linear space.*

(i) Let $Y \subset X$ be convex and $f : Y \to \mathbf{R}$. Then for all $n \geq 2$ we have

$$d^{(n)}_{CONV}(f)(Y) \leq d^{(n+1)}_{CONV}(f)(Y) \leq d^{(n)}_{CONV}(f)(Y) + d^{(2)}_{CONV}(f)(Y).$$

Also, f is convex on Y if and only if for any fixed $n \geq 2$, we have $d^{(n)}_{CONV}(f)(Y) = 0$.
For $f, g : Y \to \mathbf{R}, \alpha \geq 0$ and $n \geq 2$ we have

$$d^{(n)}_{CONV}(f+g)(Y) \leq d^{(n)}_{CONV}(f)(Y) + d^{(n)}_{CONV}(g)(Y)$$

and

$$d^{(n)}_{CONV}(\alpha f)(Y) = \alpha d^{(n)}_{CONV}(f)(Y).$$

(ii) Let $Y \subset X$ be linear subspace of X and $f : Y \to \mathbf{R}$. If $f(0) = 0$ then f is subadditive (on Y) if and only if $d_{SADD}(f)(Y) = 0$. f is absolute homogeneous (on Y) if and only if $d_{AH}(f)(Y) = 0$. Moreover, if $f, g : Y \to \mathbf{R}$ then

$$\begin{aligned}
d_{SADD}(f+g)(Y) &\leq d_{SADD}(f)(Y) + d_{SADD}(g)(Y), \\
d_{AH}(f+g)(Y) &\leq d_{AH}(f)(Y) + d_{AH}(g)(Y), \\
d_{SADD}(\alpha f)(Y) &= \alpha d_{SADD}(f)(Y), \forall \alpha \geq 0, \\
d_{AH}(\alpha f)(Y) &= |\alpha| d_{AH}(f)(Y), \forall \alpha \in \mathbf{R}.
\end{aligned}$$

For $Y \subset X$ linear subspace or convex set, let us denote

$$\begin{aligned}
B_0(Y) &= \{f : Y \to \mathbf{R}; f \text{ is bounded on } Y \text{ and } f(0) = 0\}, \\
E_{SADD}(f)(Y) &= \inf\{\|f - g\|; g \in B_0(Y), g \text{ is subadditive on } Y\},
\end{aligned}$$

$f \in B_0(Y)$ and

$$\begin{aligned}
B(Y) &= \{f : Y \to \mathbf{R}; f \text{ is bounded on } Y\}, \\
E_{CONV}(f)(Y) &= \inf\{\|f - g\|; g \in B(Y), g \text{ is convex on } Y\},
\end{aligned}$$

$f \in B(Y)$, respectively, where $\|f\| = \sup\{|f(x)|; x \in Y\}$.

Theorem 6.21 *(i) Let $Y \subset X$ be linear subspace of X and $f \in B_0(Y)$. Then*

$$E_{SADD}(f)(Y) \geq \frac{d_{SADD}(f)(Y)}{3}.$$

(ii) Let $Y \subset X$ be convex and $f \in B(Y)$. Then

$$E_{CONV}(f)(Y) \geq \frac{d_{CONV}^{(n)}(f)(Y)}{2}, \forall n \geq 2.$$

Remark. By Theorems 6.20, (i) and 6.21, (ii), there exists the limit $\lim_{n\to\infty} d_{CONV}^{(n)}(f)(Y)$ and moreover

$$\frac{1}{2}\lim_{n\to\infty} d_{CONV}^{(n)}(f)(Y) \leq E_{CONV}(f)(Y).$$

Let $(X, \langle \cdot, \cdot \rangle)$ be a Hilbert space over $\mathbf{R}$ or $\mathbf{C}$ and

$$LC(X) = \{A : X \to X; A \text{ is linear and continuous on } X\}.$$

For any $A \in LC(X)$, we define the norm $|||\cdot||| : LC(X) \to \mathbf{R}_+$ by $|||A||| = \sup\{||A(x)||; ||x|| \leq 1\}$, where $||x|| = \sqrt{\langle x, x \rangle}$.

The following concepts are well-known in functional analysis (see *e.g.* Muntean [155] and Ionescu-Tulcea [103]).

Definition 6.11 The operator $A \in LC(X)$ is called:

(i) symmetric (or Hermitean) if $\langle A(x), y \rangle = \langle x, A(y) \rangle, \forall x, y \in X$;

(ii) normal, if $AA^* = A^*A$, where A^* is the adjoint operator of A defined by $\langle A(x), y \rangle = \langle x, A^*(y) \rangle, \forall x, y \in X$ and $AA^*(x) = A(A^*(x)), \forall x \in X$;

(iii) idempotent, if $A^2 = A$, where $A^2(x) = A(A(x)), \forall x \in X$;

(iv) isometry, if $\langle A(x), A(y) \rangle = \langle x, y \rangle, \forall x, y \in X$ (or equivalently, if $||A(x)|| = ||x||, \forall x \in X$);

Also, two operators $A, B \in LC(X)$ are called permutable if $AB = BA$.

Suggested by these properties, we can introduce the following

Definition 6.12 Let $A, B \in LC(X)$.

(i) The defect of symmetry of A is given by

$$d_{SIM}(A) = \sup\{|\langle A(x), y \rangle - \langle x, A(y) \rangle|; ||x||, ||y|| \leq 1\};$$

(ii) The defect of normality of A is given by

$$d_{NOR}(A) = \sup\{||A(A^*(x)) - A^*(A(x))||; ||x|| \leq 1\} = |||AA^* - A^*A|||;$$

(*iii*) The defect of idempotency of A is defined by

$$d_{IDEM}(A) = \sup\left\{\left\|A^2(x) - A(x)\right\|; \|x\| \leq 1\right\} = \left\|\left|A^2 - A\right|\right\|;$$

(*iv*) The defect of isometry of A is given by

$$d_{ISO}(A) = \sup\left\{\|\|A(x) - x\|; \|x\| \leq 1\right\};$$

(*v*) The defect of permutability of A and B is given by

$$d_{PERM}(A, B) = \sup\left\{\|A(B(x)) - B(A(x))\|; \|x\| \leq 1\right\} = |||AB - BA|||.$$

Also, if $Y \subset LC(X)$, then we can introduce the corresponding defects for Y, by

$$\begin{aligned} D_{SIM}(Y) &= \sup\{d_{SIM}(A); A \in Y\}; \\ D_{NOR}(Y) &= \sup\{|||AA^* - A^*A|||; A \in Y\}; \\ D_{IDEM}(Y) &= \sup\left\{\left\|\left|A^2 - A\right|\right\|; A \in Y\right\}; \\ D_{ISO}(Y) &= \sup\{d_{ISO}(A); A \in Y\}; \end{aligned}$$

and

$$d_{PERM}(Y) = \sup\{|||AB - BA|||; A, B \in Y\}.$$

Theorem 6.22 *Let $A, B \in LC(X)$.*
(*i*) *A is symmetric if and only if $d_{SIM}(A) = 0$;*
(*ii*) *A is normal if and only if $d_{NOR}(A) = 0$;*
(*iii*) *A is idempotent if and only if $d_{IDEM}(A) = 0$;*
(*iv*) *A is isometric if and only if $d_{ISO}(A) = 0$;*
(*v*) *A and B are permutable if and only if $d_{PERM}(A, B) = 0$.*

Theorem 6.23 *Let $A, B \in LC(X)$. We have*
(*i*) $d_{SIM}(\lambda A) = |\lambda| d_{SIM}(A), \forall \lambda \in \mathbf{R}$;
$d_{NOR}(\lambda A) = |\lambda|^2 d_{NOR}(A), \forall \lambda \in \mathbf{R}$ *or* $\mathbf{C}$.
(*ii*) $d_{SIM}(A + B) \leq d_{SIM}(A) + d_{SIM}(B)$;
$d_{SIM}(AB) \leq 2|||AB - I|||$, *where* $I(x) = x, \forall x \in X$;
$d_{NOR}(A + B) \leq d_{NOR}(A) + d_{NOR}(B) + d_{PERM}(A, B^*)$
$+ d_{PERM}(A^*, B)$.

(iii) $d_{SIM}(A^{-1}) = d_{SIM}(A)$, *if there exists* A^{-1} *and* A *is isometry.*

(iv) $|d_{SIM}(A) - d_{SIM}(A^*)| \le 2\,|||A - A^*|||$;
$d_{NOR}(A^*) = d_{NOR}(A)$;
$d_{NOR}(A + A^*) \le d_{NOR}(A) + d_{NOR}(A^*)$;
$|d_{IDEM}(A) - d_{IDEM}(A^*)| \le |||A - A^*||| + \left|\left|\left|A^2 - (A^*)^2\right|\right|\right|$;
$d_{IDEM}(I - A) = d_{IDEM}(A)$;
$d_{PERM}(A, A^*) = d_{NOR}(A)$.

For given $A \in LC(X)$, let us introduce the following quantities of best approximation:

$$E_{SIM}(A) = \inf\{|||A - B|||\,; B \in LC(X), B \text{ is symmetric}\},$$

$$\begin{aligned} E_{NOR}(A) &= \inf\{|||A - B||| + |||A^* - B^*|||\,; B \in LC(X), \\ &\quad |||B||| \le 1, B \text{ is normal}\}, \end{aligned}$$

$$\begin{aligned} E_{IDEM}(A) &= \inf\left\{|||A - B||| + \left|\left|\left|A^2 - B^2\right|\right|\right|\,; B \in LC(X), \\ &\quad B \text{ is idempotent}\right\}, \end{aligned}$$

$$E_{ISO}(A) = \inf\{|||A - B|||\,; B \in LC(X), B \text{ is isometry}\}.$$

Theorem 6.24 *Let* $A \in LC(X)$. *We have:*

$$\begin{aligned} E_{SIM}(A) &\ge \frac{d_{SIM}(A)}{2}, \\ E_{NOR}(A) &\ge \frac{d_{NOR}(A)}{2}, \text{ if } |||A||| \le 1, \\ E_{IDEM}(A) &\ge d_{IDEM}(A), \\ E_{ISO}(A) &\ge d_{ISO}(A). \end{aligned}$$

Definition 6.13 Let (X, d) be a metric space and $M \subseteq X$. The defect of fixed point of $f : M \to X$ is defined by $e_d(f; M) = \inf\{d(x, f(x))\,; x \in M\}$. If there exists $x_0 \in M$ with $e_d(f; M) = d(x_0, f(x_0))$, then x_0 will be called best almost-fixed point for f on M.

Theorem 6.25 *Let* (X, d) *be a metric space and* $f : X \to X$ *be a non-expansive mapping, i.e.* $d(f(x), f(y)) \le d(x, y)$, *for all* $x, y \in X$. *Then* $e_d(f^n; X) \le n e_d(f; X)$ *and* $\inf\{d(f^n(x), f^{n+1}(x))\,; x \in X\} = e_d(f; X)$, $\forall n \in \{1, 2, ...\}$, *where* f^n *denotes the n-th iterate of* f.

Let us denote

$$\mathcal{F} = \{g : X \to X; g \text{ has fixed point in } X\},$$

where (X, d) is supposed to be compact. A natural question is to find the best approximation of a function $f : X \to X, f \notin \mathcal{F}$, by elements in $\mathcal{F}$. In this sense, we can define

$$E_{\mathcal{F}}(f) = \inf \{D(f, g); g \in \mathcal{F}\},$$

where $D(f, g) = \sup \{d(f(x), g(x)); x \in X\}$.

The following lower estimate for $E_{\mathcal{F}}(f)$ holds.

Theorem 6.26 *We have $e_d(f; X) \leq E_{\mathcal{F}}(f)$, for any $f : X \to X$.*

Theorem 6.30 *Let $(X, \langle , \rangle)$ be a real Hilbert space, $\|x\| = \sqrt{\langle x, x \rangle}, x \in X$, the metric generated by norm $\|\cdot\|$ denoted by d and $M \subseteq X$.*

(i) Let $f : M \to X$ be Gâteaux derivable on $x_0 \in M, x_0$ interior point of M, with $e_d(f; M) > 0$. If $\|x_0 - f(x_0)\| = e_d(f; M), x_0 \in M$, then

$$\langle x_0 - f(x_0), (1_X - \nabla f(x_0))(h) \rangle = 0,$$

for all $h \in X$.

(ii) If $F(x) = \|x - f(x)\|, x \in X$ is moreover Gâteaux derivable on $M \subset X$ open convex, then $\|x_0 - f(x_0)\| = e_d(f; M)$ if and only if

$$\langle x_0 - f(x_0), (1_X - \nabla f(x_0))(h) \rangle = 0, \forall h \in X.$$

Let us consider $C[0,1]$ endowed with the uniform metric $d(f, g) = \sup \{|f(x) - g(x)| : x \in [0,1]\}$. Denoting $A : C[0,1] \to C[0,1]$ by

$$(Au)(x) = 1 + \lambda \int_x^1 u(s - x) u(s) \mathrm{d}s,$$

where $\lambda > \frac{1}{2}$, it is known that A has no fixed points in $C[0,1]$. However, we can prove

Theorem 6.32 *Let $A : M \to C[0,1]$, where $\lambda > \frac{1}{2}, A$ is defined as above and*

$$M = \{u \in C[0,1] : u \text{ is derivable}, u'(x) \geq 0, \forall x \in [0,1], \\ 0 \leq u(0) \leq u(1) \leq 1\}.$$

Then $e_d(A;M) = \frac{4\lambda-1}{4\lambda}$ *and a best almost-fixed point for* A *on* M *is* $u:[0,1] \to \mathbf{R}, u(x) = \frac{1}{2\lambda}, \forall x \in [0,1]$.

Open problem 6.1 If $Y \subset X$ is an absorbent subset of the linear space $(X,+,\cdot)$ then the well-known Minkowski's functional attached to Y is defined by

$$p_Y(x) = \inf\{\alpha > 0; x \in \alpha Y\}, x \in X.$$

This functional characterizes the quasi-seminorms and the seminorms as follows (see *e.g.* Muntean [154], p. 43-45):

(*i*) $p : X \to \mathbf{R}$ is quasi-seminorm if and only if there exists $Y \subset X$, absorbent and convex such that $p = p_Y$;

(*ii*) $p : X \to \mathbf{R}$ is seminorm if and only if there exists $Y \subset X$, absorbent convex and balanced such that $p = p_Y$;

In the proof of (*i*), given an absorbent subset $Y \subset X$, the subadditivity of p_Y is essentially a consequence of the convexity of Y (because p_Y is always positive homogeneous if Y is absorbent). Let us suppose, in addition, that $(X, \|\cdot\|)$ is a real normed space. Then, would be natural to search for a relationship between the defect of subadditivity $d_{SADD}(p_Y)(X)$ and the defect of convexity $d_{CONV}(D)(Y)$ (where $D \equiv D_H$ or $D \equiv D^*$) in such a way that for absorbent $Y \subset X, d_{SADD}(p_Y)(X) = 0$ if and only if $d_{CONV}(D)(Y) = 0$. Similarly, in the proof of (*ii*), given an absorbent and convex subset $Y \subset X$, the absolute homogeneity of p_Y is essentially a consequence of the fact that Y is balanced. Therefore, in this case would be natural to search for a relationship between the defect of absolute homogeneity $d_{AH}(p_Y)(X)$ and the defect of balancing $d_{BAL}(D)(Y)$, (where $D \equiv D_H$ or $D \equiv D^*$) in such a way that for absorbent and convex $Y \subset X, d_{AH}(p_Y)(X) = 0$ if and only if $d_{BAL}(D)(Y) = 0$.

1.7 On Chapter 7: Defect of Property in Algebra

Let (X,d) be a metric space and $F : X \times X \to X$ be a binary operation on X. If F is not commutative, or is not associative, or is not distributive (with respect to another binary operation $G : X \times X \to X$), or has no identity element, or no every element has an inverse, so on, it is natural to look for a concept of defect of F with respect to these properties.

It is the main aim of this chapter to introduce and study the concept of defect of F with respect to the above properties.

Definition 7.1 Let (X, d) be a metric space, $Y \subseteq X$ and $F : X \times X \to X$ be a binary operation on X.

(i) The quantity

$$e_{COM}^{d}(F)(Y) = \sup\{d(F(x,y), F(y,x)); x, y \in Y\}$$

is called defect of commutativity of F on Y with respect to the metric d.

(ii) The quantity

$$e_{AS}^{d}(F)(Y) = \sup\{d(F(F(x,y),z), F(x, F(y,z))); x, y, z \in Y\}$$

is called defect of associativity of F on Y with respect to the metric d.

(iii) The quantity $\sup\{d(F(x,y), x); x \in Y\}$ is called defect of identity element at right of y (with respect to F) and

$$e_{IDR}^{d}(F)(Y) = \inf\{\sup\{d(F(x,y), x); x \in Y\}; y \in Y\}$$

is called defect of identity element at right of F on Y (with respect to d). Similarly, we can define the defect of identity element at left,

$$e_{IDL}^{d}(F)(Y) = \inf\{\sup\{d(F(y,x), x); x \in Y\}; y \in Y\}.$$

If there exists $e_R \in Y$ such that

$$e_{IDR}^{d}(F)(Y) = \sup\{d(F(x, e_R), x); x \in Y\} > 0,$$

then e_R will be called best almost-identity element at right of F on Y. Analogously, if there exists $e_L \in Y$ such that

$$e_{IDL}^{d}(F)(Y) = \sup\{d(F(e_L, x), x); x \in Y\} > 0,$$

then e_L will be called best almost-identity element at left of F on Y.

(iv) Let us suppose that there exists $a \in X$ such that $F(x,a) = F(a,x) = x, \forall x \in X$. Then

$$\inf\{d(F(x,y), a); y \in Y\}$$

and

$$\inf\{d(F(y,x), a); y \in Y\}$$

are called defects of invertibility of x at right and at left, respectively. The quantities

$$e^d_{INR}(F)(Y) = \sup\{\inf\{d(F(x,y),a); y \in Y\}; x \in Y\}$$

and

$$e^d_{INL}(F)(Y) = \sup\{\inf\{d(F(y,x),a); y \in Y\}; x \in Y\}$$

are called defects of invertibility of F on Y, at right and at left, respectively. The quantity $e^d_{IN}(F)(Y) = \max\{e^d_{INL}(F)(Y), e^d_{INR}(F)(Y)\}$ is called defect of invertibility of F on Y (with respect to d).
If there exists $x^*_R \in Y$ such that

$$0 < \inf\{d(F(x,y),a); y \in Y\} = d(F(x,x^*_R),a),$$

then x^*_R will be called best almost-inverse at right of x with respect to F. Similarly, if there is $x^*_L \in Y$ with

$$0 < \inf\{d(F(y,x),a); y \in Y\} = d(F(x^*_L,x),a),$$

then x^*_L will be called best almost-inverse at left of x with respect to F.

(v) The quantity

$$e^d_{REL}(F)(Y) = \sup\{d(x,y); x,y \in Y, F(z,x) = F(z,y)\}$$

is called defect of regularity at left of z on Y. Analogously,

$$e^d_{RER}(F)(Y) = \sup\{d(x,y); x,y \in Y, F(x,z) = F(y,z)\}$$

is called defect of regularity at right of z on Y.

(vi) The quantity

$$e^d_{IDEM}(F)(Y) = \sup\{d(F(x,x),x); x \in Y\}$$

is called defect of idempotency of F on Y (with respect to d).

(vii) If $G : X \times X \to X$ is another binary operation on X then

$$e^d_{DISL}(F;G)(Y) = \sup \quad \{d(F(x,G(y,z)), G(F(x,y),F(x,z))); \\ x,y,z \in Y\}$$

is called defect of left-distributivity of F with respect to G on Y. Similarly,

$$e^d_{DISR}(F;G)(Y) = \sup \quad \{d(F(G(y,z),x,),G(F(y,x),F(z,x))); \\ x,y,z \in Y\}$$

is called defect of right-distributivity of F with respect to G on Y. If $F \equiv G$, then $e^d_{DISL}(F;F)(Y)$ and $e^d_{DISR}(F;F)(Y)$ are called defects of autodistributivity (left and right) of F on Y. Also,

$$e^d_{AB}(F;G)(Y) = \max\{\sup\{d(F(a,G(a,b)),a); a,b \in Y\},$$

$$\sup\{d(G(a,F(a,b)),a); a,b \in Y\}\}$$

is called defect of absorption of (F,G) on Y.

Lemma 7.1 *With the notations in Definition 7.1 we have:*

(i) F is commutative on Y if and only if $e^d_{COM}(F)(Y) = 0$.

(ii) F is associative on Y if and only if $e^d_{AS}(F)(Y) = 0$.

(iii) If F has identity element at right in Y, i.e. there is $a \in Y$ such that $F(a,x) = x, \forall x \in Y$, then $e^d_{IDR}(F)(Y) = 0$. Conversely, if (Y,d) is compact and F is continuous on $Y \times Y$ (with respect to the box metric on $Y \times Y$) then $e^d_{IDR}(F)(Y) = 0$ implies that F has identity element at right in Y. Similar results hold in the case of identity element at left.

(iv) If F has identity element in Y and each $x \in Y$ has inverse at right, then $e^d_{INR}(F)(Y) = 0$. Conversely, if (Y,d) is compact and F is continuous on $Y \times Y$ (with respect to the box metric on $Y \times Y$), then $e^d_{INR}(F)(Y) = 0$ implies that each $x \in Y$ has inverse at right. Similar results hold for invertibility at left.

(v) A set $Y \subseteq X$ is called regular at left if each $z \in Y$ is regular at left (i.e. $F(z,x) = F(z,y)$ implies $x = y$). Then Y is regular at left with respect to F if and only if $e^d_{REL}(F)(Y) = 0$. Similar results hold for regularity at right.

(vi) A set $Y \subseteq X$ is called idempotent if each $x \in Y$ is idempotent (with respect to F). Then Y is idempotent with respect to F if and only if $e^d_{IDEM}(F)(Y) = 0$.

(vii) F is left-distributive (right-distributive) with respect to G on Y if and only if $e^d_{DISL}(F;G)(Y) = 0$ ($e^d_{DISR}(F;G)(Y) = 0$). Also, the pair (F,G) has the property of absorption on Y if and only if $e^d_{AB}(F,G)(Y) = 0$.

Definition 7.2 Let $f, g : X \to X$ be with $f \circ g = i$, where i is the identity function on X. If G is a binary operation on X, then $F : X \times X \to X$ given by $F = g \circ G(f,f)$ is called the (f,g)-dual of G (i.e. $F(x,y) = g(G(f(x),f(y))), \forall (x,y) \in X \times X$).

Definition 7.3 Let (X,d) be a metric space and $F, G : X \times X \to X$. We say that F is less than G (we write $F < G$) if F is the (f,g)-dual of G, where $d(g(x), g(y)) \leq kd(x,y), \forall x, y \in X$ and $0 < k < 1$ is a constant.

Theorem 7.2 *With the notations in Definitions 7.1 and 7.3, the condition $F < G$ implies:*

(i) $e^d_{COM}(F)(Y) \leq ke^d_{COM}(G)(f(Y)), \forall Y \subseteq X$. *In particular,* $e^d_{COM}(F)(X) \leq ke^d_{COM}(G)(X)$.

(ii) $e^d_{AS}(F)(Y) \leq ke^d_{AS}(G)(f(Y)), \forall Y \subseteq X$. *In particular,* $e^d_{AS}(F)(X) \leq ke^d_{AS}(G)(X)$.

(iii) Let F_i be the (f,g)-duals of $G_i, i \in \{1,2\}$, such that $F_i < G_i, i \in \{1,2\}$. Then $e^d_{DISL}(F_1;F_2)(Y) \leq ke^d_{DISL}(G_1;G_2)(f(Y)), \forall Y \subseteq X$, *and in particular* $e^d_{DISL}(F_1;F_2)(X) \leq ke^d_{DISL}(G_1;G_2)(X)$. *Similar results hold for* e^d_{DISR}.

Theorem 7.5 *Let $(X, \cdot, d)$ be a metrizable non-commutative group, N a closed normal divisor of X and d a metric that is left invariant. If $\widetilde{d}$ denotes the induced metric on X/N and $\odot$ the induced operation on X/N, then*

$$e^{\widetilde{d}}_{COM}(\odot)(X/N) \leq e^d_{COM}(\cdot)(X).$$

For $F, G : X \times X \to X$, we define the distance between F and G by

$$D(F,G) = \sup\{d(F(x,y), G(x,y)); x, y \in X\}.$$

Also, we define

$$E_{COM}(F) = \inf\{D(F,G); G \in COM(X)\}$$

and

$$E_{AS}(F) = \inf\{D(F,G); G \in AS(X)\},$$

the best approximation of a binary operation on X by commutative operations and by associative operations, respectively.

Theorem 7.6 *Let (X,d) be a metric space with d bounded on X. We have $\frac{1}{2}e^d_{COM}(F) \leq E_{COM}(F)$ and $\frac{1}{2}e^d_{AS}(F) \leq E_{AS}(F)$, for any binary operation $F : X \times X \to X$.*

In Section 7.2 we consider examples of calculations for the defects studied in Section 7.1. Section 7.3 deals with triangular norms and conorms.

Theorem 7.7 *Let T be a triangular norm and S be a triangular conorm. We have:*

(*i*) $e_{IDEM}(T) = \sup\{x - T(x,x); x \in [0,1]\}$;

(*ii*) $e_{IDEM}(S) = \sup\{S(x,x) - x; x \in [0,1]\}$;

(*iii*) *If S is the dual of T, that is* $S(x,y) = 1 - T(1-x, 1-y), \forall x, y \in [0,1]$, *then* $e_{IDEM}(T) = e_{IDEM}(S)$;

(*iv*) *If $T_1 \leq T_2$, that is* $T_1(x,y) \leq T_2(x,y), \forall x,y \in [0,1]$, *then* $e_{IDEM}(T_2) \leq e_{IDEM}(T_1)$;

(*v*) *If $S_1 \leq S_2$, that is* $S_1(x,y) \leq S_2(x,y), \forall x,y \in [0,1]$, *then* $e_{IDEM}(S_1) \leq e_{IDEM}(S_2)$;

(*vi*) *If T^* is the reverse of triangular norm T, that is* $T^*(x,y) = \max(0, x+y-1+T(1-x,1-y))$ *(see Kimberling [117] or Šabo [186]), then* $e_{IDEM}(T^*) \leq e_{IDEM}(T)$.

Theorem 7.11 (*i*) $0 \leq e_{DIS}(F;G) \leq 1$ *for every triangular norms or conorms F and G;*

(*ii*) $e_{DIS}(T;S_M) = 0$, *for every triangular norm T;*

(*iii*) $e_{DIS}(S;T_M) = 0$, *for every triangular conorm S;*

(*iv*) $e_{DIS}(F;F) = e_{IDEM}(F)$ *for every triangular norm or conorm F;*

(*v*) *If S is the dual of T then* $e_{DIS}(T;T) = e_{DIS}(S;S)$.

Chapter 7 ends with some applications.

1.8 On Chapter 8: Miscellaneous

In this chapter we study some defects of property in Complex Analysis, Geometry, Number Theory and Fuzzy Logic.

Pompeiu [163] notes that if $\int_C f(z)\,dz \neq 0$, then $\left|\int_C f(z)\,dz\right|$ can be considered as a measure of non-holomorphy of f inside of the domain bounded by the closed curve C.

Definition 8.1 Let $D \subset \mathbf{C}$ be a bounded domain and $f: D \to \mathbf{C}$, integrable on D. The number

$$d_{HOL}(f)(D) = \sup\left\{\left|\int_C f(z)\,dz\right|; C \subset D, \text{ closed rectifiable curve}\right\}$$

is called defect of holomorphy of f on D.

The following properties are immediate:

$$\begin{aligned} d_{HOL}(f+g)(D) &\leq d_{HOL}(f)(D) + d_{HOL}(g)(D), \forall f,g \in C(D) \\ d_{HOL}(\lambda f)(D) &= |\lambda| \, d_{HOL}(f)(D), \forall \lambda \in \mathbf{C}, f \in C(D). \end{aligned}$$

Continuing the ideas in Pompeiu [163], let us suppose that $f \in C^1(D)$, that is if $f = u + iv$ then u, v are of C^1 class.

Definition 8.2 (Pompeiu [163]) Let $f \in C^1(D)$, $f(z) = u(x,y) + iv(x,y)$, $z = x + iy \in D$ and $z_0 = x_0 + iy_0 \in D$. The areolar derivative of f at z_0 is given by

$$d_A(f)(z_0) = \lim_{C \to z_0} \frac{\frac{1}{2i}\int_C f(z)\,dz}{m(\Delta)},$$

where the limit is considered for all closed curves C (in D) surrounding z_0, that converge to z_0 by a continuous deformation ($m(\Delta)$ represents the area of the domain Δ closed by C).

Remark. Because for $f \in C^1(D)$ and $z_0 \in D$, it is obvious that (see Pompeiu [164])

$$d_A(f)(z_0) = 0 \text{ if and only if } f \text{ is differentiable at } z_0,$$

we can call $|d_A(f)(z_0)|$ as defect of differentiability of $f \in C^1(D)$ on z_0. In Szu-Hoa Min [209], $|d_A(f)(z_0)|$ is called deviation from analiticity.

The defect of differentiability of f on D given by

$$d_A(f)(D) = \sup\{|d_A(f)(z)|; z \in D\},$$

appears in the estimate of $f(z)$ by the Cauchy's integral:

$$\left| f(z) - \frac{1}{2\pi i}\int_C \frac{f(\xi)}{\xi - z} d\xi \right| \leq r d_A(f)(\Delta),$$

if we choose $C = \{u \in \mathbf{C}; |u-a| = r\} \subset D, \Delta = int(C) \subset D$ and $z \in D$ such that $|z - a| \geq 2r$.

In the Euclidean and non-Euclidean geometries, the concept of curvature and torsion of a curve $\mathcal{C}$ in a point M are introduced by (see *e.g.* Mihăileanu [148], p. 99-100)

$$\gamma(M) = \lim_{\Delta s \to 0} \frac{\Delta\alpha}{\Delta s}$$

and

$$\tau(M) = \lim_{\Delta s \to 0} \frac{\Delta \theta}{\Delta s},$$

respectively, where $\Delta\alpha$ is the angle of tangents in M and $M', \Delta\theta$ is the angle of binormals in M and M' and Δs is the length of the arc MM' $(M, M' \in \mathcal{C})$ when M' tends to M.

Definition 8.3 Let $\mathcal{C}$ be a curve. The quantities

$$\gamma(C) = \sup\{|\gamma(M)|; M \in \mathcal{C}\}$$

and

$$\tau(C) = \sup\{|\tau(M)|; M \in \mathcal{C}\}$$

are called the defect of right line and the defect of plane curve of $\mathcal{C}$, respectively.

Definition 8.5 Let ABC be a triangle in a geometry. The quantity

$$D_{\Delta}(ABC) = |A + B + C - \pi|$$

is called defect of Euclidean triangle.

Let us assume that the absolute of a non-Euclidean space is given by

$$q^2 x_0^2 + x_1^2 + x_2^2 + x_3^2 = 0,$$

where $q^2 \in \mathbf{R}$ and let us denote $\varepsilon = \frac{1}{q}$.

The inner product of two points X and Y with coordinates $x_i, i \in \{0,1,2,3\}$ and $y_i, i \in \{0,1,2,3\}$ is given by

$$X \cdot Y = \frac{q^2 x_0 y_0 + x_1 y_1 + x_2 y_2 + x_3 y_3}{\sqrt{q^2 x_0^2 + x_1^2 + x_2^2 + x_3^2}\sqrt{q^2 y_0^2 + y_1^2 + y_2^2 + y_3^2}}$$

and the inner product of two planes $\alpha : \alpha_0 x_0 + \alpha_1 x_1 + \alpha_2 x_2 + \alpha_3 x_3 = 0$, $\beta : \beta_0 x_0 + \beta_1 x_1 + \beta_2 x_2 + \beta_3 x_3 = 0$ is given by

$$\alpha \cdot \beta = \frac{\varepsilon^2 \alpha_0 \beta_0 + \alpha_1 \beta_1 + \alpha_2 \beta_2 + \alpha_3 \beta_3}{\sqrt{\varepsilon^2 \alpha_0^2 + \alpha_1^2 + \alpha_2^2 + \alpha_3^2}\sqrt{\varepsilon^2 \beta_0^2 + \beta_1^2 + \beta_2^2 + \beta_3^2}}.$$

Definition 8.6 (i) The quantity

$$d_{ORTH}(X, Y) = |X \cdot Y|$$

is called defect of orthogonality of the points X and Y.

(*ii*) The quantity

$$d_{ORTH}(\alpha,\beta) = |\alpha \cdot \beta|$$

is called defect of orthogonality of the planes α and β.

Theorem 8.2 (*i*) *$d_{ORTH}(X,Y) = 0$ if and only if X and Y are orthogonal; $d_{ORTH}(\alpha,\beta) = 0$ if and only if α and β are orthogonal.*

(*ii*) *$0 \le d_{ORTH}(X,Y) \le 1$ and $0 \le d_{ORTH}(\alpha,\beta) \le 1$.*

(*iii*) *$d_{ORTH}(X,Y) = d_{ORTH}(Y,X)$; $d_{ORTH}(\alpha,\beta) = d_{ORTH}(\beta,\alpha)$.*

(*iv*) *$d_{ORTH}(X,Y) = \left|\cos\frac{d}{q}\right|$, where d is the distance between X and Y; $d_{ORTH}(\alpha,\beta) = |\cos\theta|$, where θ is the angle of the planes α and β.*

Let us consider two right lines u and v with the common point X and let U, V be the orthogonals of the point X on u and v, respectively (that is $U \in u$ and $V \in v, U \cdot X = 0, V \cdot X = 0$). The angle Δ of the right lines u and v is given by (see Mihăileanu [148], p.33)

$$\cos\Delta = U \cdot V.$$

Definition 8.7 The quantity

$$d_{ORTH}(u,v) = |\cos\Delta|$$

is called defect of orthogonality of the right lines u and v.

Theorem 8.3 (*i*) *$d_{ORTH}(u,v) = 0$ if and only if u and v are orthogonal.*

(*ii*) *$0 \le d_{ORTH}(u,v) \le 1$.*

(*iii*) *$d_{ORTH}(u,v) = d_{ORTH}(v,u)$.*

(*iv*) *$d_{ORTH}(u,v) = \left|\cos\frac{\delta}{q}\right|$, where δ is the distance between u and v.*

(*v*) *If $u : ax + by + cz = 0$ and $v : a'x + b'y + c'z = 0$ are two right lines in plane, then $d_{ORTH}(u,v) = \frac{|aa'+bb'+cc'|}{\sqrt{a^2+b^2+\varepsilon^2c^2}\sqrt{a'^1+b'^2+\varepsilon^2c'^2}}$.*

Now, we consider $u : ax + by + cz = 0$ and $v : a'x + b'y + c'z = 0$ two right lines in a non-Euclidean plane. By definition, u and v are parallel if their angle is null (see Mihăileanu [148], p. 39).

Definition 8.8 The quantity

$$d_{PAR}(u,v) = |\sin\Delta|$$

is called defect of parallelness of the right lines u and v.

Theorem 8.4 *(i)* $0 \leq d_{PAR}(u,v) \leq 1$.
(ii) $d_{PAR}(u,v) = d_{PAR}(v,u)$.
(iii) $d_{PAR}(u,u) = 0$.
(iv) $d_{PAR}(u,v) = \left|\sin\frac{\delta}{q}\right| = \sqrt{\frac{(ab'-a'b)^2+\varepsilon^2(ac'-a'c)^2+\varepsilon^2(bc'-b'c)^2}{(a^2+b^2+\varepsilon^2c^2)(a'^2+b'^2+\varepsilon^2c'^2)}}$.

In what follows we deal with Number Theory.

Definition 8.12 Let $x \in \mathbf{R}$. The quantity

$$d_{\mathbf{RZ}}(x) = \min(x - [x], [x] - x + 1)$$

is called defect of integer number of x, where $[x]$ is the integer part of x.

Theorem 8.7 *(i)* $d_{\mathbf{RZ}}(x) = \min(\{x\}, 1 - \{x\})$, *where* $\{x\}$ *denotes the fractional part of* x.
(ii) $d_{\mathbf{RZ}}(x) = 0$ *if and only if* $x \in \mathbf{Z}$ *and* $d_{\mathbf{RZ}}(x) = d(x, \mathbf{Z})$.
(iii) $0 \leq d_{\mathbf{RZ}}(x) \leq \frac{1}{2}, \forall x \in \mathbf{R}$.
(iv) $d_{\mathbf{RZ}}(x) = d_{\mathbf{RZ}}(-x), \forall x \in \mathbf{R}$.

Let us denote by $X = \{x_1, ..., x_n\}$ a finite set of real numbers and let us consider the fuzzy set $A_X : \mathbf{R} \to [0,1]$ corresponding to X and defined by $A_X(x) = 0$ if $x \in \mathbf{R} \setminus X$ and $A_X(x) = \{x\}$ if $x \in X$, where $\{x\}$ is the fractional part of x. If d_c is a normalized measure of fuzziness (that is has values in $[0,1]$), then we give the following

Definition 8.13 The quantity

$$d_I(X) = d_c(A_X)$$

is called defect of integer of the set X.

Theorem 8.8 *(i)* $d_I(X) = 0$ *(i.e has the minimum value) if and only if* $x_k \in \mathbf{Z}, \forall k \in \{1, ..., n\}$.
(ii) $d_I(X) = 1$ *(i.e. has the maximum value) if and only if* $d_{\mathbf{RZ}}(x_k)$ *is maximum,* $\forall k \in \{1, ..., n\}$.
(iii) Let $Y = \{y_1, ..., y_n\}$ *be a finite set of real numbers. If* $\{x_k\} \leq \{y_k\}$ *for* $\{y_k\} \leq \frac{1}{2}$ *and* $\{x_k\} \geq \{y_k\}$ *for* $\{y_k\} \geq \frac{1}{2}, \forall k \in \{1, ..., n\}$, *then* $d_I(X) \leq d_I(Y)$.

Definition 8.14 Let p, q two natural numbers. The quantity

$$D_{div}(p,q) = \min\{r, q - 1 - r\},$$

where r is the remainder of division of p by q, is called defect of divisibility of p with respect to q.

Theorem 8.9 *Let p, p_1, p_2, q be natural numbers.*

(*i*) $D_{div}(p,q) = 0$ *if and only if* q/p *(i.e. q is divisor of p).*

(*ii*) $0 \leq D_{div}(p,q) \leq \left[\frac{q}{2}\right] + 1$, *where* $[x]$ *is the integer part of* x.

(*iii*) *If* $p_1 \equiv p_2 \,(\bmod q)$ *then* $D_{div}(p_1,q) = D_{div}(p_2,q)$ *and* $D_{div}(p_1 - p_2, q) = 0$.

(*iv*) $D_{div}(p_1 + p_2, q) = D_{div}(r_1 + r_2, q)$ *and* $D_{div}(p_1 p_2, q) = D_{div}(r_1 r_2, q)$, *where* r_1 *and* r_2 *are the remainders of division of* p_1 *and* p_2 *by* q.

Definition 8.17 Let m and n be positive integers. The quantities

$$D_{\text{perfect}}(n) = \left| n - \sum_{k/n, k \neq n} k \right|$$

and

$$D_{\text{amicable}}(m,n) = \left| \sum_{i/m, i \neq m} i - \sum_{j/n, j \neq n} j \right|$$

are called defect of perfect number of n and defect of amicable numbers of m, n, respectively.

A fuzzy logic (see *e.g.* Butnariu-Klement-Zafrany [51]) can be described as a $[0,1]$-valued logic, that is one real number $t(p) \in [0,1]$ is assigned to each proposition p.

Definition 8.18 (see Hájek-Gödel [98]) The propositional form A is a 1-tautology (or standard tautology) if $t(A) = 1$, for each evaluation.

Definition 8.19 Let A be a propositional form which is represented with propositional forms $A_1, ..., A_n$ and connectives. The quantity

$$d^1_{TAUT}(A) = 1 - \inf \quad \{t(A_1, ..., A_n, \vee_S, \wedge_T, \neg);\ A_1, ..., A_n \text{ propositional forms}\}$$

is called defect of 1-tautology of propositional form A.

Among all fuzzy logics, min – max logic is the most used in practice.

Theorem 8.12 *(see Butnariu-Klement-Zafrany [51]) A propositional form A in* $\min - \max$ *fuzzy logic is a tautology if and only if $t(A) \geq 0.5$, for every evaluation.*

Let A be a propositional form which depends on propositional forms $A_1, ..., A_n$ and connectives $\vee, \wedge, \rightarrow, \neg$.

Definition 8.20 The quantity

$$d_{TAUT}(A) = \max\{0, 0.5 - \inf\{t(A_1, ..., A_n, \vee, \wedge, \rightarrow, \neg);$$

$$A_1, ..., A_n \text{ are propositional forms}\}\}$$

is called defect of tautology in $\min - \max$ fuzzy logic (or T_M-tautology) of propositional form A.

In intuitionistic fuzzy logic, two real non-negative numbers, $\mu(p)$ and $\nu(p)$, are assigned to each proposition p, with the following constraint:

$$\mu(p) + \nu(p) \leq 1.$$

Definition 8.21 (see *e.g.* Atanassov [12]) The propositional form A is an intuitionistic fuzzy tautology if and only if

$$\mu(A) \geq \nu(A).$$

Definition 8.22 Let A be a propositional form which can be represented with arbitrary propositional forms $A_1, ..., A_n$ and connectives. The quantity

$$\begin{aligned} d_{I-TAUT}(A) &= \max\{0, \sup\{\nu(A_1, ..., A_n) - \mu(A_1, ..., A_n); \\ &\quad A_1, ..., A_n \text{ propositional forms}\}\} \end{aligned}$$

is called defect of intuitionistic fuzzy tautology of propositional form A.

Theorem 8.13 (i) $0 \leq d_{I-TAUT}(A) \leq 1$.

(ii) $d_{I-TAUT}(A) = 0$ *if and only if A is an intuitionistic fuzzy tautology.*

Chapter 2

Defect of Property in Set Theory

In this chapter we consider the measures of fuzziness as measuring the "deviation" of a fuzzy set from the concept of crisp (classical) set, that is as measuring the defect of crisp(classical) set and the intuitionistic entropies as measuring the "deviation" of an intuitionistic fuzzy set from the concept of fuzzy set, that is as measuring the defect of fuzzy set. In the last section, some applications to the determination of the degree of interference (mainly in the geography of population), to description of systems performance and to digital image processing are given.

2.1 Measures of Fuzziness

Given a set X, a fuzzy subset A of X (or a fuzzy set A on X) is defined by a function

$$\mu_A : X \to [0,1]$$

such that $\mu_A(x)$ expresses the degree of membership of x to A, i.e., the degree of compatibility of x with the concept represented by the fuzzy set A. Therefore, $\mu_A(x) = 0$ means that x is definitely not a member of A and $\mu_A(x) = 1$ means that x is definitely a member of A; if either $\mu_A(x) = 0$ or $\mu_A(x) = 1$ for every $x \in X$, then A is a crisp set. We denote by $FS(X)$ the class of all fuzzy sets on X and we identify the fuzzy sets with their membership functions, that is

$$FS(X) = \{A \,|A : X \to [0,1]\}.$$

The measures of fuzziness (see *e.g.* Ban-Fechete [20], Klir [120], Knopfmacher [122], Roventa-Vivona [180], Vivona [217]) or, in other words, the fuzzy entropies (see *e.g.* Rudas-Kaynak [181], Román Flores-Bassanezi [34]) are real functions which attach to fuzzy sets values that characterize their degree of fuzzification, that is measure the difference between fuzzy sets and classical (or crisp) sets.

A general definition of the measures of fuzziness is the following (inspired by Rudas-Kaynak [181]):

Definition 2.1 A measure of fuzziness is a positive real function d_c defined on $\mathcal{F}(X) \subseteq FS(X)$, that satisfies the following requirements:

(i) If $A \in \mathcal{F}(X), A(x) \in \{0,1\}, \forall x \in X$ then $d_c(A) = 0$.

(ii) If $A \prec B$ then $d_c(A) \leq d_c(B)$, where $A \prec B$ means that A is sharper than B(see *e.g.* relation (2.2) below).

(iii) If A is maximally fuzzy (see *e.g.* relation (2.3) below) then $d_c(A)$ assumes its maximum value.

Remarks. 1) If $(X, \mathcal{A}, \mu)$ is a measure space (all measure-theoretic terms and results used in this chapter may be found in Halmos' book [99]), we denote $\mathcal{F}_{\mathcal{A}}(X) = \{A \in FS(X); A \text{ is } \mathcal{A}\text{-measurable}\}$. In many papers (see e.g Román Flores-Bassanezi [34], Knopfmacher [122], Roventa-Vivona [180]) it is considered $\mathcal{F}(X) = \mathcal{F}_{\mathcal{A}}(X)$ and the condition (i) in Definition 2.1 is replaced by

$$d_c(A) = 0 \text{ if and only if } A(x) \in \{0,1\}, \mu\text{-}a.e.\ x \in X. \tag{2.1}$$

If the set X is finite, $\mathcal{A} = \mathcal{P}(X)$ and the measure μ is defined by $\mu(A) = cardA$, then $\mathcal{F}_{\mathcal{A}}(X) = FS(X)$ and we obtain the definition introduced by Klir [120].

2) The best known acceptations for the sharpness relation $\prec$ and for fuzzy maximality introduced by De Luca and Termini [138] and further investigated by many authors (see e.g Batle-Trillas [35], De Luca-Termini [139], Knopfmacher [122], Román Flores-Bassanezi [34], Ban-Fechete [20]), are the followings:

$$A \prec B \text{ if and only if } A(x) \leq B(x) \text{ for } B(x) \leq \frac{1}{2} \text{ and}$$

$$A(x) \geq B(x) \text{ for } B(x) \geq \frac{1}{2}. \tag{2.2}$$

and

$$A \text{ is maximally fuzzy if and only if } A(x) = \frac{1}{2}, \forall x \in X \tag{2.3}$$

$$(\text{or } A(x) = \frac{1}{2}, \mu\text{-a.e. } x \in X \text{ if } (X, \mathcal{F}(X), \mu) \text{ is a measure space}).$$

In the case when X is finite, the best known example in this sense was given by De Luca and Termini [138]:

Example 2.1 The function $d_c : FS(X) \to \mathbf{R}$ defined by

$$d_c(A) = -\sum_{x \in X} (A(x)\log_2 A(x) + (1 - A(x))\log_2 (1 - A(x)))$$

(by convention $0\log_2 0 = 0$) is a measure of fuzziness.

Theorem 2.1 *(Loo [135]) Let X be a finite set. If the function $h : \mathbf{R}_+ \to \mathbf{R}_+$ is increasing, the functions $(g_x)_{x \in X}$, $g_x : [0,1] \to \mathbf{R}_+$ are increasing on $\left[0, \frac{1}{2}\right]$, decreasing on $\left[\frac{1}{2}, 1\right]$ such that $g_x(0) = g_x(1) = 0, \forall x \in X$ and $g_x\left(\frac{1}{2}\right)$ is the unique maximum value of $g_x, \forall x \in X$, then $d_c : FS(X) \to \mathbf{R}$ defined by*

$$d_c(A) = h\left(\sum_{x \in X} g_x(A(x))\right)$$

is a measure of fuzziness.

Remarks. 1) If

$$g_x(A(x)) = -A(x)\log_2 A(x) - (1 - A(x))\log_2 (1 - A(x)),$$

for every $x \in X$ and h is the identical function on $\mathbf{R}_+$, then we get the measure of fuzziness in Example 2.1.

2) When for a given $w \in [1, \infty]$, we take $h(b) = b^{\frac{1}{w}}$ and

$$g_x(A(x)) = \begin{cases} (A(x))^w, & \text{if } A(x) \in \left[0, \frac{1}{2}\right] \\ (1 - A(x))^w, & \text{if } A(x) \in \left[\frac{1}{2}, 1\right] \end{cases}$$

for every $x \in X$, the measures of fuzziness proposed by Kaufmann [114] are obtained. In this way, a measure of fuzziness is introduced as the distance

(Hamming if $w = 1$, Euclid if $w = 2$, Minkowski if $w \in [1, \infty]$) between the fuzzy set A and its nearest crisp set C_A, that is

$$C_A(x) = \begin{cases} 0, & \text{if } A(x) \leq \frac{1}{2} \\ 1, & \text{if } A(x) > \frac{1}{2}. \end{cases}$$

Sometimes, additional conditions are required to the concept of measure of fuzziness. Thus, an usual one is the equality between the measure of fuzziness of a fuzzy set and the measure of fuzziness of its complement. In this sense, we introduce the concept of fuzzy complement.

Definition 2.2 (see Klir [120], Rudas-Kaynak [181]) A function $N : [0,1] \to [0,1]$ is a fuzzy complement, if for all $a, b \in [0,1]$, the following axioms are satisfied:

(i) $N(0) = 1$ and $N(1) = 0$, that is N gives the same results as the classical complement for crisp conditions;

(ii) If $a < b$ then $N(a) \geq N(b)$, that is N is monotonically decreasing;

(iii) N is a continuous function;

(iv) $N(N(a)) = a$, that is N is involutive.

Example 2.2 If $t \in (0,1)$ then $N_t : [0,1] \to [0,1]$ defined by

$$N_t(x) = \begin{cases} 1 - \frac{1-t}{t}x, & \text{if } x \leq t \\ \frac{t}{1-t}(1-x), & \text{if } x \geq t \end{cases}$$

is a fuzzy complement. Indeed, $N_t(0) = 1$, $N_t(1) = 0$, the continuity and the monotonicity are obvious. Because $N_t(x) \in [t,1]$ if and only if $x \in [0,t]$ and $N_t(x) \in [0,t]$ if and only if $x \in [t,1]$, we get

$$\begin{aligned} N_t(N_t(x)) &= N_t\left(1 - \frac{1-t}{t}x\right) \\ &= \frac{t}{t-1}\left(1 - 1 + \frac{1-t}{t}x\right) = x, \forall x \in [0,t], \end{aligned}$$

and

$$\begin{aligned} N_t(N_t(x)) &= N_t\left(\frac{t}{1-t}(1-x)\right) \\ &= 1 - \frac{1-t}{t}\left(\frac{t}{1-t}(1-x)\right) = x, \forall x \in [t,1]. \end{aligned}$$

Definition 2.3 Let N be a fuzzy complement and $A \in FS(X)$. The fuzzy set $\overline{A} \in FS(X)$ defined by $\overline{A}_N(x) = N(A(x))$, for every $x \in X$, is called the complement of A.

Remark. The conventional complement of a fuzzy set $A, \overline{A}(x) = 1 - A(x)$, for every $x \in X$, is obtained considering as fuzzy complement $N : [0,1] \to [0,1]$ defined by $N(a) = 1 - a$.

Another axiom considered as natural in the definition of measures of fuzziness on $\mathcal{F}(X) \subseteq FS(X)$ is

$$d_c(A) = d_c(\overline{A}_N), \tag{2.4}$$

for every $A \in \mathcal{F}(X)$, if $\overline{A}_N \in \mathcal{F}(X)$ too.

Remark. In Knopfmacher [122] and Vivona [217] the conventional complement is considered.

Also, in Knopfmacher [122] are added the following requirements for a measure of fuzziness:

(i) $d_c(A \vee B) + d_c(A \wedge B) = d_c(A) + d_c(B), \forall A, B \in \mathcal{F}_{\mathcal{A}}(X)$, where $(A \vee B)(x) = \max(A(x), B(x))$ and $(A \wedge B)(x) = \min(A(x), B(x))$;

(ii) d_c is a continuous function on $\mathcal{F}_{\mathcal{A}}(X)$ relative to uniform metric ρ on $\mathcal{F}_{\mathcal{A}}(X)$, $\rho(A, B) = \sup_{x \in X} |A(x) - B(x)|$;

(iii) The restriction of d_c to the family of constant fuzzy sets $(A_\alpha)_{\alpha \in [0, \frac{1}{2}]}$ (that is $A_\alpha(x) = \alpha, \forall x \in X$), is strictly increasing function of α.

In the same paper [122], a family of measures of fuzziness (in normalized form) which verify all these conditions is given, as follows.

Theorem 2.2 *Let $(X, \mathcal{A}, \mu)$ be a measure space with $0 < \mu(A) < +\infty$ and $\mathcal{F}_{\mathcal{A}}(X)$ the set of all fuzzy sets A on X that are measurable as real-valued functions. If Δ denotes an arbitrary real-valued function of $\alpha \in [0,1]$, such that $\Delta(0) = \Delta(1) = 0, \Delta(\alpha) = \Delta(1-\alpha), \forall \alpha \in [0,1]$ and Δ is strictly increasing on $[0, \frac{1}{2}]$, then $d_c : \mathcal{F}_{\mathcal{A}}(X) \to \mathbf{R}$ defined by*

$$d_c(A) = \frac{1}{\mu(X)} \int_X \Delta(A(x)) \, d\mu$$

is a measure of fuzziness which verifies the conditions in Definition 2.1 (in the sense of (2.2) and (2.3)), (2.4) (with the conventional complement) and the above conditions $(i) - (iii)$, where $\int$ denotes the Lebesgue integral.

Replacing the conventional interpretations (where the remarkable value is $\frac{1}{2}$) of the relation "sharper than" and of property "maximally fuzzy" in Remark after Definition 2.1, we obtain the following more general and useful definition of t-measures of fuzziness, the fixed value $t \in (0,1)$ becoming important.

Definition 2.4 Let $(X, \mathcal{A})$ be a measurable space and $t \in (0,1)$. A t-measure of fuzziness is a function $d_c^t : \mathcal{F}_{\mathcal{A}}(X) \to \mathbf{R}$ that satisfies the following conditions:

(*i*) If $A(x) \in \{0,1\}, \forall x \in X$ then $d_c^t(A) = 0$;

(*ii*) If $A \prec_t B$ then $d_c^t(A) \leq d_c^t(B)$, where $A \prec_t B$ if and only if $A(x) \leq B(x)$ for $B(x) \leq t$ and $A(x) \geq B(x)$ for $B(x) \geq t$.

(*iii*) If $A(x) = t, \forall x \in X$ then $d_c^t(A)$ is the maximum value of d_c^t.

Definition 2.5 Let $(X, \mathcal{A}, \mu)$ be a measure space and $t \in (0,1)$. A function $s_c^t : \mathcal{F}_{\mathcal{A}}(X) \to \mathbf{R}$ that satisfies the following conditions:

(*i*) $s_c^t(A) = 0$ if and only if $A(x) \in \{0,1\}$, μ-a.e $x \in X$;

(*ii*) If $A \prec_t B$ then $s_c^t(A) \leq s_c^t(B)$;

(*iii*) $s_c^t(A)$ is the maximum value of s_c^t if and only if $A(x) = t$, μ-a.e. $x \in X$,

is called strict t-measure of fuzziness with respect to μ.

Remarks. 1) In general, if X is finite then we consider $\mathcal{A} = \mathcal{P}(X)$ and $\mu(A) = cardA, \forall A \in \mathcal{P}(X)$, in this case μ-a.e meaning everywhere.

2) It is obvious that any strict t-measure of fuzziness with respect to a measure μ, is a t-measure of fuzziness.

Definition 2.6 Let N be a fuzzy complement. A t-measure of fuzziness is called symmetrical with respect to N (or N-symmetrical) if

$$d_c^t(A) = d_c^t(\overline{A}_N), \forall A \in \mathcal{F}_{\mathcal{A}}(X). \tag{2.5}$$

Remark. The most natural fuzzy complement for t-measures of fuzziness is that introduced in Example 2.2. If $t = \frac{1}{2}$ we get the conventional complement.

Based on the idea in above Theorem 2.1, we can give a family of t-measures of fuzziness. The result is more general than Theorem 2.1 in Ban-Fechete [20].

Theorem 2.3 *Let X be a finite set, $t \in (0,1)$ and $h : \mathbf{R}_+ \to \mathbf{R}_+$ increasing, such that $h(0) = 0$, $(g_x)_{x \in X}$, $g_x : [0,1] \to \mathbf{R}_+$ increasing on $[0,t]$ and* *decreasing on* $[t, 1]$, *such that* $g_x(0) = g_x(1) = 0, \forall x \in X$ *and* $g_x(t)$ *is the maximum value of all functions* $g_x, x \in X$. *The function* $d_c^t : FS(X) \to \mathbf{R}$ *defined by*

$$d_c^t(A) = h\left(\sum_{x\in X} g_x\left(A(x)\right)\right)$$

is a t*-measure of fuzziness. If, in addition,*

$$g_x(a) = g_x\left(N(a)\right), \forall a \in [0,1], \forall x \in X,$$

where N *is a fuzzy complement, then* d_c^t *is a* N*-symmetrical* t*-measure of fuzziness.*

Proof. We verify the conditions in Definition 2.4.

(i) If $A(x) \in \{0, 1\}, \forall x \in X$ then $g_x\left(A(x)\right) = 0, \forall x \in X$ and $\sum_{x\in X} g_x\left(A(x)\right) = 0$, therefore $d_c^t(A) = h(0) = 0$.

(ii) If $A \prec_t B$ then the monotonicity of functions $g_x, x \in X$ implies $g_x\left(A(x)\right) \le g_x\left(B(x)\right), \forall x \in X$. This means $\sum_{x\in X} g_x\left(A(x)\right) \le \sum_{x\in X} g_x\left(B(x)\right)$, that is $d_c^t(A) \le d_c^t(B)$ because h is increasing.

(iii) $d_c^t(A)$ assumes the maximum value if $\sum_{x\in X} g_x\left(A(x)\right)$ assumes the maximum value, that is $A(x) = t, \forall x \in X$.

Relation (2.5) is also satisfied because

$$d_c^t\left(\overline{A}_N\right) = h\left(\sum_{x\in X} g_x\left(N\left(A(x)\right)\right)\right) = h\left(\sum_{x\in X} g_x\left(A(x)\right)\right) = d_c^t(A),$$

therefore d_c^t is a N-symmetrical t-measure of fuzziness. □

Example 2.3 Let X be a finite set and $t \in (0, 1)$. If $h : \mathbf{R}_+ \to R_+, h(x) = x$ and $g_x : [0, 1] \to \mathbf{R}_+$ are defined by

$$g_x(a) = \begin{cases} a, & \text{if } a \in [0, t] \\ N(a), & \text{if } a \in [t, 1], \end{cases}$$

where N is a fuzzy complement, then the conditions in Theorem 2.3 are verified, therefore

$$d_c^t\left(A\right) = \sum_{x\in X} g_x\left(A(x)\right)$$

is a N-symmetrical t-measure of fuzziness. Indeed, $g_x(0) = 0$ and $g_x(1) = g_x\left(N(1)\right) = g_x(0) = 0, \forall x \in X, g_x$ are increasing on $[0, t]$ and decreasing on $[t,1]$, $\forall x \in X$, because N is decreasing. Also, t is the maximum value of g_x if $a \in [0,t]$ and $N(t)$ is the maximum value of g_x if $a \in [t,1]$, $\forall x \in X$, therefore $g_x(t)$ is the maximum value, $\forall x \in X$. In addition, N being involutory implies

$$\begin{aligned} g_x(N(a)) &= \begin{cases} N(a), & \text{if } N(a) \in [0,t] \\ N(N(a)), & \text{if } N(a) \in [t,1] \end{cases} \\ &= \begin{cases} N(a), & \text{if } a \in [t,1] \\ a, & \text{if } a \in [0,t] \end{cases} = g_x(a). \end{aligned}$$

Under additional conditions, we obtain a family of strict t-measures of fuzziness.

Theorem 2.4 *If the functions h and $(g_x)_{x\in X}$ satisfy the hypothesis in Theorem 2.3 and, in addition, h is strictly increasing, $(g_x)_{x\in X}$ are strictly increasing on $[0,t]$ and strictly decreasing on $[t,1]$, then the function $s_c^t : FS(X) \to \mathbf{R}$ defined by*

$$s_c^t(A) = h\left(\sum_{x\in X} g_x(A(x))\right)$$

is a strict t-measure of fuzziness which is N-symmetrical if $g_x(a) = g_x(N(a))$, $\forall a \in [0,1], \forall x \in X$, where N is a fuzzy complement.

Proof. We must prove only the necessity in conditions (i) and (iii) of Definition 2.5, the other requirements being proved by the proof of Theorem 2.3.

If $s_c^t(A) = 0$ then $\sum_{x\in X} g_x(A(x)) = 0$, which implies $A(x) \in \{0,1\}, \forall x \in X$, because the functions g_x are strictly increasing on $[0,t]$ and strictly decreasing on $[t,1]$.

If $s_c^t(A)$ is the maximum value of s_c^t, then $\sum_{x\in X} g_x(A(x))$ assumes the maximum value, that is $g_x(A(x))$ is the maximum value of g_x, $\forall x \in X$. The strict monotonicity of g_x on $[0,t]$ and on $[t,1]$, for every $x \in X$, implies $A(x) = t, \forall x \in X$. □

In what follows, we give an example of strict t-measure of fuzziness of entropy type. In the particular case $t = \frac{1}{2}$, we obtain the t-measure of fuzziness introduced by De Luca and Termini (see Example 2.1).

Example 2.4 Let X be finite, $t \in (0,1)$, $u_t : [0,1] \to [0,1]$ defined by

$$u_t(a) = \begin{cases} \frac{a}{2t}, & \text{if } a \in [0,t] \\ \frac{a}{2(1-t)} + \frac{1-2t}{2(1-t)}, & \text{if } a \in [t,1], \end{cases}$$

and $H : [0,1] \to \mathbf{R}_+$ defined by $H(y) = -y \log_2 y - (1-y)\log_2(1-y)$ (by convention, $0 \log_2 0 = 0$). We denote $g_x = H \circ u_t, \forall x \in X$. The function u_t is strictly increasing on $[0,1]$. If $a \in [0,t]$ then $u_t(a) \in \left[0, \frac{1}{2}\right]$ and H being strictly increasing on $\left[0, \frac{1}{2}\right]$, we obtain that the functions g_x are strictly increasing on $[0,t], \forall x \in X$. Analogously, if $a \in [t,1]$ then $u_t(a) \in \left[\frac{1}{2}, 1\right]$ and because the function H is strictly decreasing on $\left[\frac{1}{2}, 1\right]$, we obtain that the functions g_x are strictly increasing on $[t,1], \forall x \in X$. The monotonicity of H implies that the function g_x assumes the maximum value if and only if the function u_t has the value equal to $\frac{1}{2}$, that is, if and only if the argument of u_t and implicitely of the function g_x, $x \in X$, is t. Because

$$g_x(0) = H(u_t(0)) = H(0) = 0, \forall x \in X,$$

and

$$g_x(1) = H(u_t(1)) = H(1) = 0, \forall x \in X,$$

we get that the functions $(g_x)_{x \in X}$ verify the hypothesis in Theorem 2.4. Because $h : \mathbf{R}_+ \to \mathbf{R}_+, h(x) = x$, also verifies these hypothesis, we obtain that the function defined by

$$\begin{aligned} H_c^t(A) = \quad & -\textstyle\sum_{x \in X} \left(u_t(A(x)) \log_2 u_t(A(x)) \right. \\ & \left. + (1 - u_t(A(x))) \log_2 (1 - u_t(A(x))) \right) \end{aligned}$$

is a strict t-measure of fuzziness. We prove that this t-measure of fuzziness is N_t-symmetrical, where N_t is the fuzzy complement introduced in Example 2.2. Indeed, if $a \in [0,t]$, then $N_t(a) \in [t,1]$ and

$$\begin{aligned} u_t(N_t(a)) &= \frac{N_t(a)}{2(1-t)} + \frac{1-2t}{2(1-t)} \\ &= \frac{1 - \frac{1-t}{t}a + 1 - 2t}{2(1-t)} = 1 - \frac{a}{2t} = 1 - u_t(a). \end{aligned}$$

If $a \in [t,1]$, then $N_t(a) \in [0,t]$ and

$$u_t(N_t(a)) = \frac{N_t(a)}{2t} = \frac{\frac{t}{1-t}(1-a)}{2t} = \frac{1-a}{2(1-t)}$$

$$= 1 - \left(\frac{a}{2(1-t)} + \frac{1-2t}{2(1-t)} \right) = 1 - u_t(a).$$

Because $H(y) = H(1-y), \forall y \in [0,1]$, we obtain

$$\begin{aligned} H_c^t\left(\overline{A}_{N_t}\right) &= \sum_{x \in X} H\left(u_t\left(N_t\left(A(x)\right)\right)\right) = \sum_{x \in X} H\left(1 - u_t\left(A(x)\right)\right) \\ &= \sum_{x \in X} H\left(u_t\left(A(x)\right)\right) = H_c^t(A), \forall A \in FS(X). \end{aligned}$$

Other families of t-measures of fuzziness can be obtained by using the concept of t-norm function in the sense of Vivona [217].

Definition 2.7 Let $t \in (0,1)$ and $(X, \mathcal{A}, \mu)$ be a measure space. An $\mathcal{A}$-measurable function with respect to the first variable, $\varphi_t : X \times [0,1] \to [0,1]$ that satisfies the properties

(*i*) $\varphi_t(x,0) = \varphi_t(x,1) = 0, \forall x \in X$;

(*ii*) $\varphi_t(x,\cdot)$ is increasing on $[0,t]$ and decreasing on $[t,1]$;

(*iii*) $\varphi_t(x,t) = 1, \forall x \in X$,

is called t-norm function.

Example 2.5 The function $\varphi_t : X \times [0,1] \to [0,1]$ defined by $\varphi_t(x,a) = (H \circ u_t)(a)$, where H and u_t are defined as in Example 2.4, is a t-norm function.

Corresponding to Theorem 2.3 we prove the following result.

Theorem 2.5 *Let $t \in (0,1)$, $(X, \mathcal{A}, \mu)$ be a measure space and $\varphi_t : X \times [0,1] \to [0,1]$ a t-norm function. The function $d_c^{\varphi_t} : \mathcal{F}_{\mathcal{A}}(X) \to [0,1]$ defined by*

$$d_c^{\varphi_t}(A) = \int_X \varphi_t\left(x, A(x)\right) d\mu$$

is a t-measure of fuzziness. If $N : [0,1] \to [0,1]$ is a fuzzy complement and the t-norm function φ_t verifies

$$\varphi_t\left(x, N(a)\right) = \varphi_t\left(x, a\right), \forall x \in X, \forall a \in [0,1],$$

then the t-measure of fuzziness $d_c^{\varphi_t}(A)$, defined as above, is N-symmetrical.

Proof. It is immediate by using the properties of φ_t and the properties of Lebesgue integral. □

Remark. If the t-norm function φ_t is strictly increasing on $[0,t]$ and strictly decreasing on $[t,1]$, then the t-measures of fuzziness in Theorem 2.5 are strict.

2.2 Intuitionistic Entropies

Let X be fixed. An intuitionistic fuzzy sets (see *e.g.* Atanassov [5]) A on X is an object having the form

$$A = \{\langle x, \mu_A(x), \nu_A(x)\rangle : x \in X\},$$

where the functions $\mu_A, \nu_A : X \to [0,1]$ define the degree of membership and the degree of non-membership of the element $x \in X$ to the set $A \subseteq X$, respectively, and for every $x \in X, \mu_A(x)+\nu_A(x) \leq 1$.We denote by $IFS(X)$ the family of intuitionistic fuzzy sets on X.

We recall the following useful relations and operations on $IFS(X)$ (see Atanassov [5]-[10], Atanassov-Ban [14]):

$$A \subseteq B \text{ if and only if } \mu_A(x) \leq \mu_B(x) \text{ and } \nu_A(x) \geq \nu_B(x), \forall x \in X;$$

$$A = B \text{ if and only if } \mu_A(x) = \mu_B(x) \text{ and } \nu_A(x) = \nu_B(x), \forall x \in X;$$

$$A \preceq B \text{ if and only if } \mu_A(x) \leq \mu_B(x) \text{ and } \nu_A(x) \leq \nu_B(x), \forall x \in X;$$

$$\overline{A} = \{\langle x, \nu_A(x), \mu_A(x)\rangle : x \in X\};$$

where $A, B \in IFS(X), A = \{\langle x, \mu_A(x), \nu_A(x)\rangle : x \in X\}$ and $B = \{\langle x, \mu_B(x), \nu_B(x)\rangle : x \in X\}$. Also, if h is a triangular norm (that is, an associative and commutative binary operation $h : [0,1]\times[0,1] \to [0,1]$ such that $h(x,1) = x, \forall x \in [0,1]$ and $y \leq z$ implies $h(x,y) \leq h(x,z), \forall x \in [0,1]$) or a triangular conorm (that is, an associative and commutative binary operation $h : [0,1] \times [0,1] \to [0,1]$ such that $h(x,0) = x, \forall x \in [0,1]$ and $y \leq z$ implies $h(x,y) \leq h(x,z), \forall x \in [0,1]$), we define (see Burillo-Bustince [46])

$$A\widetilde{h}B = \{\langle x, h(\mu_A(x), \mu_B(x)), h^c(\nu_A(x), \nu_B(x))\rangle : x \in X\},$$

where $h^c(x,y) = 1 - h(1-x, 1-y), \forall x, y \in [0,1]$, is the triangular norm (conorm) corresponding to triangular conorm (norm) h. We also notice that these operations can be extended to countable case (see Ban [17]).

The intuitionistic entropy, introduced by Burillo-Bustince [47], can be considered as a measure of the level of intuitionism of an intuitionistic fuzzy set, *i.e.* it is a quantitative expression of the difference between an intuitionistic fuzzy set and a fuzzy set, similar to the fuzzy entropy (or measure of fuzziness) which measures the difference between a fuzzy set and a crisp set.

In the above cited paper, the following formal conditions are required for an intuitionistic entropy:

(*i*) to be null when the set is a fuzzy set;

(*ii*) to be maximum if the set is totally intuitionistic;

(*iii*) as in the case of fuzzy sets, the entropy of an intuitionistic fuzzy set has to be equal to the entropy of its complement;

(*iv*) If the degree of membership and the degree of non-membership of each element increase, the sum will do as well, and therefore, this set will become more fuzzy, and the entropy will decrease.

Taking into account the previous considerations, in the case when X is finite, we give the following definition.

Definition 2.8 (Burillo-Bustince [47])A real function $d_f : IFS(X) \to \mathbf{R}_+$ is called an intuitionistic entropy on $IFS(X)$, if d_f has the following properties:

(*i*) $d_f(A) = 0$ if and only if $\mu_A(x) + \nu_A(x) = 1, \forall x \in X$;

(*ii*) $d_f(A) = cardX = N$ if and only if $\mu_A(x) = \nu_A(x) = 0, \forall x \in X$;

(*iii*) $d_f(A) = d_f(\overline{A}), \forall A \in IFS(X)$;

(*iv*) If $A \preceq B$ then $I(A) \geq I(B)$.

We reformulate this definition in a more general frame, even if in Ban-Gal [23], a result of approximation of intuitionistic fuzzy sets by discrete intuitionistic fuzzy sets (that is functions whose degrees of membership and degrees of non-membership take a finite number of values) reduces, in a certain sense, the infinite case to the finite case.

Let us consider $(X, \mathcal{A}, m)$ a finite measure space and let us denote by $\mathcal{I}_{\mathcal{A}}(X)$ the family of all $\mathcal{A}$-measurable intuitionistic fuzzy sets on X, that is $A = \{\langle x, \mu_A(x), \nu_A(x)\rangle : x \in X\} \in \mathcal{I}_{\mathcal{A}}(X)$ if and only if μ_A and ν_A are $\mathcal{A}$-measurable functions.

Definition 2.9 A real function $d_f : \mathcal{I}_{\mathcal{A}}(X) \to \mathbf{R}_+$ is called an intuitionistic entropy (on $\mathcal{I}_{\mathcal{A}}(X)$) if the following properties are satisfied:

(*i*) $d_f(A) = 0$ if and only if $\mu_A(x) + \nu_A(x) = 1$, m-a.e. $x \in X$;

(ii) $d_f(A)$ is maximum if and only if $\mu_A(x) = \nu_A(x) = 0$, m-a.e. $x \in X$;
(iii) $d_f(A) = d_f(\overline{A}), \forall A \in \mathcal{I_A}(X)$;
(iv) If $A, B \in \mathcal{I_A}(X)$ and $A \preceq B$ then $d_f(A) \geq d_f(B)$.

Remark. The particular situation in Definition 2.8 is obtained if X is finite, $\mathcal{A} = \mathcal{P}(X)$ and $m(A) = cardA, \forall A \in \mathcal{A}$.

In what follows we consider only intuitionistic entropies in the normalized form, that is their maximum values are equal to 1.

Definition 2.10 Let $(X, \mathcal{A}, m)$ be a measure space and let us denote $D = \{(\mu, \nu) \in [0,1] \times [0,1] : \mu + \nu \leq 1\}$. An intuitionistic norm function is an $\mathcal{A}$-measurable function with respect to the first variable, $\Phi : X \times D \to [0,1]$, with the following properties, for every element $x \in X$:
(i) $\Phi(x, \mu, \nu) = 0$ if and only if $\mu + \nu = 1$;
(ii) $\Phi(x, \mu, \nu) = 1$ if and only if $\mu = \nu = 0$;
(iii) $\Phi(x, \mu, \nu) = \Phi(x, \nu, \mu)$;
(iv) If $\mu \leq \mu'$ and $\nu \leq \nu'$ then $\Phi(x, \mu, \nu) \geq \Phi(x, \mu', \nu')$.

We obtain families of intuitionistic entropies by using intuitionistic norm functions and Lebesgue integral, similar to Knopfmacher [122] for the case of fuzzy entropy.

Theorem 2.6 *Let $(X, \mathcal{A}, m)$ be a finite measure space and Φ an intuitionistic norm function. Then, the function $d_f^{\Phi} : \mathcal{I_A}(X) \to [0,1]$ defined by*

$$d_f^{\Phi}(A) = \frac{1}{m(X)} \int_X \Phi(x, \mu_A(x), \nu_A(x))\, dm$$

with $A = \{\langle x, \mu_A(x), \nu_A(x)\rangle : x \in X\}$, is an intuitionistic entropy.

Proof. Let $A, B \in \mathcal{I_A}(X), A = \{\langle x, \mu_A(x), \nu_A(x)\rangle : x \in X\}$ and $B = \{\langle x, \mu_B(x), \nu_B(x)\rangle : x \in X\}$. We notice that $A = B$ m-a.e., that is $\mu_A(x) = \mu_B(x)$ and $\nu_A(x) = \nu_B(x)$ m-a.e. $x \in X$, implies $\Phi(x, \mu_A(x), \nu_A(x)) = \Phi(x, \mu_B(x), \nu_B(x))$ m-a.e. $x \in X$, therefore $d_f^{\Phi}(A) = d_f^{\Phi}(B)$.

We verify the conditions $(i) - (iv)$ in Definition 2.9.

(i) If $E = \{\langle x, \mu_E(x), \nu_E(x)\rangle : x \in X\} \in \mathcal{I_A}(X)$ verifies $\mu_E(x) + \nu_E(x) = 1, \forall x \in X$, then

$$d_f^{\Phi}(E) = \frac{1}{m(X)} \int_X 0 dm = 0.$$

If $A \in \mathcal{I}_{\mathcal{A}}(X)$ is an intuitionistic fuzzy set with $d_f^{\Phi}(A) = 0$, then $\Phi(x, \mu_A(x), \nu_A(x)) = 0$ m-a.e. $x \in X$ and Definition 2.10, (i), implies $\mu_A(x) + \nu_A(x) = 1$ m-a.e. $x \in X$.

(ii) If E verifies $\mu_E(x) = \nu_E(x) = 0, \forall x \in X$, then

$$d_f^{\Phi}(E) = \frac{1}{m(X)} \int_X 1 \mathrm{d}m = 1$$

If $d_f^{\Phi}(A) = 1$ then $\int_X \Phi(x, \mu_A(x), \nu_A(x)) \mathrm{d}m = \int_X 1 \mathrm{d}m$, that is $\Phi(x, \mu_A(x), \nu_A(x)) = 1$ m-a.e. $x \in X$. By Definition 2.10, (ii), we get $\mu_A(x) = \nu_A(x) = 0$ m-a.e. $x \in X$.

(iii), (iv) Are immediate by the properties (iii) and (iv) in Definition 2.10. □

Remark. Among all the possible intuitionistic entropies previously introduced, the most natural and simple is obtained by setting $\Phi(x, \mu, \nu) = 1 - \mu - \nu$, that is

$$d_f^{\Phi}(A) = \frac{1}{m(X)} \int_X (1 - \mu_A(x) - \nu_A(x))\, dm, \forall A \in \mathcal{I}_{\mathcal{A}}(X).$$

Theorem 2.7 *Let $(X, \mathcal{A}, m)$ be a finite measure space and Φ be an intuitionistic norm function. If Φ is a continuous function with respect to the second and third variables, then the intuitionistic entropy $d_f^{\Phi} : \mathcal{I}_{\mathcal{A}}(X) \to [0,1]$ defined by*

$$d_f^{\Phi}(A) = \frac{1}{m(X)} \int_X \Phi(x, \mu_A(x), \nu_A(x))\, dm$$

is a continuous function with respect to the metric $d : \mathcal{I}_{\mathcal{A}}(X) \times \mathcal{I}_{\mathcal{A}}(X) \to [0,1]$, defined by

$$d(A, B) = \max\left(\sup_{x \in X} |\mu_A(x) - \mu_B(x)|, \sup_{x \in X} |\nu_A(x) - \nu_B(x)| \right),$$

where $A = \{\langle x, \mu_A(x), \nu_A(x)\rangle : x \in X\}, B = \{\langle x, \mu_B(x), \nu_B(x)\rangle : x \in X\}$.

Proof. Because $D = \{(\mu, \nu) \in [0,1] \times [0,1] : \mu + \nu \leq 1\}$ is a compact set, $\Phi(x, \cdot, \cdot)$ is uniformly continuous function. Let $\varepsilon > 0$. There is $\delta > 0$ such that $x \in X$ with $|\mu_A(x) - \mu_B(x)| < \delta$ and $|\nu_A(x) - \nu_B(x)| < \delta$ implies

$$|\Phi(x, \mu_A(x), \nu_A(x)) - \Phi(x, \mu_B(x), \nu_B(x))| < \varepsilon.$$

But $d(A,B) < \delta$ implies $|\mu_A(x) - \mu_B(x)| < \delta$ and $|\nu_A(x) - \nu_B(x)| < \delta, \forall x \in X$, therefore

$$|\Phi(x, \mu_A(x), \nu_A(x)) - \Phi(x, \mu_B(x), \nu_B(x))| < \varepsilon, \forall x \in X.$$

We obtain the existence of a $\delta > 0$ such that $d(A,B) < \delta$ implies

$$\left| \frac{1}{m(X)} \int_X \Phi(x, \mu_A(x), \nu_A(x))\, \mathrm{d}m - \frac{1}{m(X)} \int_X \Phi(x, \mu_B(x), \nu_B(x))\, \mathrm{d}m \right| \le \varepsilon$$

therefore $|I_\Phi(A) - I_\Phi(B)| \le \varepsilon$. □

In Burillo-Bustince [47] for the finite case, a theorem of characterization of intuitionistic entropies by using I_φ-functions is given. The idea can be extended to general case, by considering particular intuitionistic norm functions.

Definition 2.11 Let $(X, \mathcal{A}, m)$ be a measure space and $\varphi : [0,1] \to [0,1]$ be a continuous function such that if $\alpha + \beta \le 1$ then $\varphi(\alpha) + \varphi(\beta) \le 1$. The function $I_\varphi : \mathcal{I}_\mathcal{A}(X) \to [0,1]$ defined by

$$I_\varphi(A) = \frac{1}{m(X)} \int_X (1 - \varphi(\mu_A(x)) - \varphi(\nu_A(x)))\, \mathrm{d}m$$

with $A = \{\langle x, \mu_A(x), \nu_A(x)\rangle : x \in X\}$, is called I_φ-function.

Theorem 2.8 *Let $(X, \mathcal{A}, m)$ be a finite measure space, $\varphi : [0,1] \to [0,1]$ be a continuous function and $d_\varphi : \mathcal{I}_\mathcal{A}(X) \to [0,1]$. The function d_φ is an intuitionistic entropy and an I_φ-function if and only if*

$$d_\varphi(A) = \frac{1}{m(X)} \int_X (1 - \varphi(\mu_A(x)) - \varphi(\nu_A(x)))\, dm,$$

where φ satisfies the conditions:
(i) φ is increasing;
(ii) $\varphi(\alpha) = 0$ if and only if $\alpha = 0$;
(iii) $\varphi(\alpha) + \varphi(\beta) = 1$ if and only if $\alpha + \beta = 1$.

Proof. ($\Leftarrow$) Let us consider $\Phi : X \times D \to [0,1], \Phi(x, \mu, \nu) = 1 - \varphi(\mu) - \varphi(\nu)$, where $D = \{(\mu, \nu) \in [0,1] \times [0,1] : \mu + \nu \le 1\}$. The function Φ is constant with respect to x, therefore it is $\mathcal{A}$-measurable with respect to the first variable.
We verify $(i) - (iv)$ in Definition 2.10.

(i) $\Phi(x, \mu, \nu) = 1 - (\varphi(\mu) + \varphi(\nu)) = 0$ if and only if $\varphi(\mu) + \varphi(\nu) = 1$ if and only if $\mu + \nu = 1$.

(ii) $\Phi(x, \mu, \nu) = 1$ if and only if $\varphi(\mu) + \varphi(\nu) = 0$ if and only if $\varphi(\mu) = \varphi(\nu) = 0$ if and only if $\mu = \nu = 0$.

(iii) $\Phi(x, \mu, \nu) = 1 - \varphi(\mu) - \varphi(\nu) = \Phi(x, \nu, \mu)$.

(iv) If $\mu \leq \mu'$ and $\nu \leq \nu'$ then $\varphi(\mu) + \varphi(\nu) \leq \varphi(\mu') + \varphi(\nu')$, that is $\Phi(x, \mu, \nu) \geq \Phi(x, \mu', \nu')$.

Φ being an intuitionistic norm function, Theorem 2.6 implies that d_φ is an intuitionistic entropy.

If $\alpha + \beta \leq 1$ then

$$\begin{aligned} d_\varphi(\{\langle x, \alpha, \beta\rangle : x \in X\}) &= \frac{1}{m(X)} \int_X (1 - \varphi(\alpha) - \varphi(\beta))\, dm \\ &= 1 - \varphi(\alpha) - \varphi(\beta) \geq 0, \end{aligned}$$

which implies $\varphi(\alpha) + \varphi(\beta) \leq 1$, therefore d_φ is an I_φ-function.

$(\Rightarrow)$ We assume that d_φ is an intuitionistic entropy and an I_φ-function.

Being I_φ-function it has the form

$$d_\varphi(A) = \frac{1}{m(X)} \int_X (1 - \varphi(\mu_A(x)) - \varphi(\nu_A(x)))\, dm$$

if $A = \{\langle x, \mu_A(x), \nu_A(x)\rangle : x \in X\}$, where $\varphi : [0,1] \to [0,1]$ is a continuous function such that $\alpha + \beta \leq 1$ implies $\varphi(\alpha) + \varphi(\beta) \leq 1$. We prove that φ satisfies the conditions *(i)* – *(iii)*.

(i) If $\alpha \leq \alpha'$ and $\alpha, \alpha' \in [0,1]$, we construct the following intuitionistic fuzzy sets: $A = \{\langle x, \alpha, 0\rangle : x \in X\}$ and $B = \{\langle x, \alpha', 0\rangle : x \in X\}$. Because $A \preceq B$, it follows $I(A) \geq I(B)$, that is

$$\int_X (1 - \varphi(\alpha) - \varphi(0))\, dm \geq \int_X (1 - \varphi(\alpha') - \varphi(0))\, dm$$

and

$$1 - \varphi(\alpha) - \varphi(0) \geq 1 - \varphi(\alpha') - \varphi(0).$$

We obtain $\varphi(\alpha) \leq \varphi(\alpha')$.

(ii) $d_\varphi(\{\langle x, 0, 0\rangle : x \in X\}) = 1$ by Definition 2.10, *(ii)*. But

$$d_\varphi(\{\langle x, 0, 0\rangle : x \in X\})$$

$$= \frac{1}{m(X)} \int_X (1 - \varphi(0) - \varphi(0))\, dm = 1 - 2\varphi(0),$$

therefore $1 = 1 - 2\varphi(0)$ and $\varphi(0) = 0$.

Conversely, if $\varphi(\alpha) = 0$, then

$$\begin{aligned} d_\varphi(\{\langle x, \alpha, 0\rangle : x \in X\}) &= \frac{1}{m(X)} \int_X (1 - \varphi(\alpha) - \varphi(0))\, dm \\ &= 1 - \varphi(\alpha) = 1 \end{aligned}$$

and Definition 2.10, (ii), implies $\alpha = 0$.

(iii) If $\alpha + \beta = 1$, we take $A = \{\langle x, \alpha, \beta\rangle : x \in X\}$. We know by Definition 2.10, (i), that $d_\varphi(A) = 0$. But

$$d_\varphi(A) = \frac{1}{m(X)} \int_X (1 - \varphi(\alpha) - \varphi(\beta))\, dm = 1 - \varphi(\alpha) - \varphi(\beta).$$

We obtain $\varphi(\alpha) + \varphi(\beta) = 1$.

Conversely, let $\varphi(\alpha) + \varphi(\beta) = 1$. If we assume that $\alpha + \beta \neq 1$, then two cases may occur:

(a) $\alpha + \beta < 1$

Then

$$\begin{aligned} d_\varphi(\{\langle x, \alpha, \beta\rangle : x \in X\}) &= \frac{1}{m(X)} \int_X (1 - \varphi(\alpha) - \varphi(\beta))\, dm \\ &= 1 - \varphi(\alpha) - \varphi(\beta) = 0, \end{aligned}$$

therefore $\alpha + \beta = 1$ (see Definition 2.10, (i)), contradicting the hypothesis.

(b) $\alpha + \beta > 1$

Because $\alpha + (1 - \alpha) = 1$ and $\beta + (1 - \beta) = 1$, we get $\varphi(\alpha) + \varphi(1 - \alpha) = 1$ and $\varphi(\beta) + \varphi(1 - \beta) = 1$. For all α, β we can write

$$\varphi(\alpha) + \varphi(\beta) + \varphi(1 - \alpha) + \varphi(1 - \beta) = 2,$$

therefore $\varphi(1 - \alpha) + \varphi(1 - \beta) = 1$.

By $1 - \alpha + 1 - \beta = 2 - (\alpha + \beta) \leq 1$ we get

$$\begin{aligned} & d_\varphi(\{\langle x, 1 - \alpha, 1 - \beta\rangle : x \in X\}) \\ = \; & \frac{1}{m(X)} \int_X (1 - \varphi(1 - \alpha) - \varphi(1 - \beta))\, dm = 0. \end{aligned}$$

Then Definition 2.10, (i), implies $1 - \alpha + 1 - \beta = 1$, therefore $\alpha + \beta = 1$, in contradiction with the hypothesis.

We obtain the unique possibility $\alpha + \beta = 1$. □

Further we need to introduce the concept of intuitionistic fuzzy measure with respect to a triangular norm.

Definition 2.12 Let h be a triangular norm or a triangular conorm and $X \neq \emptyset$. A family $\mathcal{I} \subseteq IFS(X)$ that satisfies:

(i) $\widetilde{0}_X \in \mathcal{I}$, where $\widetilde{0}_X = \{\langle x, 0, 1\rangle : x \in X\}$;

(ii) $A \in \mathcal{I}$ implies $\overline{A} \in \mathcal{I}$;

(iii) $(A_n)_{n\in\mathbf{N}} \subseteq \mathcal{I}$ implies $\widetilde{h}_{n\in\mathbf{N}} A_n \in \mathcal{I}$,

is called intuitionistic fuzzy h-algebra and the pair $(X, \mathcal{I})$ is called intuitionistic fuzzy h-measurable space. (Here $\widetilde{h}_{n\in\mathbf{N}} A_n$ extends the operation $\widetilde{h}$ to countable case.)

Example 2.6 If $(X, \mathcal{A})$ is a measurable space then $\mathcal{I}_{\mathcal{A}}(X)$ is an intuitionistic fuzzy h-algebra, for every continuous triangular norm or conorm h.

Definition 2.13 Let $(X, \mathcal{I})$ be an intuitionistic fuzzy h-measurable space. A function $\widetilde{m} : \mathcal{I} \to \overline{\mathbf{R}}_+$ is called intuitionistic fuzzy h-measure if it satisfies the following conditions:

(i) $\widetilde{m}\left(\widetilde{0}_X\right) = 0$;

(ii) $A, B \in \mathcal{I}$ implies $\widetilde{m}(A\widetilde{h}B) + \widetilde{m}(A\widetilde{h^c}B) = \widetilde{m}(A) + \widetilde{m}(B)$;

(iii) $(A_n)_{n\in\mathbf{N}} \subseteq \mathcal{I}, A_n \subseteq A_{n+1}, \forall n \in \mathbf{N}$ and $\lim_{n\to\infty} A_n \in \mathcal{I}$ imply $\lim_{n\to\infty} \widetilde{m}(A_n) = \widetilde{m}(\lim_{n\to\infty} A_n)$.

Between intuitionistic fuzzy measures and intuitionistic entropies we can establish the following connection.

Theorem 2.9 *Let $(X, \mathcal{A}, m)$ be a measure space. If $d_\varphi : \mathcal{I}_{\mathcal{A}}(X) \to [0,1]$ is an intuitionistic entropy and an I_φ-function , then d_φ is an intuitionistic fuzzy T_M-measure on $\mathcal{I}_{\mathcal{A}}(X)$, where T_M is the triangular norm given by $T_M(x, y) = \min(x, y)$. Moreover, if φ is additive then d_φ is an intuitionistic fuzzy h-measure on $\mathcal{I}_{\mathcal{A}}(X)$, for every continuous triangular norm h which verifies $h(x, y) + h^c(x, y) = x + y, \forall x, y \in [0,1]$.*

Proof. By Theorem 2.8 we have

$$d_\varphi(A) = \frac{1}{m(X)} \int_X (1 - \varphi(\mu_A(x)) - \varphi(\nu_A(x)))\, \mathrm{d}m,$$

with φ verifying the conditions $(i) - (iii)$ in the same theorem. We obtain

$$d_\varphi\left(\widetilde{0}_X\right) = d_\varphi(\{\langle x, 0, 1\rangle : x \in X\})$$

$$= \frac{1}{m(X)} \int_X (1 - (\varphi(0) + \varphi(1)))\, \mathrm{d}m = 0.$$

Let $(A_n)_{n\in\mathbf{N}} \subseteq \mathcal{I_A}(X)$ be an increasing sequence of intuitionistic fuzzy sets, $A_n = \{\langle x, \mu_{A_n}(x), \nu_{A_n}(x)\rangle : x \in X\}$. Then

$$\lim_{n\to\infty} d_\varphi(A_n) = \lim_{n\to\infty} \frac{1}{m(X)} \int_X (1 - (\varphi(\mu_{A_n}(x)) + \varphi(\nu_{A_n}(x))))\, \mathrm{d}m$$

$$\begin{aligned} &= \frac{1}{m(X)} \int_X \left(1 - \left(\varphi\left(\lim_{n\to\infty} \mu_{A_n}(x)\right) + \varphi\left(\lim_{n\to\infty} \nu_{A_n}(x)\right)\right)\right) \mathrm{d}m \\ &= d_\varphi\left(\langle x, \lim_{n\to\infty} \mu_{A_n}(x), \lim_{n\to\infty} \nu_{A_n}(x)\rangle : x \in X\right) = d_\varphi\left(\lim_{n\to\infty} A_n\right), \end{aligned}$$

because φ is continuous (see Ban [17]).

If φ is additive and $h(x, y) + h^c(x, y) = x + y, \forall x, y \in [0, 1]$ then we have

$$\begin{aligned} & d_\varphi\left(A\tilde{h}B\right) + d_\varphi\left(A\tilde{h^c}B\right) \\ = \; & \frac{1}{m(X)} \int_X (1 - (\varphi(h(\mu_A(x), \mu_B(x))) + \varphi(h^c(\nu_A(x), \nu_B(x)))))\, \mathrm{d}m \\ & + \frac{1}{m(X)} \int_X (1 - (\varphi(h^c(\mu_A(x), \mu_B(x))) + \varphi(h(\nu_A(x), \nu_B(x)))))\, \mathrm{d}m \\ = \; & \frac{1}{m(X)} \int_X (2 - (\varphi(\mu_A(x)) + \varphi(\mu_B(x)) + \varphi(\nu_A(x)) + \varphi(\nu_B(x))))\, \mathrm{d}m \\ = \; & \frac{1}{m(X)} \int_X (1 - (\varphi(\mu_A(x)) + \varphi(\nu_A(x))))\, \mathrm{d}m \\ & + \frac{1}{m(X)} \int_X (1 - (\varphi(\mu_B(x)) + \varphi(\nu_B(x))))\, \mathrm{d}m \\ = \; & d_\varphi(A) + d_\varphi(B). \end{aligned}$$

For $f = T_M$ the proof is similar and the conclusion is true even if φ is not additive. □

Another concept of entropy for intuitionistic fuzzy sets was introduced and studied in Szmidt-Kacprzyk [206], [207]. It is similar to other considerations for ordinary fuzzy sets (see Definitions 2.1 and 2.4). In this sense, if X is a finite set, $A, B \in IFS(X), A = \{\langle x, \mu_A(x), \nu_A(x)\rangle : x \in X\}$, $B = \{\langle x, \mu_B(x), \nu_B(x)\rangle : x \in X\}$, then the entropy for intuitionistic fuzzy sets is a function which satisfies (*i*) $E(A) = 0$ if and only if A is crisp;

(*ii*) $E(A) = 1$ if and only if $\mu_A(x) = \nu_A(x)$, $\forall x \in X$;

(*iii*) $E(A) \leq E(B)$ if A is less fuzzy than B, *i.e.* if $\mu_A(x) \leq \mu_B(x)$ and $\nu_A(x) \geq \nu_B(x)$ for $\mu_B(x) \leq \nu_B(x)$ or $\mu_A(x) \geq \mu_B(x)$ and $\nu_A(x) \leq \nu_B(x)$ for $\mu_B(x) \geq \nu_B(x)$;

(*iv*) $E(A) = E(\overline{A})$.

The condition (*i*) shows that an entropy in this sense, for an intuitionistic fuzzy set can be considered as its defect of crisp set.

2.3 Applications

2.3.1 *Application to determination of degree of interference*

Let U be a finite non-empty set and $\mathcal{P}(U)$ be the family of all subsets of U. A family $X = \{C_1, ..., C_n\} \subseteq \mathcal{P}(U)$ is a partition of U if are satisfied

(*i*) $\bigcup_{i=1}^{n} C_i = U$;

(*ii*) $C_i \cap C_j = \emptyset, \forall i, j \in \{1, ..., n\}, i \neq j$.

We name atoms the elements of X and we denote by $\mathcal{T}(U)$ the class of all partitions of U.

In what follows, we introduce a normal indicator $I_{PQ}(X)$ (that is $0 \leq I_{PQ}(X) \leq 1$) of the degree of interference between the elements of U which have a certain property "P" and those with another property "Q" (in this order), with respect to the partition $X \in \mathcal{T}(U)$.

Let us denote by p_k, the number of elements in C_k which have the property "P" and by q_k, the number of elements in C_k which have property "Q". The value that presents the relation between the number of elements with the property "P" and the number of elements that have at least one of the properties "P" or "Q" on U, is

$$t = \frac{\sum_{k=1}^{n} p_k}{\sum_{k=1}^{n} (p_k + q_k)}.$$

In general, by degree of interference between the elements with property "P" and those with property "Q", we will understand a function $I_{PQ} : \mathcal{T}(U) \to [0, 1]$ which satisfies the following requirements:

($I1$) $I_{PQ}(X) = 0$ if and only if $p_k = 0$ or $q_k = 0$, for all $k \in \{1, ...n\}$, that is, the indicator I_{PQ} assumes the minimum value if and only if each atom of the partition has only elements with the property "P" or only elements with the property "Q";

($I2$) Let $k, j \in \{1, ..., n\}, k \neq j$. If $\frac{p_j}{p_j+q_j} < \frac{p_k}{p_k+q_k} \leq t$ or $\frac{p_j}{p_j+q_j} > \frac{p_k}{p_k+q_k} \geq t$ then the contribution of C_j to $I_{PQ}(X)$ is smaller than that of C_k;

($I3$) $I_{PQ}(X) = 1$ if and only if $\frac{p_k}{p_k+q_k} = t$, for every $k \in \{1, ..., n\}$, that is the indicator I_{PQ} assumes the maximum value if and only if in every atom of the partition X is fulfilled the global situation on U, between the elements which have the property "P" and those with property "Q".

These three requirements seem to be reasonable and intuitively acceptable for the characterization of degree of interference.

Now, for every $X \in \mathcal{T}(U)$ we define the fuzzy set $A : X \to [0, 1]$, $A(C_k) = \frac{p_k}{p_k+q_k}, \forall k \in \{1, ..., n\}$, where C_k, p_k, q_k are as above and we put $I_{PQ}(X) = s_c^t(A)$, where s_c^t is a strict normalized t-measure of fuzziness. Due to conditions $(i)-(iii)$ in Definition 2.5, we obtain the following result:

Theorem 2.10 *The function $I_{PQ} : \mathcal{T}(U) \to [0, 1]$ defined as above, verifies the requirements $(I1) - (I3)$.*

Proof. It is obvious. □

It is often natural to admit that the degree of interference between the elements of a set with respect to two properties "P" and "Q", does not depend on order, that is $I_{PQ} = I_{QP}$.

Theorem 2.11 *If I_{PQ} and I_{QP} are evaluated with the strict t-measure of fuzziness H_c^t in Example 2.4, given in normalized form, then $I_{PQ}(X) = I_{QP}(X), \forall X \in \mathcal{T}(U)$.*

Proof. It is obvious that the value which presents the global situation between the number of elements with the property "Q" and the number of elements that have at least one of the properties "P" or "Q" on U, is given by

$$t' = \frac{\sum_{k=1}^{n} q_k}{\sum_{k=1}^{n} (p_k + q_k)} = 1 - t.$$

We denote $\overline{A}(C_k) = \frac{q_k}{p_k+q_k} = 1 - \frac{p_k}{p_k+q_k}, \forall k \in \{1, ..., n\}$. If the strict t-measure of fuzziness H_c^t is used to calculate I_{QP}, then

$$I_{QP}(X) = H_c^{t'}(\overline{A}) = H_c^{1-t}(\overline{A})$$

$$= -\frac{1}{|X|}\sum_{x\in X} u_{1-t}\left(1 - A(C_k)\right)\log_2 u_{1-t}\left(1 - A(C_k)\right)$$
$$-\frac{1}{|X|}\sum_{x\in X}\left(1 - u_{1-t}\left(1 - A(C_k)\right)\right)\log_2\left(1 - u_{1-t}\left(1 - A(C_k)\right)\right),$$

where $|X|$ is the cardinal of the set X.
Because (see Example 2.4)

$$u_t(a) = \begin{cases} \frac{a}{2t}, & \text{if } a \in [0,t] \\ \frac{a}{2(1-t)} + \frac{1-2t}{2(1-t)}, & \text{if } a \in [t,1], \end{cases}$$

$$u_{1-t}(1-a) = \begin{cases} \frac{1-a}{2(1-t)}, & \text{if } a \in [t,1] \\ \frac{2t-a}{2t}, & \text{if } a \in [0,t], \end{cases}$$

we get $u_t(a) + u_{1-t}(1-a) = 1, \forall a \in [0,1]$, therefore

$$I_{QP}(X) = -\frac{1}{|X|}\sum_{x\in X}\left(1 - u_t\left(A(C_k)\right)\right)\log_2\left(1 - u_t\left(A(C_k)\right)\right)$$

$$-\frac{1}{|X|}\sum_{x\in X} u_t\left(A(C_k)\right)\log_2 u_t\left(A(C_k)\right) = H_c^t(A) = I_{PQ}(X).$$

□

Remarks. 1) We can use the $\frac{1}{2}$-measures of fuzziness to determinate the indicator $I_{PQ}(X)$ even if $t \neq \frac{1}{2}$. Nevertheless, for t near to 0 or 1, we get the value of the indicator near to 0, which is not eloquent.

2) We can use the formula $I_{P_1 \dots P_n}(X) = \frac{\sum_{i,j=1,i\neq j}^{n} I_{P_1 P_j}(X)}{C_n^2}$, to calculate the degree of interference of elements in U with respect to the properties "P_1", ..., "P_n", where $I_{P_i P_j}$ is the degree of interference of elements in U with the properties "P_i" and "P_j".

In what follows, we present a numerical example with applications to the geography of population.

Let U be the set of habitants of a country having n districts and let us denote by C_k, the set of habitants of the district $k, k \in \{1, ..., n\}$. Also, we consider two nationalities of this country, denoted by "P" and "Q". We desire to determinate the degree of interference of these, knowing the result of census of population (see table below). We denote by p_k and by q_k, the number of persons (expressed in tens of thousands, for example) from C_k belonging to the nationality "P" and "Q", respectively. Because it is natural to consider equality between the degree of interference of nationality "P" with "Q" and the degree of interference of nationality "Q" with "P", the above Theorem 2.11 justifies the t-measure of fuzziness given in Example 2.4 to determinate the value of indicator $I_{PQ}(X)$, where $X \in \mathcal{T}(U)$.

District	p_k	q_k	County	p_{kj}	q_{kj}
			1	8	10
			2	10	12
1	30	50	3	6	8
			4	2	7
			5	4	13
			1	8	5
2	30	20	2	3	8
			3	7	4
			4	12	3
			1	6	14
3	20	40	2	10	7
			3	4	19
			1	20	6
4	40	70	2	5	28
			3	6	22
			4	9	14

If $X = \{C_1, C_2, C_3, C_4\}$ with the values in above table, then we obtain $t = 0.4$. For this value, by using the indicated t-measure of fuzziness, we get

$$I_{PQ}(X) = 0.9723.$$

Now, let us assume that each district has more counties and let us denote by C_{kj}, the set of habitants of the county j from district k. Also, we denote by p_{kj} and q_{kj}, the number of persons (expressed in tens of thousands) from C_{kj} belonging to the nationality "P" and "Q", respectively. For this territorial structure, that is for the partition $X' = \{C_{11}, C_{12}, ..., C_{44}\}$ (see table), we obtain the degree of interference of nationality "P" with the nationality "Q",

$$I_{PQ}(X') = 0.8727.$$

The degree of interference corresponding to the districts, $I_{PQ}(X)$, is close to

1 because the ratio of the number of habitants of nationality "P" and the total number of habitants, calculated for each district $\left(\frac{p_k}{p_k+q_k}, k \in \{1,2,3,4\}\right)$ is relatively close to the same ratio calculated for the level of whole country, that is $t = 0.4$. $I_{PQ}(X')$ has a smaller value because in more counties the ratio $\frac{p_{kj}}{p_{kj}+q_{kj}}$ have remote values from $t = 0.4$, for example $\frac{p_{24}}{p_{24}+q_{24}} = 0.8$ and $\frac{p_{42}}{p_{42}+q_{42}} = 0.15$ (see also $(I2)$).

Modifications of the indicator of degree of interference (regarding more census of population or with respect to different territorial structures) leads to interesting conclusions from the geography of population viewpoint.

Finally, we observe that other indicators in geography of population can be introduced with the help of t-measures of fuzziness.

Let V be a country and let us denote by $X = \{S_1, ..., S_n\}$, a partition of V, that is $S_i, i \in \{1, ..., n\}$ represent all the districts of V. We can introduce a normal indicator $\Gamma_{U,X}$ (that is $0 \leq \Gamma_{U,X} \leq 1$) which estimates the degree of homogeneity of territorial distribution of population in V, with respect to the organization corresponding to X. We notice that, the density of population, the usual indicator used in this situation, is a global indicator which does not take account the pointwise aspects. For example, if there exists a great concentration of population in certain zones, this indicator is not significant.

Let us denote by Γ_{med}, the density of population in V, that is

$$\Gamma_{med} = \frac{\text{total population in } V}{\text{area of } V},$$

by Γ_{MAX}, the maximum possible density of population in a district of V, that is

$$\Gamma_{MAX} = \frac{\text{total population in } V}{\text{area of the smaller district } S_i}$$

and by Γ_i, the density of population in $S_i, i \in \{1, ..., n\}$, that is

$$\Gamma_i = \frac{\text{population in } S_i}{\text{area of } S_i}.$$

We define the fuzzy set $P : X \to [0,1]$ by

$$P(S_i) = \frac{\Gamma_i}{\Gamma_{MAX}},$$

and we denote $t = \frac{\Gamma_{med}}{\Gamma_{MAX}}$.

We put $\Gamma_{V,X} = s_c^t(P)$, where s_c^t is a strict t-measure of fuzziness in normalized form. Following the axioms of strict t-measures of fuzziness (see Definition 2.5), we get properties of the indicator $\Gamma_{V,X}$ which are in concordance with our intuition:

$(O1)$ $\Gamma_{V,X}$ has the minimum value (equal to 0) if and only if the entire population of V is completely situated in the district with the smallest area;

$(O2)$ If $\Gamma_j \leq \Gamma_i \leq t$ or $t \leq \Gamma_i \leq \Gamma_j$ then the contribution of the district S_i to $\Gamma_{V,X}$ is greater than the contribution of the district S_j;

$(O3)$ $\Gamma_{X,U}$ has the maximum value (equal to 1) if and only if in all districts of V the density of population is equal to Γ_{med}.

Choosing a suitable strict t-measure of fuzziness (for example, that given in Example 2.3) we can get interesting results concerning the territorial distribution of population.

2.3.2 *Application to description of the performance of systems*

Fuzzy methods in the study of the performance of systems are described in many papers (see, for example, Kaleva [111] and [112], Kaufmann [114], Kaufmann-Grouchko-Cruon [115], Misra-Sharma [150], Soman-Krishna [201]), taking into account that the performance level of a system is a value included in $[0,1]$: the value 0 corresponds to non-working and the value 1 corresponds to optimum working.

In Kaufmann [114] and Kaleva [111] a mapping indicating the performance level of a system as function of level of performance of its components is discussed. But, it is often important to know the degree of homogeneity of a system, that is how much the performance level of components is different from the performance level of system. In this application, we introduce the degree of homogeneity of a system with the help of t-measures of fuzziness, by using its performance function.

Let $X = \{x_1, ..., x_n\}$ be the set of components of a system. If to each component x_i we attach a value $a_i = A(x_i) \in [0,1]$ indicating the performance level of that component, then the working of system is described by the fuzzy set $A : X \to [0,1]$. For a coherent system (that is, a system having a series-parallel network representation) we can consider a function indicating the performance level of the system, as a function of the component performances. We denote this function by Ψ and we call it, performance function (see Kaleva [112], Kaufmann [114]).

We introduce a normal index $H_{\Psi,X}$ (that is, $0 \leq H_{\Psi,X}(A) \leq 1$, for every fuzzy set A) of the degree of homogeneity for a system with the set of its components, $X = \{x_1, ..., x_n\}$ and the performance function Ψ. We assume that $t = \Psi(a_1, ..., a_n) \in [0,1]$ is the performance level of the system X, where a_i is the performance level of component x_i, for $i \in \{1, ..., n\}$. The following four requirements appear to be reasonable and intuitively acceptable to characterize the degree of homogeneity $H_{\Psi,X}$:

$(H1)$ $H_{\Psi,X}(A) = 0$ if and only if $a_i = 0$ or $a_i = 1$, for all $i \in \{1, ..., n\}$, that is the index $H_{\Psi,X}$ assumes the minimum value if and only if all the components of the system are in two states: operating or failed.

$(H2)$ Let $i, j \in \{1, ..., n\}$. If $a_i < a_j \leq \Psi(a_1, ..., a_n)$ or $a_i > a_j \geq \Psi(a_1, ..., a_n)$ then the contribution of the component x_i to $H_{\Psi,X}(A)$ is smaller than that of the component x_j, that is the contribution of a component is greater if its working is near to the working of the system.

$(H3)$ $H_{\Psi,X}(A) = 1$ if and only if $a_i = \Psi(a_1, ..., a_n)$, for all $i \in \{1, ..., n\}$, that is the index $H_{\Psi,X}$ assumes the maximum value if and only if the performance level of every component is equal to the performance level of the entire system.

$(H4)$ The degree of homogeneity of the system is the same if the performance level of components is at equal distance from the performance level of system in any sense (from 0 or from 1).

If we compare the conditions $(i) - (iii)$ in Definition 2.5 and the additional condition (2.5) with the previous requirements, then we see that the degree of homogeneity $H_{\Psi,X}(A)$ of a system with components $\{x_1, ..., x_n\}$, can be considered as the symmetrical normalized strict $\Psi(a_1, ..., a_n)$-measure of fuzziness of the fuzzy set $A : X \to [0,1], A(x_i) = a_i$ which indicates the performance level of components, if the performance level of system is $t = \Psi(a_1, ..., a_n) \in (0,1)$.

In concordance with the situation $\Psi(a_1, ..., a_n) \in (0,1)$ and with the requirements $(H1) - (H4)$, the following statements are acceptable for the situation $t = \Psi(a_1, ..., a_n) \in \{0,1\}$.

$(h1)$ If $t = \Psi(a_1, ..., a_n) = 0$, then the degree of homogeneity of the system is equal to 1 if and only if $a_i = 0$, for all $i \in \{1, ..., n\}$, equal to 0 if and only if $a_i = 1$, for all $i \in \{1, ..., n\}$ and decreases when the performance level of components increases.

$(h2)$ If $t = \Psi(a_1, ..., a_n) = 1$, then the degree of homogeneity of the system is equal to 1 if and only if $a_i = 1$, for all $i \in \{1, ..., n\}$, equal to 0 if and only if $a_i = 0$,for all $i \in \{1, ..., n\}$ and decreases when the performance level of components decreases.

Even if any strict and symmetrical t-measure of fuzziness can be used to determine the value of the index $H_{\Psi,X}(A)$, a suitable choice is given by Example 2.3, taking the fuzzy complement in Example 2.2. If $X = \{x_1, ..., x_n\}$, then we normalize and simplify and this t-measure of fuzziness ($t \in (0,1)$) becomes

$$s^t_{c,K}(A) = \frac{1}{n}\sum_{i=1}^{n} g\left(A(x_i)\right),$$

where $g(a) = a$ if $a \in [0,t]$ and $g(a) = \frac{t}{1-t}(1-a)$ if $a \in [t,1]$.

In fact, these t-measures of fuzziness are introduced (for $t = \frac{1}{2}$) by Kaufmann [114] (see also Klir [120]).

Taking into account ($h1$) and ($h2$), a formula of calculus for the degree of homogeneity of a system is given by

$$H_{\Psi,X}(A) = \begin{cases} s^t_{c,K}(A), & \text{if } 0 < t < 1 \\ \frac{1}{n}\sum_{i=1}^{n}(1-a_i), & \text{if } t = 0 \\ \frac{1}{n}\sum_{i=1}^{n} a_i, & \text{if } t = 1, \end{cases}$$

where $t = \Psi(a_1, ..., a_n)$.

In a concrete case, we consider a system with three components $X = \{x_1, x_2, x_3\}$, having the performance function

$$\Psi(a, b, c) = \max\left(\min(a, b), \min(a, c)\right),$$

where a, b, c represent the level performance of components x_1, x_2 and x_3, respectively. The table of values of this function, corresponding to different ordering of variables is (see Kaleva [112]):

$\leq$	$\leq$		$min\,(a,b)$	$min\,(a,c)$	Ψ
a	b	c	a	a	a
a	c	b	a	a	a
b	a	c	b	a	a
b	c	a	b	c	c
c	a	b	a	c	a
c	b	a	b	c	b

We obtain the below table with the results of degrees of homogeneity corresponding to different ordering of variables ($n = 3, a_1 = a, a_2 = b, a_3 = c$):

		$\le$	$\le$	Ψ	$H_{\Psi,X}$
$a = 0$				0	$\frac{1-b-c}{3}$
$a \in (0,1)$	a	b	c	a	$\frac{a(3-a-b-c)}{3(1-a)}$
$a = 1$				1	1
$a = 0$				0	$\frac{1-b-c}{3}$
$a \in (0,1)$	a	c	b	a	$\frac{a(3-a-b-c)}{3(1-a)}$
$a = 1$				1	1
$a = 0$				0	$\frac{3-c}{3}$
$a \in (0,1)$	b	a	c	a	$\frac{b-ab+a-a^2+a-ac}{3(1-a)}$
$a = 1$				1	$\frac{b+2}{3}$
$a = 0$				0	$\frac{3-b}{3}$
$a \in (0,1)$	c	a	b	a	$\frac{c-ac+a-a^2+a-ab}{3(1-a)}$
$a = 1$				1	$\frac{c+2}{3}$
$b = 0$				0	$\frac{3-a}{3}$
$b \in (0,1)$	c	b	a	b	$\frac{c-bc+b-b^2+b-ab}{3(1-b)}$
$b = 1$				1	$\frac{c+2}{3}$
$c = 0$				0	$\frac{3-a}{3}$
$c \in (0,1)$	b	c	a	c	$\frac{c-c^2+b-bc+c-ac}{3(1-c)}$
$c = 1$				1	$\frac{b+2}{3}$

2.3.3 *Application to digital image processing*

Digital image processing is a discipline with many applications in document reading, automated assembly and inspection, radiology, hematology, meteorology, geology, land-use management, *etc.* (see Rosenfeld [175]). One of methods used in the study of digital images and, more general, of discrete arrays in two or more dimensions of whose elements which have values 0 or 1, is the digital topology (for details see Herman [101], Kong-Rosenfeld [123], Kong-Roscoe-Rosenfeld [125] and Kong-Kopperman-Meyer [124]). Although some parts of digital topology can be generalized to fuzzy digital topology, which deals with gray-scale image arrays whose elements lie in the range [0, 1] (for details, see Rosenfeld [176]-[179]), nevertheless the study of binary image is more simple.

In what follows, by using t-measures of fuzziness, we give an acceptable method to obtain a binary digital image from a fuzzy digital image.

A digital image is typically obtained quantifying the brightness values of a image in a discrete and bounded grid of points, named background and denoted in what follows by X. If the brightness values are considered in $[0,1]$, then we obtain a fuzzy digital image or a gray-scale image array. If the set of values is $\{0,1\}$ we obtain a binary digital image.

For every $x \in X$ we denote by $A(x) \in [0,1]$ the gray-level of the point x. So we obtain the fuzzy set $A : X \to [0,1]$.

Let $t \in (0,1)$. By tresholding a fuzzy digital picture with value t, that is by classifying the points $x \in X$ according to the fact that their gray-level exceed t or not exceed t, we obtain a binary digital picture (if $A(x) > t$ then the new value in $x \in X$ is 1 and if $A(x) \leq t$ then the new value in $x \in X$ is 0), but we lose a part of the information contained in the initial picture. Our purpose is to choose the value (or values) t such that the lost information to be minimum. We denote by $P^t(A)$ the lost information if the level of tresholding is $t \in (0,1)$.

Intuitively, we can accept the following natural conditions on the lost information $P^t(A)$:

$(P1)$ $P^t(A) = 0$ if and only if $A(x) \in \{0,1\}, \forall x \in X$, that is the initial digital image is binary.

$(P2)$ $a)$ If the gray-levels of two points are smaller than t then the contribution to lost information of the point with greater gray-level is more important than that of the point with smaller gray-level;
$b)$ If the gray-levels of two points are greater than t, then the contribution to lost information of the point with smaller gray-level is more important than that of the point with greater gray-level.

$(P3)$ $P^t(A)$ is maximum if and only if, for any point $x \in X$, the lost information is maximum, that is $A(x) = t, \forall x \in X$.

We notice that these conditions are verified for every strict t-measure of fuzziness s_c^t. So, if $A \in FS(X)$ is the fuzzy set which represents the gray-level of the points of background X (for a fuzzy digital image) and $\{s_c^t\}_{t\in(0,1)}$ is a family of strict t-measures of fuzziness, then

$$\mathcal{M}_{s_c^t}(A) = \inf_{t\in(0,1)} P^t(A) = \inf_{t\in(0,1)} s_c^t(A)$$

is the minimal lost information by transition from a gray-scale image array to a binary image array, with respect to the family of t-measures of fuzziness $\{s_c^t\}_{t\in(0,1)}$.

Remarks. 1) Because A is a finite fuzzy set, we can assume that

$$0 \leq A(x_0) < A(x_1) < \dots < A(x_k) < A(x_{k+1}) \leq 1$$

given up on the common values. In this case we obtain

$$\mathcal{M}_{s_c^t}(A) = \min_{p\in\{0,\dots,k\}} \left\{ s_c^{t_p}(A); A(x_p) < t_p < A(x_{p+1}) \right\}.$$

and the values of t for which is obtained this minimum represent an interval.

2) In practice, the members of family $\{s_c^t\}_{t\in(0,1)}$ are of the same kind, for any $t \in (0,1)$.

2.4 Bibliographical Remarks

Definitions 2.4 and 2.7, Example 2.5, Theorem 2.5 are in Ban [16], Definition 2.6, Theorem 2.4, Examples 2.2 and 2.3 are in Ban [21], Definitions 2.9, 2.10 are in Ban [18] and Theorems 2.6-2.9 are proved in Ban [18]. Definitions 2.12, 2.13 are in Ban [17], Theorems 2.10, 2.11 and Example 2.4 are in Ban-Fechete [20]. The application to systems performance is in Ban [21] and to the calculus of degree of interference is in Ban-Fechete [20]. Completely new are Definition 2.5 and Theorem 2.3.

Chapter 3

Defect of Property in Topology

A well-known concept in topology is that of measure of noncompactness, which in our terminology can be called defect of compactness. The main aim of this chapter is to present this concept in classical setting, in random setting and in fuzzy setting.

Also, other defects of topological properties are considered.

3.1 Measures of Noncompactness for Classical Sets

Let (X, ρ) be a metric space and let us denote

$$\mathcal{P}_b(X) = \{Y \subset X; Y \neq \emptyset, Y \text{ is bounded}\}.$$

Mainly, two functionals that measure the degree in which subsets of X fail to be compact are well-known.

Definition 3.1 (*i*) (Kuratowski [129], [130]). The Kuratowski's measure of noncompactness for $Y \in \mathcal{P}_b(X)$ is given by

$$\alpha(Y) = \inf \Big\{\varepsilon > 0; \exists n \in \mathbf{N}, A_i \in X, i = \overline{1,n} \text{ with } Y \subset \textstyle\bigcup_{i=1}^{n} A_i \text{ and } diam(A_i) < \varepsilon\Big\},$$

where $diam(A_i) = \sup\{\rho(x,y); x, y \in A_i\}, i = \overline{1,n}$.

(*ii*) (see *e.g.* Istrătescu [105]) The Hausdorff's measure of noncompactness of $Y \subset X$ is given by

$$h(Y) = \inf\left\{\varepsilon > 0; \exists n \in \mathbf{N}, y_i \in Y, i = \overline{1,n} \text{ with } Y \subset \bigcup_{i=1}^{n} B(y_i, \varepsilon)\right\}.$$

Concerning the quantities $\alpha(Y)$ and $h(Y)$ the following results are known.

Theorem 3.1 *(i) (see e.g. Rus [182], p.85-87 or Banas-Goebel [32]) Let $X, Y \in \mathcal{P}_b(X)$. We have:*
$0 \leq \alpha(A) \leq diam(A)$;
$A \subseteq B$ implies $\alpha(A) \leq \alpha(B)$;
$\alpha(\overline{A}) = \alpha(A)$ and moreover $\alpha(A) = 0$ if and only if $\overline{A}$ is compact;
$\alpha(V_\varepsilon(A)) \leq \alpha(A) + 2\varepsilon$, where $V_\varepsilon(A) = \{x \in X; \rho(x, A) < \varepsilon\}$ and $\rho(x, A) = \inf\{\rho(x, y); y \in A\}$;
$\alpha(A \cup B) = \max\{\alpha(A), \alpha(B)\}$;
$\alpha(A \cap B) \leq \min\{\alpha(A), \alpha(B)\}$;
Let $A_i \subset X, A_{i+1} \subset A_i$ be with A_i closed and nonempty, $i = 1, 2, \ldots$.
If $\lim_{n\to\infty} \alpha(A_n) = 0$ then $\bigcap_{n=1}^{\infty} A_n$ is nonempty and compact.
If, in addition, X is a Banach space, then $\alpha(A + B) \leq \alpha(A) + \alpha(B)$, $\alpha(cA) = |c|\,\alpha(A), c \in \mathbf{R}, \alpha(convA) = \alpha(A)$, where $convA$ is the convex hull of $A \in \mathcal{P}_b(X)$.

(ii) (see e.g. Beer [36], Banas-Goebel [32])
$h(A) = 0$ if and only if A is totally bounded (that is, $\forall \varepsilon > 0, \exists x_1, \ldots, x_p \in A$ such that $\forall x \in X, \exists x_i$ with $\rho(x, x_i) < \varepsilon$);
$A \subset B$ implies $h(A) \leq 2h(B)$;
$h(\overline{A}) = h(A)$;
$h(A \cup B) \leq \max\{h(A), h(B)\}$;
h is continuous on $CL(X) = \{Y \subset X; Y$ is closed, $Y \neq \emptyset\}$ with respect to Hausdorff topology, i.e. if $Y_n, Y \in CL(X), n = 1, 2, \ldots$, satisfy $D_H(Y_n, Y) \overset{n\to\infty}{\to} 0$ (where D_H is the Hausdorff-Pompeiu distance), then $\lim_{n\to\infty} h(Y_n) = h(Y)$; Also, h is upper semicontinuous on $CL(X)$ with respect to the so-called Vietoris topology.

In addition,

$$h(A) \leq \alpha(A) \leq 2h(A), \forall A \in \mathcal{P}_b(X)$$

and if moreover X is Hilbert space, then

$$\sqrt{2}h(A) \leq \alpha(A) \leq 2h(A), \forall A \in \mathcal{P}_b(X).$$

Remarks. 1) If (X, ρ) is a complete metric space, then for $A \in CL(X)$, we have

$$\alpha(A) = 0 \text{ if and only if } A \text{ is compact}$$

and

$$h(A) = 0 \text{ if and only if } A \text{ is compact.}$$

Because of these properties, we also can call $\alpha(A)$ and $h(A)$ as defects of compactness.

2) The Kuratowski's measure of noncompactness has many applications to fixed point theory (see *e.g.* Darbo [60], Sadowski [188]).

The properties in Theorem 3.1 suggest an axiomatic approach in Banach spaces $(X, \|\cdot\|)$, as follows. Firstly we need the notations:

$$\begin{aligned} RC(X) &= \{Y \subset X; Y \neq \emptyset, Y \text{ is relatively compact}\}, \\ CO(X) &= \{Y \subset X; Y \neq \emptyset, Y \text{ is compact}\}. \end{aligned}$$

Definition 3.2 (Banas-Goebel [32]). $\mathcal{K} \subset RC(X)$ is called kernel (of a measure of noncompactness) if it satisfies:

(i) $A \in \mathcal{K}$ implies $\overline{A} \in \mathcal{K}$;

(ii) $A \in \mathcal{K}, B \subset A, B \neq \emptyset$ implies $B \in \mathcal{K}$;

(iii) $A, B \in \mathcal{K}$ implies $\lambda A + (1-\lambda) B \in \mathcal{K}, \forall \lambda \in [0,1]$;

(iv) $A \in \mathcal{K}$ implies $convA \in \mathcal{K}$;

(v) The set $\mathcal{K}^c = \{A \in \mathcal{K}; A \text{ is compact}\}$ is closed in $CO(X)$ with respect to Hausdorff topology (*i.e.* the topology induced on $CO(X)$ by the Hausdorff-Pompeiu distance).

Definition 3.3 (Banas-Goebel [32]). The function $\mu : \mathcal{P}_b(X) \to [0, +\infty)$ is called measure of noncompactness (or defect of compactness, in our terminology) with the kernel $\mathcal{K}$ (denoting $\ker \mu = \mathcal{K}$) if satisfies:

(i) $\mu(A) = 0$ if and only if $A \in \mathcal{K}$;

(ii) $\mu(\overline{A}) = \mu(A)$;

(iii) $\mu(convA) = \mu(A)$;

(iv) $A \subset B$ implies $\mu(A) \leq \mu(B)$;

(v) $\mu(\lambda A + (1-\lambda) B) \leq \lambda \mu(A) + (1-\lambda) \mu(B), \forall \lambda \in [0,1]$;

(vi) If $A_n \in \mathcal{P}_b(X), \overline{A}_n = A_n$ and $A_{n+1} \subset A_n, n = 1, 2, \ldots$ and if $\lim_{n\to\infty} \mu(A_n) = 0$, then $\bigcap_{n=1}^{\infty} A_n \neq \emptyset$.

If $\mathcal{K} = RC(X)$ then μ will be called full (or complete) measure.

The following examples are in Banas-Goebel [32].

Example 3.1 α and h are full measures of noncompactness.

Example 3.2 $\mu(A) = diam(A), A \in \mathcal{P}_b(X)$, with kernel $\mathcal{K} = \{\{x\}; x \in X\}$.

Example 3.3 Let $F \subset X$ be closed and $\mathcal{K} = RC(X)$. Then $\mu(A) = h(A) + d_H(A, F), A \in \mathcal{P}_b(X)$, where

$$d_H(A, F) = \sup\{\inf\{\|a - y\|; a \in A\}; y \in F\}.$$

Example 3.4 Let us suppose that $(X, \|\cdot\|)$ is a Banach space that have a Schauder basis $\{e_i\}_{i \in \mathbf{N}}$, that is each $x \in X$ has an unique representation $x = \sum_{i=0}^{\infty} \varphi_i(x) e_i$, where $\varphi_i : X \to \mathbf{R}, i \in \mathbf{N}$. Denoting $R_n : X \to X$ by $R_n(x) = \sum_{i=n+1}^{\infty} \varphi_i(x) e_i$, the function $\mu(A) = \lim_{n\to\infty} \sup |||R_n|||$ is a regular measure of noncompactness on $\mathcal{P}_b(X)$ (that is full, and satisfies $\mu(A + B) \le \mu(A) + \mu(B), \mu(A \cup B) \le \max\{\mu(A), \mu(B)\}$), where $|||R_n|||$ denotes the norm of linear continuous operator R_n.

Example 3.5 Let (K, ρ) be a compact metric space and $X = C(K; \mathbf{R}) = \{f : K \to \mathbf{R}; f \text{ continuous on } K\}$, endowed with the uniform norm $\|f\| = \max\{|f(x)|; x \in K\}$. For any $f \in X$ we can define the modulus of continuity of f by

$$\omega(f; \varepsilon) = \sup\{|f(t) - f(s)|; t, s \in K, \rho(t, s) \le \varepsilon\}, \forall \varepsilon \ge 0$$

and for any $A \subset X$, let us define

$$\omega(A; \varepsilon) = \sup\{\omega(f; \varepsilon); f \in A\}, \forall \varepsilon \ge 0.$$

Then

$$\mu(A) = \lim_{\varepsilon \to 0} \omega(A; \varepsilon)$$

is a measure of noncompactness on $\mathcal{P}_b(X)$, which moreover satisfies

$$\mu(A) = 2h(A),$$

where $h(A)$ is the Hausdorff measure of noncompactness in Definition 3.1.

Example 3.6 Let $X = L^p[a, b], 1 \le p < +\infty$, be endowed with the norm $\|f\|_p = \left(\int_a^b |f(t)|^p \, dt\right)^{\frac{1}{p}}$ and let $\beta : [0, +\infty) \to [0, +\infty)$ be with the property $\lim_{t\to 0} \beta(t) = \beta(0) = 0$. For $f \in X$, let us define the Kolmogorov modulus of continuity of f with respect to β by

$$\omega_\beta(f; \varepsilon) = \sup\left\{\|f_h - f\|_p - \beta(|h|); |h| \le \varepsilon\right\},$$

(where $f_h(t) = f(t+h), \forall t \in [a,b]$) and for $A \in \mathcal{P}_b(X)$, let us define

$$\begin{aligned} \omega_\beta(A;\varepsilon) &= \sup\{\omega_\beta(f;\varepsilon); f \in A\}, \\ \mu_\beta(A) &= \lim_{\varepsilon\to 0}\omega_\beta(A;\varepsilon). \end{aligned}$$

Then $\mu_\beta : \mathcal{P}_b(X) \to [0,+\infty)$ is a regular measure of noncompactness.

Remark. For many other details concerning the axiomatic approach of measures of noncompactness, see the interesting book Banas-Goebel [32].

The end of this section is based on an idea used in Kupka [127] which is the following: When we want to model "metric" or "uniform" notions in pure topological case, it suffices sometimes to replace "$\varepsilon > 0$" by "open cover of X". Then, we can obtain analogous results to the metric case. In this sense, we will introduce a topological analogue of measure of noncompactness and we will show some basic properties of this kind of measure. Finally, as applications we prove the upper semicontinuity of the limit of a decreasing net of upper semicontinuous multifunctions.

Let π be the set of all open coverings of a topological space X. Let us consider the family $\mathcal{C} = 2^\pi$ of all subsets of π ordered by the relation

$$x \leq y \text{ if and only if } x \supset y, \text{ for } x, y \in \mathcal{C}.$$

Then $(\mathcal{C}, \leq)$ is a complete lattice with the minimal element $\Theta = \pi$. Now we can define the $\mathcal{C}$-measure of noncompactness as follows:

Definition 3.4 (Kupka-Toma [128]). If $A \subset X$ then the measure of noncompactness of A is the element of $\mathcal{C}$ defined by

$$m(A) = \left\{\mathcal{P} \in \pi \,\middle|\, \exists \text{ finite } \delta \subset \mathcal{P} : A \subset \bigcup \delta\right\}.$$

Remark. Lechicki in his article [132] has introduced a measure of noncompactness for uniform spaces, whose values are in a smaller system of sets than $\mathcal{C}$. But in general topological spaces we have no notion of "uniformness" as in the case of uniform spaces. The above notion of measure of noncompactness applied in the class of uniform spaces is stronger than Lechicki's one.

Theorem 3.2 *(Kupka-Toma [128]). The mapping $m : 2^X \to \mathcal{C}$ has the following properties:*

(*i*) *If A is a compact subset of X then $m(A) = \Theta$. If A is closed then the converse is true;*

(*ii*) *If $A \subset B$ then $m(A) \leq m(B)$;*

(*iii*) $m(A \cup B) = m(A) \cap m(B) = \sup\{m(A), m(B)\}$;

(*iv*) $m(A \cap B) \leq m(A) \cup m(B) = \inf\{m(A), m(B)\}$.

The next two examples prove that if A is not closed, then the equality $m(A) = \Theta$ does not imply the compactness of A and that the inequality in (*iv*) can be strict.

Example 3.7 Let X be a compact topological space, which contains a noncompact subset A. Then $m(X) = \Theta$ and consequently $m(A) = \Theta$ too, but A is not compact.

Example 3.8 Let $X = \mathbf{R}$ with the usual topology, $A = (-\infty, 0], B = [0, \infty)$. $A \cap B = \{0\}$ is compact and therefore $m(A \cap B) = \Theta$ but $\inf\{m(A), m(B)\} \neq \Theta$ because the open covering

$$\mathcal{P} = \{(n-1, n+1); n \in \mathbf{Z}\}$$

of X does not contain a finite subcovering neither of the set A nor of the set B.

In the class of complete uniform spaces, we can partially compare the measure of noncompactness m, with Lechicki's measure which is denoted by Φ (see Lechicki [132]).

Theorem 3.3 *(Kupka-Toma [128]). Let X be a complete uniform space. Then we have the following:*

(*i*) *For every closed subset $A \subset X, m(A) = \Theta$ if and only if $\Phi(A) = 0$;*

(*ii*) *For every decreasing net $(A_\gamma : \gamma \in \Gamma)$ of closed nonempty subsets of X,*

$$\inf\{\Phi(A_\gamma) : \gamma \in \Gamma\} = 0 \Rightarrow \lim_{\gamma \in \Gamma} m(A_\gamma) = \Theta.$$

The next example shows that if A is not closed, then the equivalence (*i*) of Theorem 3.3 may not be true, because in general we can have $m(A) \neq m(\overline{A})$, where $\overline{A}$ is the closure of A. This also shows that topological measure of noncompactness can distinguish such subtle differences which are indistinguishable by the classical (metric) measure of noncompactness. Moreover, one can easily see that $m(A) = \Theta$ is in topological case stronger than $m(A) = 0$ in the metric case.

Example 3.9 Consider $\mathbf{R}$ with the α-topology (open sets are the sets of the form $O \setminus N$ where O is an usual open set and N is a nowhere-dense set (see Njastad [160]). Let us take the following α-open set

$$A = (0,1) \setminus \left\{ \frac{1}{n}; n \in \mathbf{N} \right\}.$$

The α-closure of A is $\overline{A} = [0,1]$. The open covering $\mathcal{P}$ of X defined by

$$\mathcal{P} = \{A\} \cup \left\{ \left(\frac{1}{n} - \frac{1}{(n+1)^2}, \frac{1}{n} + \frac{1}{(n+1)^2} \right); n \in \mathbf{N} \right\} \cup \left\{ \mathbf{R} \setminus \left\{ \frac{1}{n}; n \in \mathbf{N} \right\} \right\}$$

contains a finite subcovering of the set A, namely $\{A\}$, and therefore $\mathcal{P} \in m(A)$. But, $\mathcal{P} \notin m(\overline{A})$ because the points $\frac{1}{n} \in \overline{A}$ cannot be covered by finite number of sets from $\mathcal{P}$.

Let $(\Gamma, \leq)$ be a upward directed set. If $(x_\gamma : \gamma \in \Gamma)$ is a net in $\mathcal{C}$, then we shall write $\lim_{\gamma \in \Gamma} x_\gamma = \Theta$ if and only if

$$\forall \{\mathcal{P}\} \in \mathcal{C}, \exists \gamma_0 \in \Gamma, \forall \gamma \geq_\Gamma \gamma_0 : x_\gamma \leq \{\mathcal{P}\}.$$

The next theorem is a topological analogue of the Theorem 2.2 in Lechicki [132]. It works with a complete uniform space which in general, of course, is not our case. Because of *(ii)* in Theorem 3.3, we can state that the theorem generalizes Theorem 2.2 of Lechicki, if the space X is supposed to be uniformly complete.

Theorem 3.4 *(Kupka-Toma [128]). Let $(A_\gamma : \gamma \in \Gamma)$ be a decreasing net of closed nonempty subsets of a topological space X. Then the following implication is true:*

$$\lim_{\gamma \in \Gamma} m(A_\gamma) = \Theta \Rightarrow A = \bigcap_{\gamma \in \Gamma} A_\gamma \text{ is nonempty and compact.}$$

The next theorem generalizes a theorem which is well-known in the metric case. The notion of upper semicontinuity being a topological one (see *e.g.* Michael [147]), it is natural to search an analogous result in topological setting.

Theorem 3.5 *(Kupka-Toma [128]). Let X, Y be topological spaces and $(F_\gamma : X \to Y)_{\gamma \in \Gamma}$ be a decreasing net of u.s.c. multifunctions with closed* *nonempty values. If for each* $x \in X, \lim_{\gamma \in \Gamma} m(F_\gamma(x)) = \Theta$*, then the multifunction*

$$F : X \to Y, F(x) = \bigcap_{\gamma \in \Gamma} F_\gamma(x)$$

is upper semicontinuous with nonempty compact values.

Remark. The net $(F_\gamma : \gamma \in \Gamma)$ is decreasing and it is easy to see that this net converges to F in the sense of the Vietoris topology (see *e.g.* Michael [147]).

The condition in Theorem 3.5 that the measure of noncompactness of $F_n(x)$ tends to zero, cannot be omitted as the next example shows.

Example 3.10 Let $X = Y = [0,1) \cup (1,+\infty)$ with the usual topology. For each $n \in \mathbf{N}$ and $x \in X$ we define the multifunctions

$$\begin{aligned} F_n(x) &= \left[0, 1 + x + \frac{1}{n}\right] \setminus \{1\} \\ F(x) &= [0, 1+x] \setminus \{1\} \text{ if } x \neq 0, F(0) = [0,1). \end{aligned}$$

Then $\forall x \in X, F_n(x) \supset F_{n+1}(x)$ and $F(x) = \bigcap_{n \in \mathbf{N}} F_n(x)$. The multifunctions F_n are upper semicontinuous but F is not upper semicontinuous at the point 0. To see this, let us remark that $F(0) \subset [0,1) = U, U$ is open in X and for each open neighborhood V of 0 we have $F(V) \not\subseteq U$.

The next example shows that in Theorem 3.5, the hypothesis that the values of multifunctions are closed, cannot be omitted.

Example 3.11 Let us consider the complet metric spaces $X = [0,+\infty)$, $Y \subset \mathbf{R}$ with the usual metric. For each $x \in X$, we define the multifunctions:

$$\begin{aligned} F_n(x) &= \left(-\infty, -\frac{n}{nx+1}\right] \cup \{1\}, \\ F(x) &= \left(-\infty, -\frac{1}{x}\right] \cup \{1\} \text{ if } x \neq 0, F(0) = \{1\}. \end{aligned}$$

Then

$$\forall x \in X, n \in \mathbf{N} : F_n(x) \supset F_{n+1}(x) \text{ and } F(x) = \bigcap_{n \in \mathbf{N}} F_n(x).$$

The multifunctions F_n are all u.s.c. but their limit F is not, because it is not u.s.c. at the point 0.

Finally, we give an example which shows why it is necessary to consider the family π of all open coverings of X and not only a subfamily of coverings of certain type, as for example ball coverings in metric space. We will construct two nice families of coverings which generate very different measures of noncompactness.

Example 3.12 Let $X = \left(-\frac{\pi}{2}, \frac{\pi}{2}\right)$ with the usual topology. We can consider two families of open coverings of X:

$$\begin{aligned} \pi_1 &= \{\mathcal{P}_\varepsilon; \varepsilon > 0\}, \mathcal{P}_\varepsilon = \{(a-\varepsilon, a+\varepsilon) \cap X; a \in X\}, \\ \pi_2 &= \{\mathcal{Q}_\varepsilon; \varepsilon > 0\}, \mathcal{Q}_\varepsilon = \{(\operatorname{arctg}(a-\varepsilon), \operatorname{arctg}(a+\varepsilon)); a \in \mathbf{R}\}. \end{aligned}$$

For ε tending to 0, the coverings $\mathcal{P}_\varepsilon$ (resp. $\mathcal{Q}_\varepsilon$) are getting finer in obvious way. The space X with the family π_1 evokes $\left(-\frac{\pi}{2}, \frac{\pi}{2}\right)$ with usual metric, X with π_2 evokes $\mathbf{R}$ with the usual metric. The second space is complete, the first one is not. Let us take the decreasing sequence of closed subsets of X

$$A_n = \left(-\frac{\pi}{2}, \frac{\pi}{2} + \frac{1}{n}\right), n \in \mathbf{N}.$$

Let $\mathcal{C}_1 = 2^{\pi_1}, \mathcal{C}_2 = 2^{\pi_2}$. If we consider m_1 (resp. m_2) -the measures of noncompactness with values in $\mathcal{C}_1$ (resp. $\mathcal{C}_2$), then

$$\begin{aligned} \lim_{n\in\mathbf{N}} m_1(A_n) &= \pi_1 \text{ (minimum of } \mathcal{C}_1), \\ \lim_{n\in\mathbf{N}} m_2(A_n) &= \emptyset \text{ (maximum of } \mathcal{C}_2). \end{aligned}$$

3.2 Random Measures of Noncompactness

In this section we present extensions of the Kuratowski's and of the Hausdorff's measures of noncompactness to probabilistic metric spaces in Menger sense.

Let us denote by Δ the set of all nondecreasing and left continuous functions $f : \mathbf{R} \to [0,1]$, such that $f(0) = 0$ and $\lim_{x\to+\infty} f(x) = 1$.

Definition 3.5 (Menger [145]). A probabilistic metric space (PM-space, shortly) is an ordered pair (S, F), where S is an arbitrary set and $F : S \times S \to \Delta$ satisfies:

(*i*) $F_{p,q}(x) = 1, \forall x > 0$ if and only if $p = q$;
(*ii*) $F_{p,q}(x) = F_{q,p}(x), \forall p, q \in S$;

(iii) If $F_{p,q}(x) = 1$ and $F_{q,r}(y) = 0, \forall p, q, r \in S$ then $F_{p,q}(x+y) = 1$. (Here $F_{p,q}(x) = F(p,q)(x)$.)

Definition 3.6 (Egbert [69]). Let $A \subset S, A \neq \emptyset$. The function $D_A(\cdot)$ defined by

$$D_A(x) = \sup_{t<x} \inf_{p,q \in A} F_{p,q}(x), x \in \mathbf{R}$$

is called the probabilistic diameter of A. If $\sup\{D_A(x); x \in \mathbf{R}\} = 1$ then A is called bounded.

Definition 3.7 (Bocşan-Constantin [41], see also Istrătescu [105], Constantin-Istrătescu [58]). Let A be a bounded subset of S. The mapping

$$\alpha_A(x) = \sup\left\{\varepsilon \geq 0; \exists n \in \mathbf{N}, A_i, i = \overline{1,n} \text{ with } A = \bigcup_{i=1}^{n} A_i \text{ and } D_{A_i}(x) \geq \varepsilon\right\}$$

where $x \in \mathbf{R}$, is called the random Kuratowski's measure of noncompactness.

Remark. In our terminology, α_A can be called random defect of compactness of A (of Kuratowski type).

Theorem 3.6 *(Bocşan-Constantin [41]). We have:*

(i) $\alpha_A \in \Delta$;

(ii) $\alpha_A(x) \geq D_A(x), \forall x \in \mathbf{R}$;

(iii) If $\emptyset \neq A \subset B$ then $\alpha_A(x) \geq \alpha_B(x), \forall x \in \mathbf{R}$;

(iv) $\alpha_{A \cup B}(x) = \min\{\alpha_A(x), \alpha_B(x)\}, \forall x \in \mathbf{R}$;

(v) $\alpha_A(x) = \alpha_{\overline{A}}(x)$, where $\overline{A}$ is the closure of A in the (ε, λ)-topology of S (where by (ε, λ)-neighborhood of $p \in S$, we understand the set $V_p(\varepsilon, \lambda) = \{q \in S; F_{p,q} > 1 - \lambda\}, \varepsilon > 0, \lambda \in [0,1]$.

The function $\alpha_A(x)$ can be calculated by

Theorem 3.7 *(Bocşan-Constantin [41]). Let K_A be the set of functions $f \in \Delta$ such that there exists a finite cover of A, $A = \bigcup_{j \in J} A_j$, J-finite, with $D_{A_j}(x) \geq f(x), \forall j \in J, x \in \mathbf{R}$. Then*

$$\alpha_A(x) = \inf\{f(x); f \in K_A\}, \forall x \in \mathbf{R}.$$

It is known that in usual metric spaces, the Kuratowski's measure is used to characterize the compactness. In this sense, for the probabilistic case, we need the following

Definition 3.8 (Bocşan-Constantin [41]). We say that the PM-space (S, F) is probabilistic precompact (or probabilistic totally bounded) if for every $\varepsilon > 0, \lambda \in (0,1)$, there exists a finite cover of S, $S = \bigcup_{i \in I} A_i$, I-finite, such that $D_{A_i}(\varepsilon) > 1 - \lambda$.

Now, let us suppose that (S, ρ) is an usual metric space. It is well-known that it generates the probabilistic metric space (S, F), with $F_{p,q}(x) = H(x - \rho(p,q))$, where H is the function given by $H(x) = 0$ if $x \leq 0$, $H(x) = 1$ if $x > 0$. We have

Theorem 3.8 *(Bocşan-Constantin [41]). Let (S, ρ) be an usual metric space and (S, F) the corresponding PM-space generated by (S, ρ).*

(i) $A \subset (S, F)$ is probabilistic precompact if and only if $\alpha_A(x) = H(x)$, $\forall x \in \mathbf{R}$;

(ii) $A \subset (S, \rho)$ is precompact if and only if A is probabilistic precompact set of (S, F);

(iii) For any bounded $A \subset (S, \rho)$ we have

$$\alpha_A(x) = H(x - \alpha(A)), \forall x \in \mathbf{R},$$

where $\alpha(A)$ is the usual Kuratowski's measure of noncompactness in (S, ρ) and $\alpha_A(x)$ is the random Kuratowski's measure of noncompactness in the generated PM-space (S, F).

Remark. For other details see *e.g.* Constantin-Istrătescu [58].

In what follows we consider the random Hausdorff's measure of noncompactness. In this sense, firstly we need the followings.

Definition 3.9 (Menger [145]). (S, F) is called PM-space of Menger-type with the t-norm T, if F satisfies the first two properties in Definition 3.5 and the third one is replaced by

$$F_{p,q}(x + y) \geq T(F_{p,r}(x), F_{r,q}(x)), \forall p, q, r \in S.$$

We denote it as the triplet (S, F, T).

Definition 3.10 (see *e.g.* Istrătescu [105]). If (S, F, T) is a PM-space of Menger-type and $A, B \subset S$, then the probabilistic (random) Hausdorff-Pompeiu distance between A and B is given by

$$D^H_{A,B}(x) = \sup_{t<x} T\left(\inf_{p \in A} \sup_{q \in B} F_{p,q}(t), \inf_{q \in B} \sup_{p \in A} F_{p,q}(t)\right), x \in \mathbf{R}.$$

Definition 3.11 (see *e.g.* Istrătescu [105]). Let (S, F, T) be a PM-space of Menger-type and $A \subset S$, bounded. The random Hausdorff's measure of noncompactness of A is given by

$$h_A(x) = \sup\{\varepsilon > 0; \exists \text{ finite } F_\varepsilon \subset A \text{ such that } D^H_{A,F_\varepsilon}(x) \geq \varepsilon\}.$$

Remark. The function $h_A(x)$ has similar properties with $\alpha_A(x)$. Other details concerning random Hausdorff's measures of noncompactness can be found in *e.g.* Constantin-Istrătescu [58].

3.3 Measures of Noncompactness for Fuzzy Subsets in Metric Space

In this section we will extend in various ways the concepts and results in Section 3.2 to fuzzy subsets of classical metric spaces.

Let (X, d) be a metric space and the box metric $d^* : X \times \mathbf{R} \to \mathbf{R}_+$ defined by $d^*(P, Q) = \max\{d(x, y), |r - s|\}$, for all $P = (x, r), Q = (y, s) \in X \times \mathbf{R}$.

For the concept of fuzzy subset of X introduced by Section 2.1, we obviously can give an equivalent definition, as a couple (A, φ_A), where $\varphi_A : X \to [0, 1]$ is the membership function and $A = \{x \in X; \varphi_A(x) > 0\}$ is the so-called support of (A, φ_A).

Let us recall that in literature both forms (notations) for the concept of fuzzy set are used. Because for the purpose of this section seems to be more convenient and to accustom the reader with it too, everywhere in Section 3.3 this second form will be used.

Firstly, let us present some concepts.

Definition 3.12 The sets $G_0(\varphi_A) = \{(x, y); 0 < y = \varphi_A(x), x \in X\}$ $= Graph(F)$ and $HG_0(\varphi_A) = \{(x, y); 0 < y \leq \varphi_A(x), x \in X\}$ $= hypo(F) \cap (A \times (0, 1])$ are called the support graph and the support hypograph of (A, φ_A) respectively, where $F : A \to (0, 1], F(x) = \varphi_A(x), \forall x \in A$.

The diameter and the hypo-diameter of the fuzzy set (A, φ_A) are given by

$$D(\varphi_A) = \sup\{d^*(a, b); a, b \in G_0(\varphi_A)\}$$

and

$$hD(\varphi_A) = \sup\{d^*(a,b); a, b \in HG_0(\varphi_A)\},$$

respectively. If $D(\varphi_A) < +\infty$ $(hD(\varphi_A) < +\infty)$ we say that (A, φ_A) is bounded (hypo-bounded).

Definition 3.13 Let (A, φ_A) be bounded (or hypo-bounded, respectively). The Kuratowski's measure and the Kuratowski's hypo-measure of noncompactness of (A, φ_A) are given by

$$\begin{aligned} K(\varphi_A) &= \inf\{\varepsilon > 0; \exists n \in \mathbf{N}, (A_i, \varphi_{A_i}) \text{ with } D(\varphi_{A_i}) \leq \varepsilon, i = \overline{1,n}, \\ &\quad \text{such that } \varphi_A(x) \leq \sup\{\varphi_{A_i}(x); i = \overline{1,n}\}, \forall x \in X\} \end{aligned}$$

and by

$$\begin{aligned} hK(\varphi_A) &= \inf\{\varepsilon > 0; \exists n \in \mathbf{N}, (A_i, \varphi_{A_i}) \text{ with } hD(\varphi_{A_i}) \leq \varepsilon, i = \overline{1,n}, \\ &\quad \text{such that } \varphi_A(x) \leq \sup\{\varphi_{A_i}(x); i = \overline{1,n}\}, \forall x \in X\}, \end{aligned}$$

respectively.
The Hausdorff's measure and the Hausdorff's hypo-measure of noncompactness of (A, φ_A) are given by

$$\begin{aligned} H(\varphi_A) &= \inf\{\varepsilon > 0; \exists n \in \mathbf{N}, \exists P_i = (x_i, r_i) \in G_0(\varphi_A), i = \overline{1,n}, \\ &\quad \text{such that } \forall P = (x, r) \in G_0(\varphi_A), \exists P_k \text{ with } d^*(P, P_k) \leq \varepsilon\} \end{aligned}$$

and by

$$\begin{aligned} hH(\varphi_A) &= \inf\{\varepsilon > 0; \exists n \in \mathbf{N}, \exists P_i = (x_i, r_i) \in HG_0(\varphi_A), i = \overline{1,n}, \\ &\quad \text{such that } \forall P = (x, r) \in HG_0(\varphi_A), \exists P_k, d^*(P, P_k) \leq \varepsilon\} \end{aligned}$$

respectively.

Concerning the above concepts can be proved the following relations.

Theorem 3.9 *(i)* $K(\varphi_A) \leq hK(\varphi_A)$.

(ii) $K(\varphi_A) \leq K^*(G_0(\varphi_A)), H(\varphi_A) = H^*(G_0(\varphi_A)), hH(\varphi_A)$ $= H^*(HG_0(\varphi_A))$, *where* K^* *and* H^* *represent the usual Kuratowski's and Hausdorff's measures of noncompactness (respectively) of the subsets in the metric space* $(X \times [0,1], d^*)$. *(Here* (A, φ_A) *is considered hypo-bounded).*

Also, we have:

Theorem 3.10 *Let $(A, \varphi_A) \subset (B, \varphi_B)$ be two bounded fuzzy subsets of (X, d). We have:*

(i) $K(\varphi_A) \leq D(\varphi_A)$.

(ii) $(A, \varphi_A) \subset (B, \varphi_B)$ *implies* $K(\varphi_A) \leq K(\varphi_B)$, *where* $(A, \varphi_A) \subset (B, \varphi_B)$ *means* $\varphi_A(x) \leq \varphi_B(x), \forall x \in X$.

(iii) $K(\varphi_A \vee \varphi_B) = \max\{K(\varphi_A), K(\varphi_B)\}$, *where* $(\varphi_A \vee \varphi_B)(x) = \max\{(\varphi_A)(x), (\varphi_B)(x)\}, \forall x \in X$.

Proof. *(i)* and *(ii)* immediately follow from Definition 3.13.

(iii) Since $(A, \varphi_A) \subset (C, \varphi_C)$ and $(B, \varphi_B) \subset (C, \varphi_C)$, where $\varphi_C(x) = \varphi_A(x) \vee \varphi_B(x), \forall x \in X$, by *(ii)* we immediately get

$$\max\{K(\varphi_A), K(\varphi_B)\} \leq K(\varphi_A \vee \varphi_B).$$

Conversely, for any fixed $\delta > 0$, by the definitions of $K(\varphi_A)$ and $K(\varphi_B)$ as infimums, there exist $\varepsilon_1, \varepsilon_2 > 0, n, m \in \mathbf{N}$ and $(A_i, \varphi_{A_i}), i = \overline{1,n}$, (B_j, φ_{B_j}) ,$j = \overline{1,m}$ with $D(\varphi_{A_i}) \leq \varepsilon_1, i = \overline{1,n}, D(\varphi_{B_j}) \leq \varepsilon_2, j = \overline{1,m}$,

$$\varphi_A(x) \leq \sup\{\varphi_{A_i}(x); i = \overline{1,n}\}, \varphi_B(x) \leq \sup\{\varphi_{B_j}(x); j = \overline{1,m}\}, x \in X,$$

such that

$$\varepsilon_1 < K(\varphi_A) + \delta, \varepsilon_2 < K(\varphi_B) + \delta.$$

This implies

$$\max\{\varepsilon_1, \varepsilon_2\} < \max\{K(\varphi_A), K(\varphi_B)\} + \delta.$$

On the other hand

$$(\varphi_A \vee \varphi_B)(x) \leq \sup\{\varphi_{C_k}(x); k = \overline{1, n+m}\},$$

where $C_i = A_i, i = \overline{1,n}, C_{n+j} = B_j, j = \overline{1,m}$ and

$$D(\varphi_{C_k}) \leq \max\{\varepsilon_1, \varepsilon_2\}, \text{ for all } k = \overline{1, n+m}.$$

We get

$$K(\varphi_A \vee \varphi_B) \leq \max\{\varepsilon_1, \varepsilon_2\} < \max\{K(\varphi_A), K(\varphi_B)\} + \delta.$$

Passing with $\delta \to 0+$ we obtain

$$K(\varphi_A \vee \varphi_B) \leq \max\{K(\varphi_A), K(\varphi_B)\},$$

which proves (iii) too. □

Theorem 3.11 *(i) $hH(\varphi_A) = +\infty$ if and only if $HG_0(\varphi_A)$ is unbounded.*

(ii) If $(A, \varphi_A) \subset (B, \varphi_B)$ then $hH(\varphi_A) \leq 2hH(\varphi_B)$.

(iii) $hH(\varphi_A) = 0$ if and only if $HG_0(\varphi_A)$ is totally bounded.

(iv) $hH(\varphi_A \vee \varphi_B) \leq \max\{hH(\varphi_A), hH(\varphi_B)\}$, where $(\varphi_A \vee \varphi_B)(x) = \max\{\varphi_A(x), \varphi_B(x)\}, \forall x \in X$.

Proof. (i) and (iii) follow immediately by Lemma 1, (a), (b) in Beer [36].

Now, let $(A, \varphi_A) \subset (B, \varphi_B)$, *i.e.* $\varphi_A(x) \leq \varphi_B(x), \forall x \in X$. It is easily seen that $HG_0(\varphi_A) \subset HG_0(\varphi_B)$, which combined with Lemma 1, (c) in Beer [36] proves (ii).

(iv) Firstly we have

$$HG_0(\varphi_A \vee \varphi_B) = HG_0(\varphi_A) \cup HG_0(\varphi_B).$$

Indeed, let $(x, r) \in HG_0(\varphi_A \vee \varphi_B)$, *i.e.*, $0 < r \leq \max\{\varphi_A(x), \varphi_B(x)\}$. This implies $0 < r \leq \varphi_A(x)$ or $0 < r \leq \varphi_B(x)$, wherefrom $(x, r) \in HG_0(\varphi_A) \cup HG_0(\varphi_B)$.
Conversely, let $(x, r) \in HG_0(\varphi_A) \cup HG_0(\varphi_B)$, which implies $0 < r \leq \varphi_A(x)$ or $0 < r \leq \varphi_B(x)$ and $0 < r \leq \varphi_A(x) \vee \varphi_B(x)$, *i.e.*, $(x, r) \in HG_0(\varphi_A \vee \varphi_B)$. By Lemma 1, (f) in Beer [36] and by Theorem 3.9, we get

$$\begin{aligned} hH(\varphi_A \vee \varphi_B) &= h^*(HG_0(\varphi_A \vee \varphi_B)) \\ &= h^*(HG_0(\varphi_A) \cup HG_0(\varphi_B)) \\ &\leq \max\{h^*(HG_0(\varphi_A)), h^*(HG_0(\varphi_B))\} \\ &= \max\{hH(\varphi_A), hH(\varphi_A)\}, \end{aligned}$$

which proves (iv) too. □

Theorem 3.12 *Let (A, φ_A) be a bounded fuzzy subset of (X, d). Then the inequality $K(\varphi_A) \geq K_0(A)$ holds, where K_0 represents the usual Kuratowski's measure of noncompactness of subsets in (X, d).*

Proof. By Definition 3.12 we can write

$$K(\varphi_A) = \inf\{\varepsilon > 0; \exists n \in \mathbf{N}, (A_i, \varphi_{A_i}) \text{ with } D^*(G_0(\varphi_{A_i})) \leq \varepsilon,$$

$$i = \overline{1, n}, \text{ such that } 0 < \varphi_A(x) \leq \sup\{\varphi_{A_i}(x); i = \overline{1, n}\}, \forall x \in X\},$$

where D^* represents the usual diameter of subsets in $(X \times [0,1], d^*)$.

Firstly we will prove that

$$K(\varphi_A) = M = \inf\{\varepsilon > 0; \exists n \in \mathbf{N}, (A_i, \varphi_{A_i}) \text{ with } D^*(G_0(\varphi_{A_i})) \le \varepsilon, \\ i = \overline{1,n}, \text{ such that } G_0(\varphi_A) \subset \cup_{i=1}^n HG_0(\varphi_{A_i})\}. \quad (3.1)$$

Indeed, let $\varepsilon > 0$ be such that $\exists n \in \mathbf{N}, (A_i, \varphi_{A_i})$ with $D^*(G_0(\varphi_{A_i})) \le \varepsilon$ and $0 < \varphi_A(x) \le \sup\{\varphi_{A_i}(x); i = \overline{1,n}\}, \forall x \in A$. Then if $(x,y) \in G_0(\varphi_A)$, *i.e.*, $0 < y = \varphi_A(x)$, we get that there exists $j \in \{1,...,n\}$ such that $0 < y = \varphi_A(x) \le \varphi_{A_j}(x)$, *i.e.*, $(x,y) \in HG_0(\varphi_{A_j}) \subset \cup_{i=1}^n HG_0(\varphi_{A_i})$, which immediately proves that $K(\varphi_A) \ge M$.
Conversely, let $\varepsilon > 0$ be such that $\exists n \in \mathbf{N}, (A_i, \varphi_{A_i})$ with $D^*(G_0(\varphi_{A_i})) \le \varepsilon, i = \overline{1,n}$ and $G_0(\varphi_A) \subset \cup_{i=1}^n HG_0(\varphi_{A_i})$. Suppose that $x \in A$. We have $0 < \varphi_A(x) = y$ and by $(x,y) \in G_0(\varphi_A)$ there exists $j \in \{1,...,n\}$ such that $(x,y) \in HG_0(\varphi_{A_j})$, *i.e.* $0 < \varphi_A(x) \le \varphi_{A_j}(x) \le \sup\{\varphi_{A_i}(x); i = \overline{1,n}\}$. This implies that $M \ge K(\varphi_A)$ and as a conclusion we get (3.1).

Further we will prove that

$$M \ge \inf\{\varepsilon > 0; \exists n \in \mathbf{N}, \exists A_i \subset X \text{ with } D_0(A_i) \le \varepsilon, i = \overline{1,n},$$

$$\text{such that } A \subset \cup_{i=1}^n A_i\} = K_0(A), \quad (3.2)$$

where D_0 represents the usual diameter of subsets in (X,d). Indeed, let $\varepsilon > 0$ be such that $\exists n \in \mathbf{N}, (A_i, \varphi_{A_i})$ with $D^*(G_0(\varphi_{A_i})) \le \varepsilon, i = \overline{1,n}$ and $G_0(\varphi_A) \subset \cup_{i=1}^n HG_0(\varphi_{A_i})$. By $D^*(G_0(\varphi_{A_i})) \le \varepsilon$ we immediately get $d^*[(x_1,y_1),(x_2,y_2)] \le \varepsilon$ for all $(x_k,y_k) \in G_0(\varphi_{A_i}), k = \overline{1,2}$. This implies

$$d(x_1,x_2) \le \varepsilon, |y_1 - y_2| \le \varepsilon, \text{ for all } x_1, x_2 \in A, 0 < y_k = \varphi_{A_i}(x_k), k = \overline{1,2}.$$

Therefore, $D^*(G_0(\varphi_{A_i})) \le \varepsilon$ implies $D_0(A_i) \le \varepsilon$. Then, let $(x,y) \in G_0(\varphi_A)$, *i.e.*, $0 < y = \varphi_A(x)$. There exists $j \in \{1,...,n\}$ such that $(x,y) \in HG_0(\varphi_{A_j})$, *i.e.*, $0 < y \le \varphi_{A_j}(x)$, which implies $x \in A_j$. As a conclusion, $A \subset \cup_{i=1}^n A_i$, which immediately implies (3.2).

Now by (3.1) and (3.2) the theorem is proved. □

Remarks. 1) Because in general the converse inequality in (3.2) does not hold, this means that in general we have $K(\varphi_A) > K_0(A)$.

2) By Gal [84], Theorem 3.5, (*ii*), $hH(\varphi_A) = H^*(HG_0(\varphi_A)) = 0$, if and only if $HG_0(\varphi_A)$ is totally bounded. Now we want to prove that $HG_0(\varphi_A)$ cannot be closed in the metric space $(X \times [0,1], d^*)$.

Indeed,let $x \in A$, *i.e.* $\varphi_A(x) > 0$. Obviously that $(x, 0) \notin HG_0(\varphi_A)$. On the other hand, $(x, 0) \notin \overline{HG_0(\varphi_A)}$, since for $r_n = \frac{1}{n+n_0}, n \in \mathbf{N}$, where $0 < \frac{1}{n} \leq \varphi_A(x), \forall n \geq n_0$, we have $(x, r_n) \in HG_0(\varphi_A), \forall n \in \mathbf{N}$ and $d^*[(x, r_n), (x, 0)] = \max\{d(x, x), r_n\} \xrightarrow{n} 0$.
As a conclusion, $HG_0(\varphi_A) \subset \overline{HG_0(\varphi_A)}$, strictly. But because there exist totally bounded sets which are not closed (even in a compact metric space), this means that in general can exist fuzzy sets (A, φ_A) with $hH(\varphi_A) = 0$.

By using the level sets method, we can introduce

Definition 3.14 The (α)-Kuratowski measure of noncompactness of (A, φ_A) is given by

$$\alpha K(\varphi_A) = \sup\{K_0(A_\lambda); \lambda \in (0, 1]\},$$

where $A_\lambda = \{x \in X; \varphi_A(x) \geq \lambda\}$ and K_0 is the usual Kuratowski's measure of noncompactness of usual subsets in (X, d).
The (α)-Hausdorff measure of noncompactness of (A, φ_A) is given by

$$\alpha H(\varphi_A) = \sup\{H_0(A_\lambda); \lambda \in (0, 1]\},$$

where H_0 represents the usual Hausdorff's measure of noncompactness of usual subsets in (X, d). Obviously, $\alpha H(\varphi_A)$ can take the $+\infty$ value.

We also need the following

Definition 3.15 (see *e.g.* Weiss [220]) If (X, d) is a metric space, then the induced fuzzy topology on (X, d) is the collection of all fuzzy subsets of (X, d) with $\varphi_A : X \to [0, 1]$ lower semicontinuous on X.
A fuzzy set (A, φ_A) is called (α)-bounded if for each $\lambda \in (0, 1], A_\lambda$ is bounded in the metric space (X, d).
A fuzzy set (A, φ_A) is called (α)-compact if for each $\lambda \in (0, 1], A_\lambda$ is compact in the metric space (X, d).

Now we are in position to prove the

Theorem 3.13 *Let $(A, \varphi_A), (B, \varphi_B)$ be (α)-bounded.*

(i) $\alpha K(\varphi_A) = 0$ if and only if each $\overline{A_\lambda}, \lambda \in (0, 1]$ is compact in (X, d), where $\overline{A_\lambda}$ denotes the closure of A_λ in (X, d). Also, if (A, φ_A) is (α)-compact then $\alpha K(\varphi_A) = 0$.

(ii) If $(A, \varphi_A) \subset (B, \varphi_B)$ (i.e. $\varphi_A(x) \leq \varphi_B(x), \forall x \in X$) then $\alpha K(\varphi_A) \leq \alpha K(\varphi_B)$.

(iii) $\alpha K(\varphi_A) \leq K_0(A)$, *where* K_0 *represents the usual Kuratowski's measure for subsets in* (X,d), *and in general we have not equality.*

(iv) $\alpha K(\varphi_A \vee \varphi_B) = \max\{\alpha K(\varphi_A), \alpha K(\varphi_B)\}$, *where* $(\varphi_A \vee \varphi_B)(x) = \max\{\varphi_A(x), \varphi_B(x)\}, \forall x \in X$.

(v) $\alpha H(\varphi_A) = 0$ *if and only if each* $A_\lambda, \lambda \in (0,1]$, *is totally bounded.*

(vi) If $(A, \varphi_A) \subset (B, \varphi_B)$ *then* $\alpha H(\varphi_A) \leq 2 \cdot \alpha H(\varphi_B)$.

(vii) $\alpha H(\varphi_A \vee \varphi_B) \leq \max\{\alpha H(\varphi_A), \alpha H(\varphi_B)\}$.

Proof. *(i)* We have $\alpha K(\varphi_A) = 0$ if and only if $K_0(A_\lambda) = 0, \forall \lambda \in (0,1]$, which immediately implies the desired conclusion.

(ii) Since $(A, \varphi_A) \subset (B, \varphi_B)$ implies $A_\lambda = \{x \in X; \varphi_A(x) \geq \lambda\} \subset B_\lambda = \{x \in X; \varphi_B(x) \geq \lambda\}, \forall \lambda \in (0,1]$, the proof is immediate.

(iii) Firstly it is easy to check that

$$A = \{x \in X; \varphi_A(x) > 0\} = \bigcup_{\lambda \in (0,1]} A_\lambda = \bigcup_{\lambda \in (0,1]} \{x \in X; \varphi_A(x) \geq \lambda\}.$$

Since $A_\lambda \subset A$, we have $K_0(A_\lambda) \leq K_0(A), \forall \lambda \in (0,1]$, and passing to supremum we get $\alpha K(\varphi_A) \leq K_0(A)$.

Now if for example $X = \mathbf{R}, A = [0, +\infty)$, $A = \cup_{n=1}^{\infty} [0,n]$, then $\sup\{K_0([0,n]); n \in \mathbf{N}\} = 0$ but $K_0(A) \neq 0$, which means that in general $\alpha K(\varphi_A) < K_0(A)$.

(iv) By *(ii)* we easily get

$$\max\{\alpha K(\varphi_A), \alpha K(\varphi_B)\} \leq \alpha K(\varphi_A \vee \varphi_B).$$

Conversely, for all $\lambda \in (0,1]$ we have

$$\begin{aligned} & \{x \in X; \max\{\varphi_A(x), \varphi_B(x)\} \geq \lambda\} \\ = \; & \{x \in X; \varphi_A(x) \geq \lambda\} \cup \{x \in X; \varphi_B(x) \geq \lambda\}. \end{aligned}$$

Applying the usual Kuratowski's measure K_0, we get

$$K_0(\{x \in X; \max\{\varphi_A(x), \varphi_B(x)\} \geq \lambda\}) = \max\{K_0(A_\lambda), K_0(B_\lambda)\}$$

$$\leq \max\{\alpha K(\varphi_A), \alpha K(\varphi_B)\}, \forall \lambda \in (0,1],$$

wherefrom passing to supremum with $\lambda \in (0,1]$, we obtain $\alpha K(\varphi_A \vee \varphi_B) \leq \max\{\alpha K(\varphi_A), \alpha K(\varphi_B)\}$.

(v), (vi) and (vii) are immediate consequences of Definition 3.14 and of Beer [36], Lemma 1, $(a), (c), (f)$. □

Corollary 3.1 *For all (A, φ_A) hypo-bounded, we have*

$$\alpha K(\varphi_A) \leq K(\varphi_A) \leq K^*(G_0(\varphi_A)) \leq K^*(HG_0(\varphi_A)) \leq hK(\varphi_A).$$

Proof. Firstly we will prove that $hK(\varphi_A) \geq K^*(G_0(\varphi_A))$, *i.e.* that

$$\begin{aligned} hK(\varphi_A) &= \inf\{\varepsilon > 0; \exists n \in \mathbf{N}, (A_i, \varphi_{A_i}) \text{ with } D^*(HG_0(\varphi_{A_i})) \leq \varepsilon, i = \\ &\quad \overline{1,n}, \text{and } 0 < \varphi_A(x) \leq \sup\{\varphi_{A_i}(x); i = \overline{1,n}\}, \forall x \in A\} \\ &\geq \inf\Big\{\varepsilon > 0; \exists M_i \subset X \times (0,1] \text{ with } D^*(M_i) \leq \varepsilon, i = \overline{1,n}, \\ &\quad \text{such that } HG_0(\varphi_A) \subset \bigcup_{i=1}^{n} M_i\Big\}. \end{aligned}$$

So, let $\varepsilon > 0$ be for which there exist $n \in \mathbf{N}, (A_i, \varphi_{A_i})$ with $D^*(HG_0(\varphi_{A_i})) \leq \varepsilon, i = \overline{1,n}$, such that $0 < \varphi_A(x) \leq \sup\{\varphi_{A_i}(x); i = \overline{1,n}\}$, for all $x \in A$. Let us denote $M_i = HG_0(\varphi_{A_i}), i = \overline{1,n}$. We have $D^*(M_i) \leq \varepsilon$. On the other hand, let $(x, y) \in HG_0(\varphi_A)$, *i.e.*. $x \in A, 0 < y \leq \varphi_A(x)$. Then by $\varphi_A(x) \leq \sup\{\varphi_{A_i}(x); i = \overline{1,n}\}$, there exists $j \in \{1, ..., n\}$ such that $\varphi_A(x) \leq \varphi_{A_j}(x)$, *i.e.*, $0 < y \leq \varphi_{A_j}(x)$, which means $(x, y) \in M_j \subset \bigcup_{i=1}^{n} M_i$. As a conclusion, $hK(\varphi_A) \geq K^*(G_0(\varphi_A))$, which together with the obvious inequality $K^*(HG_0(\varphi_A)) \geq K^*(G_0(\varphi_A))$ and with the Theorems 3.9, 3.12 and 3.13, prove the corollary. □

Remarks. 1) In the axiomatic definition of the measure of noncompactness (denoted by *e.g.*, μ) for usual subsets of a Banach space, among others must be checked that $\mu(A) = \mu(\overline{A})$ and $\mu(A) = \mu(conv(A))$, where $\overline{A}$ is the closure of A and $conv(A)$ is the convex hull of A (see *e.g.*, Banas [31]). From this viewpoint, similar questions can be considered for the above measures of noncompactness $\alpha K(\varphi_A), K(\varphi_A), hK(\varphi_A), \alpha H(\varphi_A), H(\varphi_A), hH(\varphi_A)$. The difficulty consists in proper definitions for the closure and for the convex hull of a fuzzy set.

For example, for (A, φ_A) we can consider the closure $\overline{(A, \varphi_A)}$ in the induced topology in Definition 3.15, *i.e.*, $\overline{(A, \varphi_A)} = (B, \overline{\varphi}_A)$, where

$$\overline{\varphi}_A(x) = \inf \quad \{\varphi(x); \varphi_A \leq \varphi \text{ on } X, \varphi : X \to [0,1], \text{ upper semicontinuous on } X\},$$

for every $x \in X$. Also, the convex hull of (A, φ_A) denoted by $(B, conv\varphi_A)$ can be defined by

$$(conv\varphi_A)(x) = \inf\{\varphi(x); \varphi_A \geq \varphi \text{ on } X, \varphi\text{-fuzzy convex on } X\}, \forall x \in X.$$

2) By using the above measures of noncompactness, it would be interesting to extend the known Darbo's [60] and Sadovski's [187] results of fixed point to fuzzy-set-valued mappings.

3) By Corollary 3.1 it follows that if (A, φ_A) has the support A relatively compact in (X, d) then $\alpha K(\varphi_A) = 0$ and that if $G_0(\varphi_A)$ is relatively compact in $(X \times [0,1], d^*)$ then $\alpha K(\varphi_A) = K(\varphi_A) = 0$.

4) By Theorem 3.9, (ii), the measures of noncompactness of Hausdorff's type for fuzzy sets are completely characterized by means of the usual Hausdorff's measure of noncompactness for some usual sets. Then, from the inequalities in Corollary 3.1 arise the question:
If we denote any measure of noncompactness of Kuratowski-type of a fuzzy set (A, φ_A) by $K(\varphi_A)$, then find an usual set (in X or in $X \times [0,1]$, respectively) such that $K(\varphi_A) = K^*(M)$, where $K^*(M)$ represents the usual Kuratowski's measure of noncompactness of M.

While the above question seems to be difficult, we will solve it by introducing other measures of noncompactness of Kuratowski-type for fuzzy sets.

These new measures are suggested by the following remark.

Let (X, d) be a metric space and $A \subset X$. As it is well-known, the Kuratowski's measure of noncompactness of A denoted by $K_0(A)$ is given by

$$\begin{aligned} K_0(A) &= \inf\{\varepsilon > 0; \exists n \in \mathbf{N}, A_i \in X \text{ with } D_0(A_i) \leq \varepsilon, \\ &\qquad i = \overline{1,n} \text{ such that } A \subset \bigcup_{i=1}^{n} A_i \right\}. \end{aligned}$$

But if we denote $B_i = A_i \cap A$, then obviously get $D_0(B_i) \leq D_0(A_i) \leq \varepsilon$ and $A = \bigcup_{i=1}^{n} B_i$. Therefore we immediately obtain

$$\begin{aligned} K_0(A) &= \inf\{\varepsilon > 0; \exists n \in \mathbf{N}, A_i \in X \text{ with } D_0(A_i) \leq \varepsilon, \\ &\qquad i = \overline{1,n} \text{ such that } A = \bigcup_{i=1}^{n} A_i \right\}. \end{aligned}$$

(Here $D_0(A_i)$ is the usual diameter of A_i in (X, d)).

As a consequence we can introduce

Definition 3.16 Let (A, φ_A) be a bounded (or hypo-bounded, respectively) fuzzy subset of (X, d). Then we define

$$K_1(\varphi_A) = \inf\{\varepsilon > 0; \exists n \in \mathbf{N}, (A_i, \varphi_{A_i}) \text{ with } D(\varphi_{A_i}) \leq \varepsilon, i = \overline{1,n}, \text{ such that } \varphi_A(x) = \sup\{\varphi_{A_i}(x); i = \overline{1,n}\}, \forall x \in X\},$$

and

$$hK_1(\varphi_A) = \inf\{\varepsilon > 0; \exists n \in \mathbf{N}, (A_i, \varphi_{A_i}) \text{ with } hD(\varphi_{A_i}) \leq \varepsilon, i = \overline{1,n}, \text{ such that } \varphi_A(x) = \sup\{\varphi_{A_i}(x); i = \overline{1,n}\}, \forall x \in X\},$$

respectively.

Theorem 3.14 *(i)* $hK(\varphi_A) \leq hK_1(\varphi_A), \forall (A, \varphi_A)$ *hypo-bounded.*
(ii) $K(\varphi_A) \leq K_1(\varphi_A) = K^*(G_0(\varphi_A)), \forall (A, \varphi_A)$ *bounded.*

Proof. *(i)* It is immediate by definitions.

(ii) The first inequality is immediate by definitions. On the other hand, the converse inequality $K_1(\varphi_A) \leq K(\varphi_A)$ seems that does not hold in general, because if we follow the idea in the usual case by defining $\varphi^*_{A_i}(x) = \min\{\varphi_A(x), \varphi_{A_i}(x)\}, x \in X, i = \overline{1,n}$, then although we have

$$\varphi_A(x) = \sup\{\varphi^*_{A_i}(x); i = \overline{1,n}\}, x \in X \tag{3.3}$$

and $\varphi^*_{A_i}(x) \leq \varphi_A(x), \forall x \in X$, however $G_0(\varphi^*_{A_i}) \not\subseteq G_0(\varphi_{A_i})$ and $D(\varphi^*_{A_i}) \not\leq D(\varphi_{A_i}) \leq \varepsilon$.

In order to obtain $K_1(\varphi_A) = K^*(G_0(\varphi_A))$, firstly we will prove the inequalities

$$K_1(\varphi_A) \geq \inf\{\varepsilon > 0; \exists n \in \mathbf{N}, (A_i, \varphi_{A_i}) \text{ with } D^*(G_0(\varphi_{A_i})) \leq \varepsilon, i = \overline{1,n}, \text{such that } G_0(\varphi_A) \subset \bigcup_{i=1}^{n} G_0(\varphi_{A_i})\Big\} \geq K^*(G_0(\varphi_A)).$$

Indeed, the second inequality is obvious. Now, let $\varepsilon > 0$ be for which there exist $n \in \mathbf{N}, (A_i, \varphi_{A_i})$ with $D^*(G_0(\varphi_{A_i})) = D(\varphi_{A_i}) \leq \varepsilon, i = \overline{1,n}$, such that $\varphi_A(x) = \sup\{\varphi_{A_i}(x); i = \overline{1,n}\}, \forall x \in X$ and let $(x, y) \in G_0(\varphi_A)$, *i.e.*, $0 < y = \varphi_A(x)$. Then there exists $j \in \{1, ..., n\}$ such that $0 < y = \varphi_A(x) = \varphi_{A_j}(x)$, *i.e.*, $(x, y) \in G_0(\varphi_{A_j})$. As a conclusion, $G_0(\varphi_A) \subset \bigcup_{i=1}^{n} G_0(\varphi_{A_i})$ which proves the first inequality too.

Conversely, by the proof of Theorem 3.3 in Gal [84] we have

$$K^*(G_0(\varphi_A)) = \inf\{\varepsilon > 0; \exists n \in \mathbf{N}, (A_i, \varphi_{A_i}) \text{ with } D(\varphi_{A_i}) \leq \varepsilon, i = \overline{1,n},$$

$$\left.\text{such that } G_0(\varphi_A) \subset \bigcup_{i=1}^{n} G_0(\varphi_{A_i})\right\}.$$

Therefore, let $\varepsilon > 0$ be for which there exist $n \in \mathbf{N}, (A_i, \varphi_{A_i})$ with $D^*(G_0(\varphi_{A_i})) \leq \varepsilon, i = \overline{1,n}$ and

$$G_0(\varphi_A) = \bigcup_{i=1}^{n} G_0(\varphi_{A_i}). \tag{3.4}$$

Let $x \in X$. There exist two possibilities:

$a)$ $\varphi_A(x) > 0$;

$b)$ $\varphi_A(x) = 0$.

Case $a)$. If we denote $y = \varphi_A(x)$, we get $(x, y) \in G_0(\varphi_A)$ and by relation (3.4) there exists $j \in \{1, ..., n\}$ such that $y = \varphi_A(x) = \varphi_{A_j}(x)$. Let us denote $J = \{j \in \{1, ..., n\}; \varphi_A(x) = \varphi_{A_j}(x)\}$. Obviously we have $\varphi_A(x) = \sup\{\varphi_{A_j}(x); j \in J\}$. If $\varphi_A(x) < \sup\{\varphi_{A_j}(x); j = \overline{1,n}\}$ then there exists $k \in \{1, ..., n\} \setminus J$ such that $0 < \varphi_A(x) < \varphi_{A_k}(x) = y'$, which implies that $(x, y') \in G_0(\varphi_{A_k})$. By (3.4) we get $(x, y') \in G_0(\varphi_A)$, *i.e.*, $y' = \varphi_A(x)$, which implies the contradiction $k \in J$. As a conclusion, $\varphi_A(x) = \sup\{\varphi_{A_j}(x); j \in \{1, ..., n\}\}$.

Case $b)$. Let $\varphi_A(x) = 0$ and suppose that there exists $j \in \{1, ..., n\}$ with $0 < y = \varphi_{A_j}(x)$. Then by (3.4) we get $(x, y) \in G_0(\varphi_A)$ and the contradiction $\varphi_A(x) > 0$. Therefore, $\varphi_{A_i}(x) = 0$, for all $i = \overline{1,n}$ and $0 = \varphi_A(x) = \sup\{\varphi_{A_i}(x); i = \overline{1,n}\}$.

As a conclusion, $K^*(G_0(\varphi_A)) \geq K_1(\varphi_A)$, which combined with the inequality $K_1(\varphi_A) \geq K^*(G_0(\varphi_A))$ proves (ii). □

Corollary 3.2 (i) $K_1(\varphi_A \vee \varphi_B) \leq \max\{K_1(\varphi_A), K_1(\varphi_B)\}$.

(ii) $hK_1(\varphi_A \vee \varphi_B) \leq \max\{hK_1(\varphi_A), hK_1(\varphi_B)\}$.

Proof. (i) For any fixed $\delta > 0$, by the definition of $K_1(\varphi_A)$ and $K_1(\varphi_B)$ as infimums, there exist $\varepsilon_1, \varepsilon_2 > 0, n, m \in \mathbf{N}$ and $(A_i, \varphi_{A_i}), i = \overline{1,n}$, $(B_j, \varphi_{B_j}), j = \overline{1,m}$ with $D(\varphi_{A_i}) \leq \varepsilon_1, i = \overline{1,n}$, $D(\varphi_{B_j}) \leq \varepsilon_2, j = \overline{1,m}$, $\varphi_A(x) = \sup\{\varphi_{A_i}(x); i = \overline{1,n}\}, \varphi_B(x) = \sup\{\varphi_{B_j}(x); j = \overline{1,m}\}, x \in X$, such that

$$\varepsilon_1 < K_1(\varphi_A) + \delta, \varepsilon_2 < K_1(\varphi_B) + \delta.$$

This implies

$$\max\{\varepsilon_1, \varepsilon_2\} < \max\{K_1(\varphi_A), K_1(\varphi_B)\} + \delta.$$

On the other hand, we easily have

$$(\varphi_A \vee \varphi_B)(x) = \sup\left\{\varphi_{C_k}(x) ; k = \overline{1, n+m}\right\},$$

where $C_i = A_i, i = \overline{1,n}, C_{n+j} = B_j, j = \overline{1,m}$ and

$$D(\varphi_{C_k}) \leq \max\{\varepsilon_1, \varepsilon_2\} \text{ for all } k = \overline{1, n+m}.$$

We have

$$K_1(\varphi_A \vee \varphi_B) \leq \max\{\varepsilon_1, \varepsilon_2\} < \max\{K_1(\varphi_A), K_1(\varphi_B)\} + \delta$$

and passing with $\delta \to 0+$ we obtain

$$K_1(\varphi_A \vee \varphi_B) \leq \max\{K_1(\varphi_A), K_1(\varphi_B)\}.$$

(*ii*) The proof is entirely analogous. □

Remark. The converse inequalities in (*i*) and (*ii*), in general are not valid because it is easy to check that K_1 and hK_1 lose the property of monotony, *i.e.*, $(A, \varphi_A) \subset (B, \varphi_B)$ does not imply that $K_1(\varphi_A) \leq K_1(\varphi_B)$ and that $hK_1(\varphi_A) \leq hK_1(\varphi_B)$.

The fuzzy metric concepts for fuzzy sets can be obtained by replacing the usual metric space with a fuzzy metric space. At the end of this section we introduce some concepts concerning this idea.

First, we need some known concepts, *i.e.*, those of fuzzy real number and of fuzzy metric space (see *e.g.*, Mashhour-Allam-El Saady [142]).

Definition 3.17 Let $\varphi, \psi : \mathbf{R} \to [0,1]$ be non-ascending functions.
We say that φ is equivalent with ψ if $\varphi_-(t) = \psi_-(t), \forall t \in \mathbf{R}$ and we write $\varphi \sim \psi$, where $\varphi_-(t) = \lim_{s \nearrow t} \varphi(s)$.
A fuzzy real number is an equivalence class $\widehat{\varphi}$ under the above relation $\sim$, such that if $\xi \in \widehat{\varphi}$ then $\xi_+(-\infty) = 1$ and $\xi_-(+\infty) = 0$, where $\xi_+(t) = \lim_{s \searrow t} \xi(s)$ and $\xi_-(t) = \lim_{s \nearrow t} \xi(s)$.

The set of all fuzzy real numbers will be denoted by $\mathbf{R}([0,1])$, while by $\mathbf{R}^*([0,1])$ one denotes the class of all fuzzy real numbers satisfying $\varphi_-(0) = 1$. If $\varphi, \psi \in \mathbf{R}([0,1])$ then $(\varphi + \psi)(s) = \sup\{\varphi(a) + \psi(b) ; a + b = s\}$.

Definition 3.18 (see *e.g.*, Mashhour-Allam-El Saady [142], p.59). (X, d_f) is called fuzzy pseudo-metric space if $d_f : X \times X \to \mathbf{R}^*([0,1])$ satisfies:

(*i*) $d_f(x,x) = \widetilde{0}, \forall x \in X$, where $\widetilde{0} : \mathbf{R} \to [0,1]$ is given by $\widetilde{0}(s) = 1$, if $s < 0$ and $\widetilde{0}(s) = 0$, if $s \geq 0$.

(ii) $d_f(x,y) = d_f(y,x), \forall x,y \in X$.

(iii) $d_f(x,z) \leq d_f(x,y) + d_f(y,z), \forall x,y,z \in X$,

where if $\widehat{\varphi}, \widehat{\psi} \in \mathbf{R}([0,1])$ then $\widehat{\varphi} \leq \widehat{\psi}$ means $\varphi'(t) \leq \psi'(t), \forall t \in \mathbf{R}, \forall \varphi' \in \widehat{\varphi}, \psi' \in \widehat{\psi}$.

Now, let (X, d_f) be a fuzzy pseudo-metric space and (A, φ_A) be a fuzzy subset of X. We can introduce the following

Definition 3.19 The fuzzy diameter of the fuzzy set (A, φ_A) will be

$$D_f(\varphi_A)(s) = \sup\left\{d_f^*(a,b)(s); a,b \in G_0(\varphi_A)\right\}, s \in \mathbf{R},$$

where $d_f^*(a,b)(s) = \max\{d_f(a_1,b_1)(s), |a_2 - b_2|\}, a = (a_1,a_2)$, $b = (b_1,b_2) \in G_0(\varphi_A)$.

If we replace $G_0(\varphi_A)$ by $HG_0(\varphi_A)$, then we obtain the concept of fuzzy hypo-diameter denoted by $hD_f(\varphi_A)(s)$. If $D_f(\varphi_A)(s)$ $(hD_f(\varphi_A)(s)) < +\infty, \forall s \in \mathbf{R}$, we say that (A, φ_A) is of finite fuzzy diameter (hypo-diameter).

For simplicity, let us denote by $D_f^*(\varphi_A)$ any between $D_f(\varphi_A)$ and $hD_f(\varphi_A)$.

Definition 3.20 Let (A, φ_A) be with $D_f^*(\varphi_A)(s) < +\infty, \forall s \in \mathbf{R}$. The fuzzy Kuratowski's measure of noncompactness of (A, φ_A) will be

$$\begin{aligned} \alpha_{\mathcal{F}}(\varphi_A)(s) &= \inf\{\varepsilon > 0; \exists n \in \mathbf{N}, (A_i, \varphi_{A_i}), i = \overline{1,n} \text{ with } D_f^*(\varphi_A)(s), \\ &\leq \varepsilon, \text{such that } \varphi_A(x) \leq \sup\left\{\varphi_{A_i}(x); i = \overline{1,n}\right\}, \forall x \in X\} \end{aligned}$$

Analogously, the fuzzy Hausdorff's measure of noncompactness of (A, φ_A) will be

$$h_{\mathcal{F}}(\varphi_A)(s) = \inf\{\varepsilon > 0; \exists n \in \mathbf{N}, \exists P_i = (x_i, r_i) \in M(\varphi_A), i = \overline{1,n},$$

$$\text{such that } \forall P = (x,r) \in M(\varphi_A), \exists P_k \text{ with } d^*(P, P_k)(s) \leq \varepsilon\}, s \in \mathbf{R},$$

where for simplicity, $M(\varphi_A)$ denotes any between $G_0(\varphi_A)$ and $HG_0(\varphi_A)$.

Remarks. 1) In Definition 3.20, we therefore have defined four fuzzy measures of noncompactness for fuzzy sets.

2) By analogy with the introduction of the so-called concept of abstract fuzzy measure (for usual sets), it would be interesting to use the axiomatic definition of the measure of noncompactness in Banach spaces in Banas [31], p. 133-134, in order to introduce a concept of abstract fuzzy measure of noncompactness for (usual) sets.

3.4 Measures of Noncompactness for Fuzzy Subsets in Topological Space

The main aim of this section is to generalize the results in Kupka-Toma [128] to fuzzy subsets in (fuzzy) topological spaces.

Let m and $\mathcal{C}$ be as in Definition 3.4.

By using the weak α-cuts of a fuzzy set, we can introduce the following crisp-fuzzy kind of definition.

Definition 3.21 Let (X, T) be a classical topological space and A a fuzzy subset of X (*i.e.* $A : X \to [0,1]$). Then, the (α)-measure of noncompactness of the fuzzy set A is given by

$$\alpha M(A) = \bigcap_{\alpha \in (0,1]} m(A_\alpha),$$

where $A_\alpha = \{x \in X; A(x) \geq \alpha\}$ and $\alpha M : I^X \to \mathcal{C}, I^X$ being the class of all fuzzy subsets of X.

Remarks. 1) If $\alpha, \beta \in (0,1]$ satisfy $\alpha \leq \beta$, then obviously $A_\beta \subseteq A_\alpha$ and by Kupka-Toma [128], Proposition 1, (b), we get $m(A_\beta) \leq m(A_\alpha)$ (*i.e.* $m(A_\alpha) \subseteq m(A_\beta)$).

2) If A is a classical set, then $A_\alpha = A$ for all $\alpha \in (0,1]$ and $\alpha M(A) = m(A)$. Therefore, in this case αM one reduces to the classical m in Definition 3.4.

In what follows we introduce the concept of $\widetilde{s}$-compactness for fuzzy sets. We recall that a triangular norm is a function $t : [0,1] \times [0,1] \to [0,1]$ which is commutative, associative, monotone in each component and satisfies the condition $t(x, 1) = x$. Also, $s : [0,1] \times [0,1] \to [0,1]$ defined by $s(x, y) = 1 - t(1 - x, 1 - y)$ is called the corresponding conorm of t. We extend t to I^X pointwise, *i.e.* $(A\widetilde{t}B)(x) = t(A(x), B(x))$. Then $\widetilde{t}$ can be considered as "intersection" of fuzzy subsets. Similarly, $\widetilde{s}$ correspond to the "union" of fuzzy subsets. Also, finite (or countable) "intersections" and "unions" are defined in the obvious way (see *e.g.* Mesiar [146]).

Let (X, T_F) be a quasi-fuzzy topological space (Lowen [136]), *i.e.* a fuzzy topological space in Chang's sense and let s be a triangular conorm.

We can introduce the following.

Definition 3.22 Let $A_i \in T_F, i \in I$ and $A^1, ..., A^n \in T_F$. The family $\mathcal{K} = \{(A_i)_{i \in I}, A^1, ..., A^n\}$ of fuzzy sets is an open $\widetilde{s}$-cover of a fuzzy set A

if

$$A \subset (\vee_{i\in I} A_i)\,\widetilde{s}A^1\widetilde{s}...\widetilde{s}A^n,$$

where $(\vee_{i\in I} A_i)(x) = \sup\{A_i(x); i \in I\}, x \in X$ and $A \subset B$ means $A(x) \leq B(x), \forall x \in X$.

We say that $\left\{(A_j)_{j\in J}, A^{k_1}, ..., A^{k_p}\right\}$ is a finite open $\widetilde{s}$-subcover in $\mathcal{K}$ of the fuzzy set A, if $A \subset (\vee_{j\in J} A_j)\,\widetilde{s}A^{k_1}\widetilde{s}...\widetilde{s}A^{k_p}$, where $J \subset I$ is finite and $\{k_1, ..., k_p\} \subset \{1, ..., n\}$.

Definition 3.23 A fuzzy set $A \in I^X$ is called $(C, \widetilde{s})$-compact if each open $\widetilde{s}$-cover of A has a finite $\widetilde{s}$-subcover of A. A fuzzy set A is called $(L, \widetilde{s})$-compact if for all open $\widetilde{s}$-cover of A and for all $\varepsilon > 0$, there exists a finite $\widetilde{s}$-subcover of B, where B is the fuzzy set defined by $B(x) = \max(0, A(x) - \varepsilon), \forall x \in X$.

Because $B(x) \leq A(x), \forall x \in X$, it follows that if A is $(C, \widetilde{s})$-compact, then A also is $(L, \widetilde{s})$-compact.

Remarks. 1) If the triangular conorm is $s(x, y) = s_0(x, y) = \max(x, y)$, then $(C, \widetilde{s}_0)$-compactness becomes the compactness in Chang's sense (see Chang [54]) and $(L, \widetilde{s}_0)$-compactness becomes the compactness in Lowen's sense (see Lowen [136] and Lowen-Lowen [137]).

2) If s_1 and s_2 are triangular conorm and $s_1 \leq s_2$, then any open $\widetilde{s}_1$-cover is an open $\widetilde{s}_2$-cover. In particular, any classical open cover (*i.e.* $\widetilde{s}_0$-cover) is an open $\widetilde{s}$-cover because $\max(x, y) \leq s(x, y), \forall x, y \in [0, 1]$, for every triangular conorm s.

3) Let $s \neq s_0$. If $\mathcal{K} = \{(A_i)_{i\in I}, A^1, ..., A^n\}$ is an open $\widetilde{s}$-cover of a fuzzy set, then in general $(\vee_{i\in I} A_i)\,\widetilde{s}A^1\widetilde{s}...\widetilde{s}A^n$ is not an open fuzzy set in (X, T_F).

Now, let π^S be the set of all open $\widetilde{s}$-covers of $(X, T), \mathcal{C}^S = \mathcal{P}(\pi^S)$ the family of all subsets of π^S, ordered by the relation $\Delta \leq \Omega$ if and only if $\Omega \subset \Delta, \forall \Delta, \Omega \in \mathcal{C}^S$, where $\subset$ means the classical inclusion. Then, obviously $(\mathcal{C}^S, \leq)$ is a complete lattice with the minimal element $0^S_{\mathcal{C}} = \pi^S$.

We can introduce the following.

Definition 3.24 Let $A \in I^X, (X, T_F)$ be a quasi-fuzzy topological space and s a triangular conorm. The (C, s)-measure of noncompactness of A is the element of $\mathcal{C}^S$ given by

$$m^S_{\mathcal{C}}(A) = \left\{\mathcal{D} \in \pi^S; \exists\delta \subset \mathcal{D}, \text{ finite open } \widetilde{s}\text{-subcover of } A\right\}.$$

The (L, s)-measure of noncompactness of A is the element of $\mathcal{C}^S$ given by

$$m_L^S(A) = \Big\{\mathcal{D}_* \in \pi^S; \forall \varepsilon > 0, \exists \delta_* \subset \mathcal{D}_*, \text{ finite}, \ \delta_* = \Big\{(A_j)_{j\in J}, A^1, ..., A^p\Big\}$$

$$\text{with } \max\{0, A(x) - \varepsilon\} \le \big((\vee_{j\in J} A_j)\, \tilde{s} A^1 \tilde{s} ... \tilde{s} A^p\big)(x), \forall x \in X\Big\}.$$

Remarks. 1) Obviously, m_C^S and m_L^S are similarly defined if (X,T) is a fuzzy topological space in the Lowen's sense (see *e.g.* Lowen [136], Definition 1.1).

2) The (L, s)-measure m_L^S is suggested by Lowen [136], Definition 4.1 and means in fact s-measure of noncompactness in Lowen's sense.

3) By Definition 3.24 we obviously get

$$m_L^S(A) \le m_C^S(A) \, (i.e. \ m_C^S(A) \subset m_L^S(A)).$$

4) If s_1 and s_2 are triangular conorms and $s_1 \le s_2$, then

$$m_C^{S_1}(A) \subset m_C^{S_2}(A) \ \text{ and } \ m_L^{S_1}(A) \subset m_L^{S_2}(A) \ \text{ for all } A \in I^X.$$

In what follows we study the measure αM. Thus, we present

Theorem 3.15 *Let (X,T) be a topological space. The mapping $\alpha M : I^X \to \mathcal{C}$ has the following properties:*

(i) If $A \in I^X$ is compact in the sense of Weiss [220], Definition 3.5, then $\alpha M(A) = 0_{\mathcal{C}}$. Conversely, if A is closed (in the sense of Proposition 3.3 in Weiss [220]) and $\alpha M(A) = 0_{\mathcal{C}}$, then A is compact (in the sense of Definition 3.5 in Weiss [220]).

(ii) If $A \subset B$ then $\alpha M(A) \le \alpha M(B)$.

(iii) $\alpha M(A \vee B) = \alpha M(A) \cap \alpha M(B)$.

(iv) $\alpha M(A \wedge B) \le \alpha M(A) \cup \alpha M(B)$.

(where $(A \vee B)(x) = \max\{A(x), B(x)\}, (A \wedge B)(x) = \min\{A(x), B(x)\}$ $\forall x \in X$).

Proof. (i) Let $A \in I^X$ be compact as in Weiss [220], Definition 3.5. It follows that all $A_\alpha, \alpha \in (0,1]$ are compact in the topological space (X,T) and then by Kupka-Toma [128], Proposition 1, (a), we get $m(A_\alpha) = 0_{\mathcal{C}}$, for all $\alpha \in (0,1]$. This immediately implies

$$\alpha M(A) = \bigcap_{\alpha\in(0,1]} m(A_\alpha) = 0_{\mathcal{C}}.$$

Conversely, suppose that $\alpha M(A) = \bigcap_{\alpha\in(0,1]} m(A_\alpha) = 0_{\mathcal{C}}$. If $\alpha \in (0,1]$ then $m(A_\alpha) = 0_{\mathcal{C}}$, since $0_{\mathcal{C}} = \bigcap_{\alpha\in(0,1]} m(A_\alpha) \subset m(A_\alpha) \subset 0_{\mathcal{C}}$. Because A_α is closed, by Kupka-Toma [128], Proposition 1, (a), we obtain that A_α is compact. Hence A is compact (according to Definition 3.5 in Weiss [220]).

(ii) If $A(x) \le B(x), \forall x \in X$, then $A_\alpha \subset B_\alpha, \forall \alpha \in (0,1]$, where A_α, B_α are the corresponding level sets of A and B, hence $m(A_\alpha) \supset m(B_\alpha), \forall \alpha \in (0,1]$ (see Kupka-Toma [128], Proposition 1, (b)).

We get

$$\alpha M(A) = \bigcap_{\alpha\in(0,1]} m(A_\alpha) \supset \bigcap_{\alpha\in(0,1]} m(B_\alpha) = \alpha M(B).$$

(iii) Because $A_\alpha \cup B_\alpha = (A \vee B)_\alpha$, by using Proposition 1, (c), in Kupka-Toma [128] we get

$$m(A_\alpha) \cap m(B_\alpha) = m(A_\alpha \cup B_\alpha) = m((A \vee B)_\alpha).$$

Therefore,

$$\begin{aligned} \alpha M(A) \cap \alpha M(B) &= \left(\bigcap_{\alpha\in(0,1]} m(A_\alpha)\right) \cap \left(\bigcap_{\alpha\in(0,1]} m(B_\alpha)\right) \\ &= \bigcap_{\alpha\in(0,1]} (m(A_\alpha) \cap m(B_\alpha)) = \bigcap_{\alpha\in(0,1]} m((A \vee B)_\alpha) \\ &= \alpha M(A \vee B). \end{aligned}$$

(iv) The proof is immediate by (ii). □

The next example proves that the inequality in (iv) can be strict.

Example 3.13 Let $X = \mathbf{R}$ be endowed with the usual topology and

$$\begin{aligned} A(x) &= \frac{1}{2} \text{ if } x \in [0,+\infty) \text{ and } A(x) = 0 \text{ if } x \in (-\infty, 0), \\ B(x) &= \frac{1}{2} \text{ if } x \in (-\infty, 0] \text{ and } B(x) = 0 \text{ if } x \in (0,+\infty), \end{aligned}$$

two fuzzy subsets of X. Then $(A \wedge B)_\alpha = \{0\}$ for $\alpha \in \left(0, \frac{1}{2}\right]$ and $(A \wedge B)_\alpha = \emptyset$ for $\alpha \in \left[\frac{1}{2}, 1\right]$, hence $\alpha M(A \wedge B) = 0_{\mathcal{C}}$.

On the other hand,

$$\alpha M(A) = \left(\bigcap_{\alpha \in \left(0, \frac{1}{2}\right]} m(A_\alpha) \right) \cap \left(\bigcap_{\alpha \in \left(\frac{1}{2}, 1\right]} m(A_\alpha) \right) = m([0, +\infty))$$

and

$$\alpha M(B) = \left(\bigcap_{\alpha \in \left(0, \frac{1}{2}\right]} m(B_\alpha) \right) \cap \left(\bigcap_{\alpha \in \left(\frac{1}{2}, 1\right]} m(B_\alpha) \right) = m((-\infty, 0]),$$

but $\alpha M(A) \cup \alpha M(B) \neq 0_{\mathcal{C}}$, because the open cover $\mathcal{P} = \{(n-1, n+1); n \in \mathbf{Z}\}$ of X does not contain a finite subcover neither of the set $[0, +\infty)$, nor of the set $(-\infty, 0]$.

Remark. Theorem 3.15 is an extension of Proposition 1 in Kupka-Toma [128] (see also Theorem 3.2).

Now, for the (C, s)-measure of noncompactness we get

Theorem 3.16 *Let (X, T_F) be a quasi-fuzzy topological space. The mapping $m_C^S : I^X \to \mathcal{C}^S$ has the properties:*

(i) If $A \in I^X$ is $(C, \widetilde{s})$-compact then $m_C^S(A) = 0_C^S$.
(ii) If $A \subset B$ then $m_C^S(A) \leq m_C^S(B)$.
(iii) $m_C^S(A \vee B) = m_C^S(A) \cap m_C^S(B)$.
(iv) $m_C^S(A \wedge B) \leq m_C^S(A) \cup m_C^S(B)$.

Proof. (i) Let us suppose that $A \in I^X$ is $(C, \widetilde{s})$-compact and $\mathcal{D}$ is an open $\widetilde{s}$-cover of X. Then evidently $\mathcal{D}$ also is an open $\widetilde{s}$-cover of A, so that there exists δ, a finite $\widetilde{s}$-subcover of A such that $\delta \subset \mathcal{D}$. Thus $m_C^S(A) = 0_C^S$.

(ii) If $\mathcal{D} \in \pi^S$ contains a finite $\widetilde{s}$-subcover of the fuzzy set B, then the same $\widetilde{s}$-subcover also covers A and this proves the inclusion $m_C^S(B) \subset m_C^S(A)$, which is equivalent to $m_C^S(A) \leq m_C^S(B)$.

(iii) Because of (ii) it is evident that $m_C^S(A \vee B) \geq m_C^S(A)$ and $m_C^S(A \vee B) \geq m_C^S(B)$, hence $m_C^S(A \vee B) \geq m_C^S(A) \cap m_C^S(B)$. In order to prove the converse inequality, which is equivalent to the inclusion $m_C^S(A) \cap m_C^S(B) \subset m_C^S(A \vee B)$, let $\mathcal{D} \in m_C^S(A) \cap m_C^S(B), \mathcal{D} = \left\{(A_i)_{i \in I}, A^1, ..., A^n\right\}$. Then there exists $K \subset I, K$ finite and $\{k_1, ..., k_p\} \subset \{1, ..., n\}$ such that $\mathcal{K} = \left\{(A_k)_{k \in K}, A^{k_1}, ..., A^{k_p}\right\}$ is a finite $\widetilde{s}$-subcover of

A, that is

$$A \subset (\vee_{k\in K} A_k)\, \widetilde{s} A^{k_1} \widetilde{s} ... \widetilde{s} A^{k_p}.$$

Also, there exists a finite subset $L \subset I$ and $\{l_1, ..., l_r\} \subset \{1, ..., n\}$ such that $\mathcal{L} = \{(A_l)_{l\in L}, A^{l_1}, ..., A^{l_r}\}$ is a finite $\widetilde{s}$-subcover of B, that is $B \subset (\vee_{l\in L} A_l)\, \widetilde{s} A^{l_1} \widetilde{s} ... \widetilde{s} A^{l_r}$. Then

$$A \vee B \subset (\vee_{j\in K\cup L} A_j)\, \widetilde{s} A^{k_1} \widetilde{s} ... \widetilde{s} A^{k_p} \widetilde{s} A^{l_1} \widetilde{s} ... \widetilde{s} A^{l_r},$$

which implies that $\mathcal{K} \cup \mathcal{L} = \left\{(A_j)_{j\in K\cup L}, A^{k_1}, ..., A^{k_p}, A^{l_1}, ..., A^{l_r}\right\}$ is a finite $\widetilde{s}$-subcover (in $\mathcal{D}$) of $A \vee B$. This proves that $\mathcal{D} \in m_C^S(A \vee B)$.

(*iv*) The proof is evident by (*ii*). □

The next examples prove that for $s = s_0$ and $s = s_\infty$ the converse statement in (*i*) can be false even if A is C-closed (*i.e.* in the Chang's sense, see Chang [54]).

Example 3.14 Let $X = \mathbf{R}$ be endowed with the quasi-fuzzy topology $T_F = \{D; D(x) = c \in [0,1], \forall x \in \mathbf{R}\}$. If $s = s_0$, where $s_0(x, y) = \max(x, y)$ is the triangular conorm corresponding to the usual union, then the fuzzy set $A(x) = \frac{1}{2}, \forall x \in X$ is a C-closed fuzzy set and obviously $m_C^{S_0}(A) = 0_C^{S_0}$. But A is not $(C, \widetilde{s}_0)$-compact (*i.e.* compact in Chang's sense) because the open $\widetilde{s}_0$-cover of A, $\{P_i; i = 1, 2, 3, ...\}$ where $P_i(x) = \frac{1}{2} - \frac{1}{i}, \forall x \in X$, does not contain a finite $\widetilde{s}_0$-subcover of A.

Example 3.15 For arbitrary X, let us consider

$$\begin{aligned} T_F &= \left\{f_n : X \to [0,1]; f_n(x) = 1 - \frac{1}{n}, n \in \mathbf{N}^*, n\text{-even}\right\} \cup \\ &\cup \left\{g_n : X \to [0,1]; g_n(x) = \frac{1}{2} - \frac{1}{2n}, n \in \mathbf{N}, n \geq 3\right\} \cup \{\emptyset, X\}, \end{aligned}$$

where $\emptyset, X \in I^X, \emptyset(x) = 0$ and $X(x) = 1, \forall x \in X$.
It is easy to see that T_F is a quasi-fuzzy topology on X. We will prove that the fuzzy set $A(x) = \frac{1}{2}, \forall x \in X$ is a C-closed set with $m_C^{S_\infty} = 0_C^{S_\infty}$, but A is not $(C, \widetilde{s}_\infty)$-compact.
Indeed, by $\frac{1}{2} = 1 - \left(1 - \frac{1}{2}\right)$ we get that A is C-closed. Also, A is not $(C, \widetilde{s}_\infty)$-compact, because by $A(x) = \vee_{n\geq 3} g_n(x), \forall x \in X$, we cannot choose a finite number of $g_n(x)$ with $A(x) = \vee g_n(x)$.
On the other hand, if $X = (\vee_{k\in I} h_k)\, \widetilde{s}_\infty A^1 \widetilde{s}_\infty ... \widetilde{s}_\infty A^m$, we have the following possibilities: (*i*) I is finite; (*ii*) I is infinite and $h_k = f_k$ for an infinite

number of k; (iii) I is infinite and $h_k = g_k$, for any $k \geq k_0$.
The case (i) obviously implies that A is covered by a finite $\widetilde{s}_\infty$-subcover.
In the case (ii), we can choose a $k_0 \in I$ with $\frac{1}{2} < h_{k_0}(x) = f_{k_0}(x), \forall x \in X$ and $A \subset h_{k_0}\widetilde{s}_\infty A^1 \widetilde{s}_\infty ... \widetilde{s}_\infty A^m$.
Finally, in the case (iii) we get $\vee_{k \in I} h_k(x) = \frac{1}{2}, \forall x \in X$, which implies

$$\frac{1}{2} = A^1(x)\, s_\infty ... s_\infty A^m(x), \forall x \in X, \ \ i.e.\ A \subset A^1 \widetilde{s}_\infty ... \widetilde{s}_\infty A^m.$$

As a conclusion, taking into account the definition of $m_C^{S_\infty}$, we get $m_C^{S_\infty}(A) = 0_C^{S_\infty}$.

Note that any fuzzy set A satisfying $A(x) \leq c < \frac{1}{2}, \forall x \in X$ is $(C, \widetilde{s}_\infty)$-compact.
However, a natural question arises: if A is C-closed and $m_C^{S_\infty}(A) = 0_C^{S_\infty}$, then in what conditions A still has some properties of compactness ?

The following remarks give answers to this question.

Remarks. 1) Firstly, we will prove that if (X, T_F) is $(C, \widetilde{s}_\infty)$-compact (which implies $m_C^{S_\infty}(X) = 0_C^{S_\infty}$ and therefore $m_C^{S_\infty}(A) = 0_C^{S_\infty}$), then each C-closed fuzzy set $A \subset X$ also is $(C, \widetilde{s}_\infty)$-compact.
Indeed, let $A \subset (\vee_{i \in I} A_i)\, \widetilde{s}_\infty A^1 \widetilde{s}_\infty ... \widetilde{s}_\infty A^m$ be given with I finite. Since A is C-closed and $A \widetilde{s}_\infty (1 - A) = X$, we obtain that $1 - A$ is open and

$$X = (\vee_{i \in I} A_i)\, \widetilde{s}_\infty A^1 \widetilde{s}_\infty ... \widetilde{s}_\infty A^m \widetilde{s}_\infty (1 - A).$$

Taking into account that X is $(C, \widetilde{s}_\infty)$-compact, there is $J \subset I$, J-finite with

$$1 = \max\{A_i(x); i \in J\}\, s_\infty A^{k_1}(x)\, s_\infty ... s_\infty A^{k_p}(x), \forall x \in X$$

or

$$1 = \max\{A_i(x); i \in J\}\, s_\infty A^{k_1}(x)\, s_\infty ... s_\infty A^{k_p}(x)\, s_\infty (1 - A(x)), \forall x \in X,$$

where $k_1, ..., k_p \in \{1, 2, ..., m\}, k_i \neq k_j$.
In the first case, we obviously can cover A with a finite $\widetilde{s}_\infty$-subcover. In the second case, denoting $B(x) = \max\{A_i(x); i \in J\}\, s_\infty A^{k_1}(x)\, s_\infty ... s_\infty A^{k_p}(x)$ we therefore can write

$$1 = B(x)\, s_\infty (1 - A(x)), \forall x \in X,$$

i.e.

$$1 = \min\{B(x) = 1 - A(x), 1\} = \frac{B(x) - A(x) + 2}{2} - \frac{|B(x) - A(x)|}{2},$$

for every $x \in X$, which means

$$0 = (B(x) - A(x)) - |B(x) - A(x)|, \forall x \in X.$$

This immediately implies $A(x) \leq B(x), \forall x \in X$, where $B(x)$ is a finite $\widetilde{s}_\infty$-subcover.

2) Let (X, T_F) be an arbitrary quasi-fuzzy topological space (not necessarily $(C, \widetilde{s}_\infty)$-compact) and $A \subset X, A$ being C-closed and satisfying $m_C^{S_\infty}(A) = 0_{\mathcal{C}}^{S_\infty}$. If moreover $A(x) \geq 1 - \varepsilon_0, \forall x \in X$ (where $\varepsilon_0 \in (0, \frac{1}{2})$) then for any $\varepsilon \in [\varepsilon_0, 1]$ and any $\widetilde{s}_\infty$-cover of A, there is a finite $\widetilde{s}_\infty$-subcover of $A - \varepsilon$.
Indeed, let $A \subset (\vee_{i \in I} A_i) \widetilde{s}_\infty A^1 \widetilde{s}_\infty ... \widetilde{s}_\infty A^m = B$. Reasoning exactly as the above remark, we get $X = B\widetilde{s}_\infty (1 - A)$, and taking into account that $m_C^{S_\infty}(A) = 0_{\mathcal{C}}^{S_\infty}$, we can write

$$A \subset (\vee_{i \in J} A_i) \widetilde{s}_\infty A^{k_1} \widetilde{s}_\infty ... \widetilde{s}_\infty A^{k_p}$$

or

$$A \subset (\vee_{i \in J} A_i) \widetilde{s}_\infty A^{k_1} \widetilde{s}_\infty ... \widetilde{s}_\infty A^{k_p} \widetilde{s}_\infty (1 - A),$$

where $J \subset I$ is finite and $k_1, ..., k_p \in \{1, 2, ..., m\}, k_i \neq k_j$.
In the first case it is obvious that $A - \varepsilon \subset (\vee_{i \in J} A_i) \widetilde{s}_\infty A^{k_1} \widetilde{s}_\infty ... \widetilde{s}_\infty A^{k_p}$, for all $\varepsilon \in [\varepsilon_0, 1]$.
In the second case, we easily obtain

$$2A - 1 \subset (\vee_{i \in J} A_i) \widetilde{s}_\infty A^{k_1} \widetilde{s}_\infty ... \widetilde{s}_\infty A^{k_p}.$$

Taking now into account that

$$0 < A(x) - \varepsilon \leq A(x) - \varepsilon_0 \leq 2A(x) - 1, \forall x \in X, \forall \varepsilon \in [\varepsilon_0, 1],$$

we get our conclusion.

Theorem 3.17 *Let (X, T_F) be a quasi-fuzzy topological space. The mapping $m_L^S : I^X \to \mathcal{C}^S$ has the properties:*
(i) If $A \in I^X$ is $(L, \widetilde{s})$-compact then $m_L^S(A) = 0_{\mathcal{C}}^S$.
(ii) If $A \subset B$ then $m_L^S(A) \leq m_L^S(B)$.
(iii) $m_L^S(A \vee B) = m_L^S(A) \cap m_L^S(B)$.

(iv) $m_L^S(A \wedge B) \leq m_L^S(A) \cup m_L^S(B)$.

Proof. (i) If $A \in I^X$ is $(L, \widetilde{s})$-compact and $\mathcal{D}_*$ is an open $\widetilde{s}$-cover of X, then $\mathcal{D}_*$ is an open $\widetilde{s}$-cover of A and for all $\varepsilon > 0$ there exists $\delta \subset \mathcal{D}_*$, a finite $\widetilde{s}$-subcover of B, where $B(x) = \max(0, A(x) - \varepsilon), \forall x \in X$. Thus $m_L^S(A) = 0_{\mathcal{C}}^S$.

(ii) Let $\mathcal{D}_* \in \pi^S$ and $\varepsilon > 0$. If $\mathcal{D}_*$ contains a finite $\widetilde{s}$-subcover of the fuzzy set defined by $\max(0, B(x) - \varepsilon), \forall x \in X$, then the same $\widetilde{s}$-subcover also covers the fuzzy set defined by $\max(0, A(x) - \varepsilon), \forall x \in X$. The inequality $\max(0, A(x) - \varepsilon) \leq \max(0, B(x) - \varepsilon), \forall x \in X$ implies the inclusion $m_L^S(B) \subset m_L^S(A)$, which is equivalent to $m_L^S(A) \leq m_L^S(B)$.

(iii) The inequality $m_L^S(A \vee B) \geq m_L^S(A) \cap m_L^S(B)$ is evident because of (ii). In order to prove the inclusion $m_L^S(A) \cap m_L^S(B) \subset m_L^S(A \vee B)$, let $\mathcal{D}_* \in m_L^S(A) \cap m_L^S(B), \mathcal{D}_* = \left\{(A_i)_{i \in I}, A^1, ..., A^n\right\}$ and $\varepsilon > 0$. Then there exist $K \subset I, K$ finite and $\{k_1, ..., k_p\} \subset \{1, ..., n\}$ such that

$$\max(A(x) - \varepsilon, 0) \leq (\vee_{k \in K} A_k)(x)\, sA^{k_1}(x)\, s ... sA^{k_p}(x), \forall x \in X.$$

Also, there exist a finite subset $L \subset I$ and $\{l_1, ..., l_r\} \subset \{1, ..., n\}$ such that

$$\max(B(x) - \varepsilon, 0) \leq (\vee_{l \in L} A_l)(x)\, sA^{l_1}(x)\, s ... sA^{l_r}(x), \forall x \in X.$$

The equality

$$\max(\max(A(x), B(x)) - \varepsilon, 0) = \max(\max(A(x) - \varepsilon, 0),$$

$$\max(B(x) - \varepsilon, 0)), \forall x \in X$$

implies

$$\max((A(x) \vee B(x)) - \varepsilon, 0)$$

$$\leq (\vee_{j \in K \cup L} A_j)(x)\, sA^{k_1}(x)\, s ... sA^{k_p}(x)\, sA^{l_1}(x)\, s ... sA^{l_r}(x),$$

for every $x \in X$. This proves that $\mathcal{D}_* \in m_L^S(A \vee B)$.

(iv) It is evident from (ii). □

In what follows we deal with the connections between upper semicontinuity and measures of noncompactness.

Firstly we recall the following:

Definition 3.25 (Kupka-Toma [128], p.458). Let $(\Gamma, \leq_\Gamma)$ be an upward directed set. If $\left(x^{(\gamma)}, \gamma \in \Gamma\right)$ is a net in $\mathcal{C}$, we will write $\lim_{\gamma\in\Gamma} x^{(\gamma)} = 0_{\mathcal{C}}$, if

$$\forall \{\mathcal{B}\} \in \mathcal{C}, \exists \gamma_0 \in \Gamma \text{ such that } x^{(\gamma)} \leq \{\mathcal{B}\}, \forall \gamma \geq_\Gamma \gamma_0.$$

Let us introduce some new concepts for fuzzy sets.

Definition 3.26 Let (X, T) be a topological space and $(\Gamma, \leq_\Gamma)$ an upward directed set. Then $\left(A^{(\gamma)} \in I^X; \gamma \in \Gamma\right)$ is called decreasing net if $\gamma_1 \leq_\Gamma \gamma_2$ implies $A^{(\gamma_2)} \subset A^{(\gamma_1)}$. If (Y, T_F) is a quasi-fuzzy (or fuzzy) (see *e.g.* Lowen [136], p.622) topological space, then a mapping $F : X \to I^Y$ will be called fuzzy set-valued mapping (see *e.g.* Gal [85]).

Definition 3.27 We say that F is u.s.c. (upper semicontinuous) at $x_0 \in X$, if for any $U \in T_F$ with $F(x_0) \subset U$, there exists a neighborhood $V(x_0)$ of x_0 (in the classical topology T) such that $F(t) \subset U, \forall t \in V(x_0)$.

On $\mathbf{R}$ we consider the topology $\tau = \{(\alpha, \infty); \alpha \in \mathbf{R}\} \cup \{\emptyset\}$. The topology which one obtains by taking on $I = [0,1]$ the induced topology will be denoted by $\mathcal{T}$. If (Y, T_F) is a fuzzy (or quasi-fuzzy) topological space, then we denote by T^I the initial topology on Y for the family of "functions" T_F and the topological space $\mathcal{T}$. T^I is the topology associated in a natural way to the fuzzy topology T_F (see Lowen [136]).

Definition 3.28 Let (X, T_F) be a quasi-fuzzy topological space. The fuzzy set-valued mapping $F : X \to I^Y$ is called (α)-u.s.c. at $x_0 \in X$ if the usual set-valued mappings $F_\alpha : X \to \mathcal{P}(Y)$ are u.s.c. at x_0 for all $\alpha \in (0,1]$, where $F_\alpha(x) = A_\alpha^{(x)}$ if $F(x) = A^{(x)} \in I^Y$ and the topology considered on Y is T^I. We say that $\left(F^{(\gamma)} : X \to I^Y\right)_{\gamma\in\Gamma}$ is a decreasing net if $\left(F^{(\gamma)}(x) \in I^Y\right)_{\gamma\in\Gamma}$ is a decreasing net for all $x \in X$.

We can present the following two results which are extensions of the Theorems 1 and 2 in Kupka-Toma [128].

Theorem 3.18 *Let (X, T) be a topological space and $(\Gamma, \leq_\Gamma)$ an upward directed set. If $\left(A^{(\gamma)}; \gamma \in \Gamma\right)$ is a decreasing net of closed subsets in I^X (in the sense of Proposition 3.3 in Weiss [220]) which satisfy the condition*

$$\forall \gamma \in \Gamma, \exists x_0 \in X \text{ (depending on } \gamma) \text{ with } A^{(\gamma)}(x_0) = 1, \tag{3.5}$$

then the following implication is true:

$$\lim_{\gamma\in\Gamma} \alpha M\left(A^{(\gamma)}\right) = 0_{\mathcal{C}} \Rightarrow A = \wedge_{\gamma\in\Gamma} A^{(\gamma)} \text{ is nonempty and compact}$$

(in the sense of Definition 3.5 in Weiss [220]).

Proof. By the definition of α-measure of noncompactness, we have

$$\lim_{\gamma\in\Gamma}\alpha M\left(A^{(\gamma)}\right)=0_{\mathcal{C}}\Leftrightarrow\lim_{\gamma\in\Gamma}\bigcap_{\alpha\in(0,1]}m\left(A_\alpha^{(\gamma)}\right)=0_{\mathcal{C}}.$$

Using Definition 3.25 we obtain

$$\forall\{\mathcal{B}\}\in\mathcal{C},\exists\gamma_0\in\Gamma:\mathcal{B}\subset\bigcap_{\alpha\in(0,1]}m\left(A_\alpha^{(\gamma)}\right),\forall\gamma\geq_\Gamma\gamma_0$$

or equivalently

$$\forall\{\mathcal{B}\}\in\mathcal{C},\exists\gamma_0\in\Gamma:\mathcal{B}\subset m\left(A_\alpha^{(\gamma)}\right),\forall\gamma\geq_\Gamma\gamma_0,\forall\alpha\in(0,1],$$

that is

$$\lim_{\gamma\in\Gamma}m\left(A_\alpha^{(\gamma)}\right)=0_{\mathcal{C}},\forall\alpha\in(0,1].$$

For $\gamma\in\Gamma,A^{(\gamma)}$ is closed and satisfies condition (3.5), therefore $A_\alpha^{(\gamma)}$ is closed (and nonempty by (3.5)), for every $\alpha\in(0,1]$.

Moreover, $\left(A_\alpha^{(\gamma)};\gamma\in\Gamma\right)$ is a decreasing net of subsets in $\mathcal{P}(X)$, for every $\alpha\in(0,1]$. By Kupka-Toma [128], Theorem 1, we get that $\bigcap_{\gamma\in\Gamma}A_\alpha^{(\gamma)}$ is nonempty and compact for every $\alpha\in(0,1]$, therefore $A=\wedge_{\gamma\in\Gamma}A^{(\gamma)}$ is nonempty and compact, because $\bigcap_{\gamma\in\Gamma}A_\alpha^{(\gamma)}=\left(\wedge_{\gamma\in\Gamma}A^{(\gamma)}\right)_\alpha$. □

The next example proves that condition (3.5) in Theorem 3.18 is essential.

Example 3.16 Let $X=\mathbf{R}$ be endowed with the usual topology and $\Gamma=\mathbf{N}\setminus\{0,1\}$. Then, $\left(A^{(k)}:k\in\Gamma\right)$, where $A^{(k)}:X\to[0,1],A^{(k)}(x)=\frac{1}{k}$ if $x\in\left[0,\frac{1}{k}\right]$ and $A^{(k)}(x)=0$ contrariwise, is a decreasing net of closed fuzzy subsets, because $A_\alpha^{(k)}=\emptyset$ if $\alpha>\frac{1}{k}$ and $A_\alpha^{(k)}=\left[0,\frac{1}{k}\right]$ if $\alpha\leq\frac{1}{k}$.
Also

$$\begin{aligned}\lim_{k\in\Gamma}\alpha M\left(A^{(k)}\right)&=\lim_{k\in\Gamma}\bigcap_{\alpha\in(0,1]}m\left(A_\alpha^{(k)}\right)\\&=\lim_{k\in\Gamma}\left(\bigcap_{\alpha\in\left(0,\frac{1}{k}\right]}m\left(A_\alpha^{(k)}\right)\right)\cap\left(\bigcap_{\alpha\in\left(\frac{1}{k},1\right]}m\left(A_\alpha^{(k)}\right)\right)\end{aligned}$$

$$= \lim_{k\in\Gamma} m\left(\left[0,\frac{1}{k}\right]\right) \cap m(\emptyset) = 0_{\mathcal{C}}.$$

Moreover, $A^{(k)}$ is a nonempty fuzzy set for all $k \in \Gamma$. However

$$\left(\wedge_{k\in\Gamma} A^{(\gamma)}\right)(x) = \inf_{k\in\Gamma} A^{(k)}(x) = 0, \forall x \in X.$$

Theorem 3.19 *Let (X,T) be a topological space, $(\Gamma, \leq_\Gamma)$ an upward directed set and (Y, T_F) a quasi-fuzzy topological space. If $\left(F^{(\gamma)} : X \to I^Y;\ \gamma \in \Gamma\right)$ is a decreasing net of (α)-u.s.c. fuzzy set-valued mappings on X such that each $F^{(\gamma)}(x) = A_x^{(\gamma)}$ is a closed fuzzy set (in the sense of Proposition 3.3 in Weiss [220]) which satisfies condition* (3.5) *in Theorem 3.18 and if for each $x \in X, \lim_{\gamma\in\Gamma} \alpha M\left(A_x^{(\gamma)}\right) = 0_{\mathcal{C}}$, then the fuzzy set-valued mapping $F : X \to I^Y$ given by $F(x) = \wedge_{\gamma\in\Gamma} F^{(\gamma)}(x), x \in X$, is (α)-u.s.c. on X, with nonempty and compact values (in the sense of the Definition* 3.5 *in Weiss [220]).*

Proof. $\lim_{\gamma\in\Gamma} \alpha M\left(A_x^{(\gamma)}\right) = 0_{\mathcal{C}}, \forall x \in X$ implies (see the proof of Theorem 3.18),

$$\lim_{\gamma\in\Gamma} m\left(\left(A_x^{(\gamma)}\right)_\alpha\right) = \lim_{\gamma\in\Gamma} m\left(\left(F^{(\gamma)}(x)\right)_\alpha\right) = 0_{\mathcal{C}}, \forall x \in X, \forall \alpha \in (0,1].$$

Let $\alpha \in (0,1]$. If $\left(F^{(\gamma)}; \gamma \in \Gamma\right)$ is a decreasing net (see Definition 3.28), then $\left(F^{(\gamma)}(x); \gamma \in \Gamma\right)$ is a decreasing net for all $x \in X$. Because $F^{(\gamma)}$ is (α)-u.s.c. $\forall \gamma \in \Gamma, F_\alpha^{(\gamma)} : X \to \mathcal{P}(Y)$ is u.s.c. $\forall \gamma \in \Gamma, \forall \alpha \in (0,1]$, if we consider on Y the topology T^I (see Definition 3.28). Moreover, $\left(F^{(\gamma)}(x)\right)_\alpha$ is closed for every $\gamma \in \Gamma, \alpha \in (0,1]$ and $x \in X$, and the condition (3.5) implies that $\left(A_x^{(\gamma)}\right)_\alpha = \left(F^{(\gamma)}(x)\right)_\alpha$ is nonempty for every $\gamma \in \Gamma, \alpha \in (0,1]$ and $x \in X$. By Kupka-Toma [128], Theorem 2, applied for the net of classical multifunctions $\left(F_\alpha^{(\gamma)} : \gamma \in \Gamma\right)_\alpha$, we obtain that the multifunction

$$\begin{aligned} F_\alpha &: X \to \mathcal{P}(Y), \\ F_\alpha(x) &= \bigcap_{\gamma\in\Gamma} \left(F^{(\gamma)}(x)\right)_\alpha \end{aligned}$$

is u.s.c. with nonempty compact values of T^I. But

$$\bigcap_{\gamma\in\Gamma} \left(F^{(\gamma)}(x)\right)_\alpha = \bigcap_{\gamma\in\Gamma} \left(A_x^{(\gamma)}\right)_\alpha = \left(\wedge_{\gamma\in\Gamma} A_x^{(\gamma)}\right)_\alpha = \left(\wedge_{\gamma\in\Gamma} F^{(\gamma)}(x)\right)_\alpha.$$

Hence F_α is u.s.c. and $\left(\bigcap_{\gamma\in\Gamma} F^{(\gamma)}(x)\right)_\alpha$ is compact and nonempty for every $\alpha \in (0,1]$, if we consider on Y the topology T^I. Therefore, the fuzzy set valued mapping F is (α)-u.s.c. on X with nonempty and compact values (in the sense of Definition 3.5 in Weiss [220]). $\square$

In what follows, we consider the measures of noncompactness m_C^S and m_L^S introduced by Definition 3.24. Firstly, we need two concepts.

Definition 3.29 (Lowen [136]). Let (X, T_F) be a quasi-fuzzy topological space (or a fuzzy topological space in the sense of Lowen in [136]) and $A \in I^X$. We say that A is C-closed (L-closed, respectively) if $B \in T_F$, where $B(x) = 1 - A(x), \forall x \in X$.

Definition 3.30 Let $(\Gamma, \leq_\Gamma)$ be an upward directed set. If $\left(x^{(\gamma)} : \gamma \in \Gamma\right)$ is a net in $\mathcal{C}^S$, we will write $\lim_{\gamma\in\Gamma} x^{(\gamma)} = 0_{\mathcal{C}}^S$, if $\forall \{\mathcal{B}\} \in \mathcal{C}^S, \exists \gamma_0 \in \Gamma$ such that $x^{(\gamma)} \leq \{\mathcal{B}\}, \forall \gamma \geq_\Gamma \gamma_0$.

We present

Theorem 3.20 *Let (X, T_F) be a quasi-fuzzy (or fuzzy) topological space and $(\Gamma, \leq_\Gamma)$ an upward directed set. If $\left(A^{(\gamma)} : \gamma \in \Gamma\right)$ is a decreasing net of nonempty and C-closed fuzzy subsets in I^X which satisfy the condition (3.5) in Theorem 3.18, then the following hold:*

$$\lim_{\gamma\in\Gamma} m_C^{S_\infty}\left(A^{(\gamma)}\right) = 0_{\mathcal{C}}^{S_\infty} \Rightarrow A = \wedge_{\gamma\in\Gamma} A^{(\gamma)} \text{ is nonempty,}$$

$$C\text{-closed and } m_C^{S_\infty}(A) = 0_{\mathcal{C}}^{S_\infty}.$$

Proof. Suppose that $A = \emptyset$. Then $X = X \setminus A = X \setminus \left(\wedge_{\gamma\in\Gamma} A^{(\gamma)}\right)$ $= \vee_{\gamma\in\Gamma}\left(X \setminus A^{(\gamma)}\right)$. The hypothesis implies $\forall \mathcal{P} \in \pi^{S_\infty}, \exists \gamma_0 \in \Gamma : \forall \gamma \geq_\Gamma \gamma_0, \mathcal{P} \in m_C^{S_\infty}\left(A^{(\gamma)}\right)$. If we apply this condition to the open $\widetilde{s}_\infty$-cover $\mathcal{P} = \left\{X \setminus A^{(\gamma)}; \gamma \in \Gamma\right\}$, we can state the existence of a finite $\widetilde{s}_\infty$-subcover of the fuzzy set $A^{(\gamma_0)}$, *i.e.* $A^{(\gamma_0)} \subset \vee_{i=1}^n \left(X \setminus A^{(\gamma_i)}\right)$. Let us choose $\gamma \geq_\Gamma \gamma_i, i \in \{0, 1, ..., n\}$. Then $A^{(\gamma)} \subset A^{(\gamma_i)}, \forall i \in \{0, 1, ..., n\}$, therefore $X \setminus A^{(\gamma)} \supset X \setminus A^{(\gamma_i)}, \forall i \in \{0, 1, ..., n\}$ and

$$A^{(\gamma)} \subset A^{(\gamma_0)} \subset \vee_{i=1}^n \left(X \setminus A^{(\gamma_i)}\right) \subset \vee_{i=1}^n \left(X \setminus A^{(\gamma)}\right) = X \setminus A^{(\gamma)},$$

which is a contradiction with the condition (3.5). As a conclusion, we obtain $A \neq \emptyset$. Since $\left(A^{(\gamma)} : \gamma \in \Gamma\right)$ is a decreasing net, we get

$$0_{\mathcal{C}}^{S_\infty} = \lim_{\gamma \in \Gamma} m_C^{S_\infty}\left(A^{(\gamma)}\right) = \inf\left\{m_C^{S_\infty}\left(A^{(\gamma)}\right); \gamma \in \Gamma\right\}$$

$$\text{(here inf is in } (\mathcal{C}^{S_\infty}, \leq)\text{)} = \bigcup_{\gamma \in \Gamma} m_C^{S_\infty}\left(A^{(\gamma)}\right)$$

$$\geq m_C^{S_\infty}\left(\wedge_{\gamma \in \Gamma} A^{(\gamma)}\right) = m_C^{S_\infty}(A) \geq 0_{\mathcal{C}}^{S_\infty},$$

i.e. $m_C^{S_\infty}(A) = 0_{\mathcal{C}}^{S_\infty}$.
Obviously then A is C-closed. □

Theorem 3.21 *Let (X, T_F) be a fuzzy topological space and $(\Gamma, \leq_\Gamma)$ an upward directed set. If $\left(A^{(\gamma)} : \gamma \in \Gamma\right)$ is a decreasing net of nonempty and L-closed fuzzy subsets in I^X which satisfy the condition (3.5) in Theorem 3.18, then the following hold:*

$$\lim_{\gamma \in \Gamma} m_L^{S_\infty}\left(A^{(\gamma)}\right) = 0_{\mathcal{C}}^{S_\infty} \Rightarrow A = \wedge_{\gamma \in \Gamma} A^{(\gamma)} \text{ is}$$

$$\text{nonempty, } L\text{-closed and } m_L^{S_\infty}(A) = 0_{\mathcal{C}}^{S_\infty}.$$

Proof. Similar to the proof of Theorem 3.20. □

Remarks. 1) The condition (3.5) in the hypothesis of Theorems 3.20 and 3.21 can be replaced by the following:

$$\forall \gamma \in \Gamma, \exists x_0 \in X \text{ (depending on } \gamma\text{) with } A^{(\gamma)}(x_0) > \frac{1}{2}. \tag{3.6}$$

2) Because of the examples given after Theorem 3.16, in the conclusions of Theorems 3.20 and 3.21 we cannot state that A is $(C, \widetilde{s}_\infty)$-compact and $(L, \widetilde{s}_\infty)$-compact, respectively. In fact, the compactness cannot be derived neither for other triangular norms, as can be seen in the following example: Let $X = \mathbf{R}, T_F = \{\alpha \,|\alpha : X \to [0,1], \alpha(x) = \alpha \in [0,1]\}, s(x,y) = s_0(x,y)$ $= \max(x,y)$. Choose $\Gamma = \{3, 4, ...\}$ and $A^{(\gamma)}(x) = \frac{1}{2} + \frac{1}{\gamma}, \forall x \in X$. Then $\left(A^{(\gamma)} : \gamma \in \Gamma\right)$ is a decreasing net of nonempty and C-closed (or L-closed) fuzzy subsets which satisfies the condition (3.6). Also, $\lim_{\gamma \in \Gamma} m_C^S\left(A^{(\gamma)}\right) = 0_{\mathcal{C}}^S$, because $m_C^S\left(A^{(\gamma)}\right) = 0_{\mathcal{C}}^S$ for all $\gamma \in \Gamma$. Finally, $A \in I^X, A(x) =$

$\left(\wedge_{\gamma\in\Gamma}A^{(\gamma)}\right)(x) = \frac{1}{2}, \forall x \in X$ and A is not $(C, \widetilde{s})$-compact (see also the example after Theorem 3.16).

At the end of this section, we briefly present another kind of measure of noncompactness in fuzzy topological spaces.

Let (X, T_F) be an arbitrary fuzzy topological space.

Definition 3.31 (Lowen-Lowen [137]). We say that X is (α, ε)-compact (where $\alpha \in (0,1], \varepsilon \in (0, \alpha)$) if each open α-cover has a finite sub-(α, ε)-cover.
We say that X is α^{+}-compact (where $\alpha \in (0,1]$) if it is (α, ε)-compact for all $\varepsilon \in (0, \alpha)$.
We say that X is ε^{-}-compact if it is (α, ε)-compact for all $\alpha \in (\varepsilon, 1)$.

Remark. (Lowen-Lowen [137]) X is compact in the sense in Lowen [136] if and only if it is (α, ε)-compact for all $\alpha \in (0,1]$ and $\varepsilon \in (0, \alpha)$, if and only if it is α^{+}-compact for all $\alpha \in (0,1]$, if and only if it is ε^{-}-compact for all $\varepsilon \in (0,1)$.

Definition 3.32 (Lowen-Lowen [137]). The degree of compactness of X is defined by

$$c(X) = \sup\left\{1 - \varepsilon; X \text{ is } \varepsilon^{-}\text{-compact}\right\}.$$

Theorem 3.22 *(Lowen-Lowen [137]). (X, T_F) is compact (in Lowen [136] sense) if and only if $c(X) = 1$.*

Remarks. 1) The quantity $d_c(X) = 1 - c(X)$ can be called defect of compactness (or measure of noncompactness) for X.

2) For other details see Lowen [136].

3.5 Defects of Opening and of Closure for Subsets in Metric Space

Let $(X, \mathcal{T})$ be a topological space, *i.e.* $X \neq \emptyset$ and $\mathcal{T}$ a topology on X.

The following concepts are well-known.

Definition 3.33 A subset $Y \subset X$ is called:

(i) open, if $Y \in \mathcal{T}$;

(ii) closed, if $X \setminus Y \in \mathcal{T}$, *i.e.* if $Y = X \setminus A$ with $A \in \mathcal{T}$.

Remark. Let (X, ρ) be a metric space and $\mathcal{T}_\rho$ be the topology generated by the metric ρ. The following characterizations are well-known:

$Y \subset X$ is open if and only if $Y = intY$, where

$$intY = \{y \in Y; \exists r > 0 \text{ such that } B(y; r) \subset Y\}$$

and

$$B(y; r) = \{z \in X; \rho(y, z) < r\}.$$

$Y \subset X$ is closed if and only if $Y = \overline{Y}$, where

$$\overline{Y} = \left\{z \in X; \exists y_n \in Y, n \in \mathbf{N} \text{ such that } \rho(y_n, z) \overset{n \to \infty}{\to} 0\right\}.$$

If $Y \subset X$ is not open or closed, it is natural to introduce the following

Definition 3.34 Let (X, ρ) be a metric space and $D(A, B)$ be a certain distance between the subsets $A, B \subset X$. For $Y \subset X$, we call:

(i) defect of opening of Y, the quantity $d_{OP}(D)(Y) = D(Y, intY)$;

(ii) defect of closure of Y, the quantity $d_{CL}(D)(Y) = D\left(Y, \overline{Y}\right)$.

Remark. A natural candidate for D might be the Hausdorff-Pompeiu distance

$$D_H(Y_1, Y_2) = \max\{\rho^*(Y_1, Y_2), \rho^*(Y_2, Y_1)\},$$

where $\rho^*(Y_1, Y_2) = \sup\{\rho(y_1, Y_2); y_1 \in Y_1\}, \rho(y_1, Y_2) = \inf\{\rho(y_1, y_2); y_2 \in Y_2\}$. However, D_H seems to be not suitable for d_{CL} and d_{OP} introduced by Definition 3.34. Indeed, we have

$$d_{CL}(D_H)(Y) = D_H\left(Y, \overline{Y}\right) = D_H\left(\overline{Y}, \overline{\overline{Y}}\right) = D_H\left(\overline{Y}, \overline{Y}\right) = 0,$$

for all $Y \subset X$ and

$$d_{OP}(D)(Y) = 0,$$

for all convex $Y \subset X$, when $(X, \|\cdot\|)$ is a normed space and $\rho(x, y) = \|x - y\|$ (by using the relation $\overline{Y} = \overline{intY}$ for convex Y, see *e.g.* Popa [165], p. 13).

That is why instead of D_H we have to use another distance, as for example, $D^* : \mathcal{P}_b(X) \times \mathcal{P}_b(X) \to \mathbf{R}_+$ given by

$$D^*(Y_1, Y_2) = \begin{cases} 0, & \text{if } Y_1 = Y_2 \\ \sup\{\rho(y_1, y_2); y_1 \in Y_1, y_2 \in Y_2\}, & \text{if } Y_1 \neq Y_2, \end{cases}$$

where $\mathcal{P}_b(X) = \{Y \subset (X, \rho); Y \neq \emptyset, Y \text{ is bounded}\}$ (we recall that by definition Y is bounded if $diam(Y) = \sup\{\rho(x, y); x, y \in Y\} < +\infty$).

Remark. It is not difficult to prove that D^* is a metric on $\mathcal{P}_b(X)$.

As an immediate consequence, it follows

Theorem 3.23 *If (X, ρ) is a metric space and $Y \in \mathcal{P}_b(X)$, then:*
(i) $d_{OP}(D^*)(Y) = 0$ if and only if Y is open;
(ii) $d_{CL}(D^*)(Y) = 0$ if and only if Y is closed.
In addition,

$$\begin{aligned} d_{CL}(D^*)(\lambda Y) &= \lambda d_{CL}(D^*)(Y), \forall \lambda > 0; \\ d_{OP}(D^*)(\lambda Y) &= \lambda d_{OP}(D^*)(Y), \forall \lambda > 0. \end{aligned}$$

Proof. While the first two relations are obvious, the last two relations follows from $\overline{\lambda Y} = \lambda \overline{Y}$ and $int(\lambda Y) = \lambda int(Y), \forall \lambda > 0$. □

3.6 Bibliographical Remarks and Open Problems

Definitions 3.12-3.14, 3.16, 3.19, 3.20, Theorems 3.9, 3.12-3.14 and Corollaries 3.1, 3.2 are in Gal [84], Theorems 3.10, 3.11 are in Gal [85], Definitions 3.21-3.24, 3.26-3.28, 3.30 and Theorems 3.15-3.21 are in Ban-Gal [22]. New are Definition 3.34 and Theorem 3.23.

Open problem 3.1 An open problem is to introduce and study concepts of measures of noncompactness for fuzzy subsets of a PM-space. In what follows we point out some helpful details. Firstly we need to recall some concepts in random analysis.
It is known (see *e.g.* Istrătescu [105], p.25) that to any usual metric space (S, ρ) can be attached the so-called simple probabilistic metric space of Menger-type, for any t-norm T, denoted by (S, F^ρ, T), with $F^\rho : S \times S \to \Delta$ defined by $F^\rho_{r,s}(x) = G(x/\rho(r, s))$, if $r \neq s$, $F^\rho_{r,s}(x) = H(x)$, if $r = s$, where $G \in \Delta$ is fixed and satisfies $\lim_{n\to\infty} G(x) = 1$ and $H(x) = 0$, if $x \leq 0, H(x) = 1$, if $x > 0$. Also, if (S_1, F, T) and (S_2, G, T)

are two PM-spaces, by the T-product of them we mean the ordered triplet $(S_1 \times S_2, T(F,G), T)$, where $T(F,G) : (S_1 \times S_2) \times (S_1 \times S_2) \to \Delta$ is defined for all $r, s \in S_1 \times S_2, r = (r_1, r_2), s = (s_1, s_2)$ by

$$T(F,G)(r,s)(x) = T\left(F_{r_1,s_1}(x), G_{r_2,s_2}(x)\right).$$

(see *e.g.* Egbert [69], Xavier [226], Istrătescu [105], p.81-82). Given (S, F, T) a PM-space of Menger-type, by a fuzzy subset of it we mean a mapping $A : S \to [0,1]$. Let us attach to the metric space $([0,1], \rho), \rho^*(x,y) = |x-y|$, its PM-space $\left([0,1], F^{\rho^*}, T\right)$ and let us consider its T-product with (S, F, T), that is $\left(S \times [0,1], T\left(F, F^{\rho^*}\right), T\right)$.
In Gal [83] were introduced the concept of probabilistic Hausdorff distance between two fuzzy subsets A, B of (S, F, T), by

$$D_H^{(F)}(A,B)(x) = \sup_{t<x} T\left(\inf_{r \in G_0(A)} \sup_{s \in G_0(B)} T\left(F, F^{\rho^*}\right)(r,s)(t),\right.$$

$$\left.\inf_{s \in G_0(B)} \sup_{r \in G_0(A)} T\left(F, F^{\rho^*}\right)(r,s)(t)\right), x \in \mathbf{R},$$

where

$$\begin{aligned} G_0(A) &= \{(x,y) \in S \times [0,1]; 0 < y = A(x), x \in S\}, \\ G_0(B) &= \{(x,y) \in S \times [0,1]; 0 < y = B(x), x \in S\}, \end{aligned}$$

and the concept of probabilistic diameter of a fuzzy subset A of (S, F, T), by

$$D_P(A)(x) = \sup_{t<x}\left(\inf_{r,s \in G_0(A)} T\left(F, F^{\rho^*}\right)(r,s)(t)\right), x \in \mathbf{R}.$$

If $\sup\{D_P(A)(x); x \in \mathbf{R}\} = 1$ then we say that the fuzzy set A is probabilistically bounded.
Then we can introduce the random probabilistic Kuratowski's and Hausdorff's measures of noncompactness of the fuzzy set A, by

$$\alpha_P(A)(x) = \sup\left\{\varepsilon > 0; \exists n \in \mathbf{N}, A_i : S \to [0,1], i = \overline{1,n},\right.$$

$$\left.\text{with } D_P(A_i)(x) \geq \varepsilon \text{ and } A(t) \leq \sup\left\{A_i(t); i = \overline{1,n}\right\}, \forall t \in S\right\}$$

and

$$h_P(A)(x) = \sup \{ \varepsilon > 0; \exists A_\varepsilon \text{ finite fuzzy subset of } A$$

$$\text{such that } D_H^{(\mathcal{F})}(A, A_\varepsilon)(x) \geq \varepsilon \Big\},$$

respectively, where by a finite fuzzy subset of A, denoted here A_ε, we mean that there exist $n \in \mathbf{N}$ and $a_1, ..., a_n \in A$, such that $A_\varepsilon : S \to [0,1], A_\varepsilon(s) = 0$ if $s \neq a_i, \forall i = \overline{1,n}$ and $A_\varepsilon(a_i) = A(a_i), \forall i = \overline{1,n}$.
The above concepts that combine fuzziness with randomness, obviously extend the corresponding random (probabilistic) concepts in Section 3.2. Therefore, would be interesting to extend/study the results mentioned in Section 3.2 for the above $\alpha_P(A)(x)$ and $h_P(A)(x)$.

Chapter 4

Defect of Property in Measure Theory

If a set function (particularly, a fuzzy measure or a non-monotonic fuzzy measure) is non-additive or non-monotonic, it is natural to look for a concept which measure the deviation of that set function from additivity or from monotonicity. In Section 4.1 and Section 4.3 we introduce and study these deviations, called by us defects. The non-additivity implies that the sum between the measure of a set and the measure of its complement is not equal to the measure of entire space. As a consequence, in Section 4.2 we introduce and discuss the defect of complementarity. Defects of other properties (subadditivity, superadditivity, submodularity, supermodularity, k-monotonicity, k-alternativity)are discussed in Section 4.4. Finally, in Section 4.5 we discuss two kinds of defects of measurability for sets.

4.1 Defect of Additivity: Basic Definitions and Properties

The measure is one of the most important concepts in mathematics and so is the integral with respect to measure. In applications, the property of additivity is very convenient, but it is considered too rigid. For example, the experience of artificial intelligence researches shows that the use of probability functions for describing subjective judgements is unjustified and that it leads to many erroneous results (see Szolovitz-Pauker [208], Wierzchon [221]). As a solution, the concept of fuzzy measure was proposed by Sugeno [204], generalizing the usual definition of a measure by replacing the additivity property by a weaker requirement.

The concepts of fuzzy measure and their corresponding fuzzy integrals constitute an important topic in fuzzy mathematics (see *e.g.* Butnariu-

Klement [52], Congxin-Minghu [57], Dennenberg [61], [62], Dubois-Prade [64]-[67], Kruse [126], Kwon-Sugeno [131], Lee-Leekwang [134], De Luca-Termini [138], Murofushi-Sugeno [156]-[159], Ralescu-Adams [168], Schmeidler [189], Suarez Diaz-Suarez Garcia [202], Sugeno [204], Sugeno-Murofushi [205], Wang-Klir [219], Wierzchon [221], [222], Yoneda-Fukami-Grabisch [228]). In this section we consider the following definition of fuzzy measure:

Definition 4.1 (see *e.g.* Ralescu-Adams [168]) Let X be a set and $\mathcal{A}$ be a σ-algebra of subsets of X. A set function $\mu : \mathcal{A} \to [0, \infty)$ is said to be a fuzzy measure if the followings hold:

(*i*) $\mu(\emptyset) = 0$;

(*ii*) $A, B \in \mathcal{A}$ and $A \subseteq B$ implies $\mu(A) \leq \mu(B)$.

If moreover we have

(*iii*) $\{A_n\} \subseteq \mathcal{A}, A_n \nearrow A$ implies $\mu(A_n) \nearrow \mu(A)$ (continuity from below);

(*iv*) $\{A_n\} \subseteq \mathcal{A}, A_n \searrow A$ implies $\mu(A_n) \searrow \mu(A)$ (continuity from above),

then μ is called continuous fuzzy measure.

Comparing it with a classical measure, the main difference is that the fuzzy measure is not additive. Consequently, appears as natural the following question: how can be evaluated the degree of additivity of a fuzzy measure ?

In the sequel we give an answer to this question, by introducing the concept of defect of additivity for a fuzzy measure. Also, for λ-additive fuzzy measures we are able to calculate and estimate this defect. Then, we estimate the defect of additivity for some fuzzy measures generated by a solution of the general functional equation

$$f(x+y) - f(x) - f(y) = \varphi(x, y).$$

Finally, we give applications to some questions like: the approximative calculation of some fuzzy integrals, the introduction of a metric on the family of fuzzy measures and the best approximation of a fuzzy measure.

Let $\mu : \mathcal{A} \to [0, \infty)$ be a fuzzy measure, where $\mathcal{A} \subseteq \mathcal{P}(X)$ is a σ-algebra.

Definition 4.2 The defect of additivity of order $n, n \in \mathbf{N}, n \geq 2$, for the measure μ is given by

$$a_n(\mu) = \sup\left\{\left|\mu\left(\bigcup_{i=1}^{n} A_i\right) - \sum_{i=1}^{n} \mu(A_i)\right| :\right.$$

$$A_i \in \mathcal{A}, A_i \cap A_j = \emptyset, i \neq j\}$$

The defect of countable additivity for the measure μ is given by

$$a_\infty(\mu) = \sup\left\{\left|\mu\left(\bigcup_{i=1}^{\infty} A_i\right) - \sum_{i=1}^{\infty} \mu(A_i)\right| :\right.$$

$$\left. A_i \in \mathcal{A}, \forall i \in \mathbf{N}, A_i \cap A_j = \emptyset, i \neq j\right\}.$$

We have

Theorem 4.1 (i) $a_n(\mu) \leq a_{n+1}(\mu), \forall n \geq 2;$

(ii) If μ is a continuous from below fuzzy measure then $a_\infty(\mu)$ $= \lim_{n\to\infty} a_n(\mu)$;

(iii) $a_n(\mu) \leq a_{n-1}(\mu) + a_2(\mu), \forall n \geq 3;$

(iv) $a_n(\mu) \leq (n-1)\, a_2(\mu), \forall n \geq 3;$

(v) $a_n(\mu) \leq (n-1)\, \mu(X), \forall n \geq 2;$

(vi) If μ is a superadditive fuzzy measure then $a_n(\mu) \leq \mu(X), \forall n \geq 2$.

Proof. (i) It is immediate by taking $A_{n+1} = \emptyset$ in the definition of $a_{n+1}(\mu)$.

(ii) Using the continuity of μ on increasing sequences of sets (see Definition 4.1, (iii)), we have

$$\left|\mu\left(\bigcup_{i=1}^{\infty} A_i\right) - \sum_{i=1}^{\infty} \mu(A_i)\right| = \lim_{n\to\infty}\left(\left|\mu\left(\bigcup_{i=1}^{n} A_i\right) - \sum_{i=1}^{n} \mu(A_i)\right|\right) \leq \lim_{n\to\infty} a_n(\mu)$$

for every sequence $(A_i)_{i\in\mathbf{N}} \subseteq \mathcal{A}, A_i \cap A_j = \emptyset$ if $i \neq j$. Passing to supremum we obtain $a_\infty(\mu) \leq \lim_{n\to\infty} a_n(\mu)$.

For the proof of the converse inequality, let $(A_i)_{i\in\mathbf{N}} \subseteq \mathcal{A}$ be a disjoint sequence such that $A_m = \emptyset, \forall m \geq n+1$. We obtain

$$\left|\mu\left(\bigcup_{i=1}^{\infty} A_i\right) - \sum_{i=1}^{\infty} \mu(A_i)\right| = \left|\mu\left(\bigcup_{i=1}^{n} A_i\right) - \sum_{i=1}^{n} \mu(A_i)\right| \leq a_\infty(\mu),$$

therefore $a_n(\mu) \leq a_\infty(\mu), \forall n \geq 2$. Passing to limit with $n \to \infty$, we get $\lim_{n\to\infty} a_n(\mu) \leq a_\infty(\mu)$.

As a conclusion,

$$a_\infty(\mu) = \lim_{n\to\infty} a_n(\mu).$$

(*iii*) We get

$$\begin{aligned}
&\left|\mu\left(\bigcup_{i=1}^{n} A_i\right)-\sum_{i=1}^{n}\mu\left(A_i\right)\right| \\
&= \left|\mu\left(\bigcup_{i=1}^{n} A_i\right)-\mu\left(\bigcup_{i=1}^{n-1} A_i\right)-\mu\left(A_n\right)+\mu\left(\bigcup_{i=1}^{n-1} A_i\right)-\sum_{i=1}^{n-1}\mu\left(A_i\right)\right| \\
&\leq \left|\mu\left(\bigcup_{i=1}^{n-1} A_i\right)-\sum_{i=1}^{n-1}\mu\left(A_i\right)\right|+\left|\mu\left(\bigcup_{i=1}^{n} A_i\right)-\mu\left(\bigcup_{i=1}^{n-1} A_i\right)-\mu\left(A_n\right)\right| \\
&\leq a_{n-1}\left(\mu\right)+a_2\left(\mu\right),
\end{aligned}$$

for every disjoint family $\{A_1, ..., A_n\} \subseteq \mathcal{A}$, which immediately implies $a_n(\mu) \leq a_{n-1}(\mu) + a_2(\mu)$.

(*iv*) From (*iii*), by recurrence we obtain $a_3(\mu) \leq a_2(\mu) + a_2(\mu)$, afterwards $a_4(\mu) \leq a_3(\mu) + a_2(\mu) \leq 3a_2(\mu)$, $a_5(\mu) \leq a_4(\mu) + a_2(\mu) \leq 4a_2(\mu)$ and finally $a_n(\mu) \leq (n-1)\, a_2(\mu)$.

(*v*) By

$$|\mu(A \cup B) - \mu(A) - \mu(B)| \leq \mu(X), \forall A, B \in \mathcal{A},$$

we get the inequality $a_2(\mu) \leq \mu(X)$, that is (using (*iv*)) $a_n(\mu) \leq (n-1)\,\mu(X)$.

(*vi*) If μ is superadditive, that is

$$\mu(A \cup B) \geq \mu(A) + \mu(B), \forall A, B \in \mathcal{A}, A \cap B = \emptyset,$$

then we obtain

$$\left|\mu\left(\bigcup_{i=1}^{n} A_i\right)-\sum_{i=1}^{n}\mu\left(A_i\right)\right| = \mu\left(\bigcup_{i=1}^{n} A_i\right)-\sum_{i=1}^{n}\mu\left(A_i\right) \leq \mu(X)$$

for every sequence $\{A_1, ..., A_n\} \subseteq \mathcal{A}, A_i \cap A_j = \emptyset$ if $i \neq j$. Passing to supremum we get the desired inequality. □

Example 4.1 If μ is additive then $a_n(\mu) = 0$, for all $n \geq 2$.

Example 4.2 If μ_0 is the fuzzy measure which models the total ignorance (see Dubois-Prade [64], p.128), that is $\mu_0 : \mathcal{P}(X) \to [0,1], \mu_0(A) = 0$ if $A \neq X$ and $\mu_0(A) = 1$ if $A = X$, then $a_n(\mu_0) = 1$ for all $n \geq 2$. Indeed, because $\mu_0(X) - \mu_0(A) - \mu_0(X \setminus A) = 1, \forall A \in \mathcal{P}(X), A \neq X, A \neq \emptyset$ we have $a_2(\mu_0) \geq 1$, that is (see Theorem 4.1, (*i*)) $a_n(\mu_0) \geq 1, \forall n \geq 2$.

On the other hand, μ_0 is superadditive, therefore (see Theorem 4.1, (vi)) $a_n(\mu_0) \le 1, \forall n \ge 2$.
(Here X is supposed to have at least two distinct elements).

Example 4.3 If $\mu : \mathcal{P}(X) \to [0,1]$ is the fuzzy measure defined by $\mu(A) = 0$ if $A = \emptyset$ and $\mu(A) = 1$ if $A \ne \emptyset$, then taking $(A_i)_{i\in\{1,\dots,n\}}$, $A_i \ne \emptyset, A_i \cap A_j = \emptyset$, $\forall i,j, i \ne j$ we get

$$\left|\mu\left(\cup_{i=1}^{n} A_i\right) - \sum_{i=1}^{n} \mu(A_i)\right| = n-1,$$

that is (by using Theorem 4.1, (v)) $a_n(\mu) = n-1$.
(Here X is supposed to have at least n distinct elements).

Example 4.4 Let $\mathcal{A}$ be a σ-algebra of subsets of X and $m : \mathcal{A} \to [0,1]$ be of exponential-type (see Choquet [55]), i. e. $m(A \cup B) = m(A)\, m(B)$ for all $A, B \in \mathcal{A}, A \cap B = \emptyset$. If we define $\mu : \mathcal{A} \to [0,1]$ by $\mu(A) = 1 - m(A)$, we easily get $\mu(\emptyset) = 0, A \subseteq B$ implies $\mu(A) \le \mu(B)$ and $\mu(A \cup B) \le \mu(A) + \mu(B)$ for all $A, B \in \mathcal{A}$. Also, by using the previous theorem we obtain $a_n(\mu) \le n-1$ for all $n \ge 2$.

Example 4.5 We consider the Lebesgue measure induced on the interval $[a,b] \subset \mathbf{R}$, denoted by m, and the induced usual topology on $[a,b]$, denoted by $\mathcal{T}_{[a,b]}$. It is well-known that the outer measure m^* (where, by definition, $m^*(A) = \inf\{m(G) : A \subseteq G, G \in \mathcal{T}_{[a,b]}\}$) and the inner measure m_* (where, by definition, $m_*(A) = \sup\{m(F) : F \subseteq A, [a,b] \setminus F \in \mathcal{T}_{[a,b]}\}$) induced by m are monotone, $m^*(\emptyset) = m_*(\emptyset) = 0, m^*$ is subadditive and m_* is superadditive. We can prove that $a_n(m^*) = (n-1)(b-a)$ and $a_n(m_*) = b-a$ that is the defect of additivity for m^* and m_* is maximum possible. Indeed, for every $n \in \mathbf{N}^*$ there exists a disjoint family $A_1, \dots, A_n \subset [a,b]$ such that $m^*(A_i) = b-a$ and $m_*(A_i) = 0, \forall i \in \{1,\dots,n\}$ (see Halmos [99], p.70 and Gelbaum-Olmsted [88], p.147). Then

$$a_n(m^*) \ge m^*(A_1) + \dots + m^*(A_n) - m^*\left(\bigcup_{i=1}^{n} A_i\right) = (n-1)(b-a)$$

because $m^*\left(\bigcup_{i=1}^{n} A_i\right) = b-a$. By using Theorem 4.1, (v) we obtain $a_n(m^*) = = (n-1)(b-a)$. Also, there exists a set $A \subset [a,b]$ such that $m^*(A) = b-a$ and $m_*(A) = 0$. Because $m^*(A) + m_*([a,b] \setminus A) = b-a$ (see Halmos [99], p.61) we obtain $m_*([a,b]\setminus A)=0$. We have

$$a_2(m_*)\geq m_*([a,b])-m_*(A)-m_*([a,b]\setminus A)=b-a$$

and by using Theorem 4.1, (i) and (vi) we get $a_n(m_*)=b-a$.

Example 4.6 It is well-known (see Congxin-Minghu [57]) that for a null-additive fuzzy measure, that is a fuzzy measure $\mu:\mathcal{A}\to[0,\infty)$ with $\mu(A\cup B)=\mu(A),\forall A,B\in\mathcal{A}$ with $\mu(B)=0$, we can repeat the classical construction of completion. We obtain the complete fuzzy measure $\widetilde{\mu}:\widetilde{\mathcal{A}}\to[0,\infty)$, where

$$\widetilde{\mathcal{A}}=\{A\cup N : A\in\mathcal{A},\exists B\in\mathcal{A}\text{ such that } N\subseteq B,\mu(B)=0\}$$

and

$$\widetilde{\mu}(A\cup N)=\mu(A).$$

By completion, the defect of additivity is not modified, that is $a_n(\mu)=a_n(\widetilde{\mu})$, $\forall n\geq 2$. Indeed, the inequality $a_n(\mu)\leq a_n(\widetilde{\mu})$ is obvious and we have to prove only the converse inequality. We obtain

$$a_n(\widetilde{\mu})=\sup_{(\widetilde{A}_i)\subseteq\widetilde{\mathcal{A}},\text{ disjoint}}\left\{\left|\widetilde{\mu}\left(\bigcup_{i=1}^n\widetilde{A}_i\right)-\sum_{i=1}^n\widetilde{\mu}\left(\widetilde{A}_i\right)\right|\right\}$$

$$\begin{aligned}
&= \sup_{(A_i\cup N_i)\subseteq\widetilde{\mathcal{A}},\text{ disjoint}}\left\{\left|\widetilde{\mu}\left(\left(\bigcup_{i=1}^n A_i\right)\cup\left(\bigcup_{i=1}^n N_i\right)\right)-\sum_{i=1}^n\widetilde{\mu}(A_i\cup N_i)\right|\right\}\\
&= \sup_{(A_i\cup N_i)\subseteq\widetilde{\mathcal{A}},\text{ disjoint}}\left\{\left|\mu\left(\bigcup_{i=1}^n A_i\right)-\sum_{i=1}^n\mu(A_i)\right|\right\}\\
&\leq \sup_{(A_i)\subseteq\mathcal{A},\text{ disjoint}}\left\{\left|\mu\left(\bigcup_{i=1}^n A_i\right)-\sum_{i=1}^n\mu(A_i)\right|\right\}\\
&= a_n(\mu).
\end{aligned}$$

In what follows, we deal with the class of λ-additive measures. Firstly we recall the following.

Definition 4.3 (see *e.g.* Kruse [126]) Let $\lambda\in(-1,\infty)$, $\lambda\neq 0$, and $\mathcal{A}$ be a σ-algebra on the set X. A continuous fuzzy measure μ with $\mu(X)=1$ is called λ-additive if, when $A, B \in \mathcal{A}, A \cap B = \emptyset$, then

$$\mu(A \cup B) = \mu(A) + \mu(B) + \lambda\mu(A)\mu(B).$$

Theorem 4.2 *(Wierzchon [222]) Let m be a classical finite measure, $m : \mathcal{A} \to [0, \infty)$. Then $\mu = t \circ m$ is a λ-additive fuzzy measure if and only if the function $t : [0, \infty) \to [0, \infty)$ has the form $t(x) = \frac{c^x - 1}{\lambda}, c > 0, c \neq 1$, $\lambda \in (-1, 0) \cup (0, \infty)$.*

Concerning the defect $a_{n,\lambda}(\mu)$ of the above measures, we present

Theorem 4.3 *If $\mu = t \circ m$ (with t and m as above) then we have*

$$a_{n,\lambda}(\mu) \leq \frac{\lambda + n - n\sqrt[n]{\lambda + 1}}{|\lambda|}, \forall n \geq 2, \forall \lambda \in (-1, 0) \cup (0, \infty)$$

and

$$a_{\infty,\lambda}(\mu) \leq \left|1 - \frac{\ln(\lambda + 1)}{\lambda}\right|, \forall \lambda \in (-1, 0) \cup (0, \infty).$$

Proof. Let $\lambda \in (-1, 0) \cup (0, \infty)$. The connection between c and λ is given by $c^{m(X)} = \lambda + 1$. If $\lambda \in (-1, 0)$ then μ is subadditive and using the mean inequality we obtain

$$\left|\mu\left(\bigcup_{i=1}^{n} A_i\right) - \sum_{i=1}^{n} \mu(A_i)\right| = \sum_{i=1}^{n} \mu(A_i) - \mu\left(\bigcup_{i=1}^{n} A_i\right)$$

$$= \sum_{i=1}^{n} \frac{c^{m(A_i)} - 1}{\lambda} - \frac{c^{m(\cup_{i=1}^{n} A_i)} - 1}{\lambda} = \frac{1 - n}{\lambda} + \frac{1}{\lambda}\sum_{i=1}^{n} c^{m(A_i)} - \frac{1}{\lambda} c^{m(\cup_{i=1}^{n} A_i)}$$

$$\leq -\frac{1}{\lambda}\left(c^{m(\cup_{i=1}^{n} A_i)} - n\sqrt[n]{c^{m(\cup_{i=1}^{n} A_i)}} + n - 1\right),$$

for every disjoint family $(A_i)_{i \in \{1, \ldots, n\}} \subseteq \mathcal{A}$. In addition, $c \in (0, 1)$ implies $c \leq c^{m(\cup_{i=1}^{n} A_i)} \leq 1$. Because the function $h : \left[\sqrt[n]{c^{m(X)}}, 1\right] \to \mathbf{R}$ defined by $h(t) = t^n - nt + n - 1$ is decreasing and

$$h\left(\sqrt[n]{c^{m(X)}}\right) = c^{m(X)} - n\sqrt[n]{c^{m(X)}} + n - 1 = \lambda + n - n\sqrt[n]{\lambda + 1}$$

we have

$$\left|\mu\left(\bigcup_{i=1}^{n} A_i\right) - \sum_{i=1}^{n} \mu(A_i)\right| \leq \frac{\lambda + n - n\sqrt[n]{\lambda+1}}{-\lambda}$$

for every disjoint family $(A_i)_{i\in\{1,\ldots,n\}} \subseteq \mathcal{A}$. This means that $a_{n,\lambda}(\mu) \leq$ $\leq \frac{\lambda+n-n\sqrt[n]{\lambda+1}}{-\lambda}$ for all $n \geq 2$.

Similarly, we can prove that $a_{n,\lambda}(\mu) \leq \frac{\lambda+n-n\sqrt[n]{\lambda+1}}{\lambda}$ for all $n \geq 2, \lambda \in (0,\infty)$.

Passing to limit with $n \to \infty$ and using Theorem 4.1, (ii), we obtain the second inequality. □

Remarks. 1) Passing to limit with $\lambda \to 0$ in the inequalities of Theorem 4.3, we obtain $\lim_{\lambda\to 0} a_{n,\lambda}(\mu) = \lim_{\lambda\to 0} a_{\infty,\lambda}(\mu) = 0$. The results are in concordance with the fact that for $\lambda = 0$, the defect (of any order) must be 0.

2) Inequalities given by Theorem 4.3 become equalities in some situations. For example, if $m : \mathcal{A} \to [0,\infty)$ is the induced Lebesgue measure on the interval $X = [a, a+\alpha]$, where $\mathcal{A} \subseteq \mathcal{P}(X), a, \alpha \in \mathbf{R}, \alpha > 0$ and $A_i = \left[a + \frac{(i-1)\alpha}{n}, a + \frac{i\alpha}{n}\right), \forall i \in \{1,\ldots,n\}$ then

$$\begin{aligned} a_{n,\lambda}(\mu) &\geq \left|\frac{c^{m(\cup_{i=1}^{n} A_i)} - 1}{\lambda} - \sum_{i=1}^{n} \frac{c^{m(A_i)} - 1}{\lambda}\right| \\ &= \left|\frac{c^{\alpha} - 1 - nc^{\frac{\alpha}{n}} + n}{\lambda}\right| = \frac{\lambda + n - n\sqrt[n]{\lambda+1}}{|\lambda|}, \end{aligned}$$

because $c^{\alpha} = \lambda+1$. Also by using Theorem 4.1, (ii), $a_{\infty,\lambda}(\mu) = \left|1 - \frac{\ln(\lambda+1)}{\lambda}\right|$.

It is known that if $f : [0,\infty) \to [0,\infty)$ is increasing and $f(0) = 0$, then for any classical measure $m : \mathcal{A} \to [0,\infty)$, the function $\mu(A) = (f \circ m)(A)$ becomes a fuzzy measure (see *e.g.* Wierzchon [222]). In this case we obtain

$$\mu(A \cup B) - \mu(A) - \mu(B) = f(m(A) + m(B)) - f(m(A)) - f(m(B)),$$

$\forall A, B \in \mathcal{A}, A \cap B = \emptyset$. This relation suggests us to consider the general functional equation

$$f(x+y) - f(x) - f(y) = \varphi(x,y) \tag{4.1}$$

Concerning this functional equation it is known the following

Theorem 4.4 *(see Ghermănescu [89], p.260, Th. 4.9) The measurable solutions of the functional equation* (4.1) *are given by*

$$\begin{aligned} f(x) &= B(x) + ax \\ \varphi(x,y) &= B(x+y) - B(x) - B(y), \end{aligned}$$

where B is an arbitrary measurable function and a is an arbitrary real constant.

By using Theorem 4.4 we can generate the following classes of fuzzy measures.

Theorem 4.5 *Let $B(x)$ be such that $B(0) = 0$, $f(x) = B(x) + ax$ is strictly increasing and $f : [0,\infty) \to [0,\infty)$. Then $\mu(A) = (f \circ m)(A)$ (where m is a classical finite measure) is a fuzzy measure which satisfies the functional equation*

$$\mu(A \cup B) = \mu(A) + \mu(B) + \varphi\left(f^{-1}(\mu(A)), f^{-1}(\mu(B))\right), \tag{4.2}$$

$\forall A, B \in \mathcal{A}, A \cap B = \emptyset$,*where φ and f are those in Theorem 4.4 and f^{-1} represents the inverse function of f (here we assume $f(0) = 0$ and* $\lim_{x \to +\infty} f(x) = +\infty$*).*

Proof. Take in (4.1), $x = m(A), y = m(B)$ and we take into account that $m = f^{-1} \circ \mu$. □

Example 4.7 Take in Theorem 4.5, $f(x) = B(x) = \frac{c^x - 1}{\lambda}, \lambda \neq 0, \lambda > -1, c > 0, c \neq 1$. We get $\varphi(x,y) = B(x+y) - B(x) - B(y)$ $= \frac{1}{\lambda}(c^x c^y - c^x - c^y + 1)$, $f^{-1}(x) = \log_c(\lambda x + 1)$ and (4.2) becomes the functional equation of λ-additive fuzzy measures

$$\mu(A \cup B) = \mu(A) + \mu(B) + \lambda \mu(A) \mu(B), \forall A, B \in \mathcal{A}, A \cap B = \emptyset.$$

Example 4.8 Take in Theorem 4.5, $f(x) = B(x) = \lambda x^p, \lambda > 0, p \in \mathbf{N}, p \geq 2$. Then (4.2) becomes the functional equation

$$\mu(A \cup B) = \left(\sqrt[p]{\mu(A)} + \sqrt[p]{\mu(B)}\right)^p, \forall A, B \in \mathcal{A}, A \cap B = \emptyset.$$

In this way we obtain the example of fuzzy measure proposed by Ralescu-Adams [168].

Example 4.9 Take in Theorem 4.5, $f(x) = B(x) = \frac{\ln(\lambda x+1)}{\ln(\lambda+1)}, \lambda > 0$. Then $\varphi(x,y) = \frac{\ln \frac{\lambda x+\lambda y+1}{\lambda^2 xy+\lambda x+\lambda y+1}}{\ln(\lambda+1)}$, $f^{-1}(x) = \frac{(\lambda+1)^x-1}{\lambda}$ and (4.2) becomes

$$\mu(A \cup B) = \frac{\ln\left((\lambda+1)^{\mu(A)} + (\lambda+1)^{\mu(B)} - 1\right)}{\ln(\lambda+1)},$$

for every $A, B \in \mathcal{A}, A \cap B = \emptyset$. In fact, $\mu(A) = \frac{\ln(\lambda m(A)+1)}{\ln(\lambda+1)}, \forall A \in \mathcal{A}$ where m is a classical measure and we can prove (by induction) the relation

$$\mu\left(\cup_{i=1}^{n} A_i\right) = \frac{\ln\left(\sum_{i=1}^{n}(\lambda+1)^{\mu(A_i)} - (n-1)\right)}{\ln(\lambda+1)},$$

if $(A_i)_{i\in\{1,\dots,n\}} \subseteq \mathcal{A}$ is disjoint.

Of course that would be interesting to estimate and calculate the defect of additivity for such of fuzzy measures too. For example, concerning the fuzzy measure introduced by Example 4.8, we can prove the following.

Theorem 4.6 *If $\mu : \mathcal{A} \to [0, \infty)$ is the fuzzy measure defined by $\mu(A) = \lambda m^2(A)$, where m is a classical finite measure, then $a_n(\mu) \le \frac{n-1}{n}\mu(X)$, for all $n \ge 2$ and $\lambda > 0$.*

Proof. By Definition 4.2 we have,

$$a_n(\mu) = 2\lambda \sup\left\{ \sum_{1\le i<j\le n} m(A_i)\, m(A_j) : A_i \in \mathcal{A}, A_i \cap A_j = \emptyset, i \ne j \right\}.$$

Because the function $h : [0, m(X)]^n \to \mathbf{R}$ defined by

$$h(t_1, \dots, t_n) = \sum_{1\le i<j\le n} t_i t_j$$

with the condition $\sum_{i=1}^{n} t_i = \alpha \in [0, m(X)]$ has the maximum value $\frac{\alpha^2(n-1)}{2n}$ obtained for $t_1 = \dots = t_n = \frac{\alpha}{n}$ (see *e.g.* Udrişte-Tănăsescu [212], p.95), we get

$$a_n(\mu) \le \frac{\lambda(n-1)}{n} m^2(X) = \frac{n-1}{n}\mu(X). \tag{4.3}$$

□

Remark. If we consider $X = \{x_1, ..., x_n\}$ and $m : \mathcal{P}(X) \to \mathbf{R}$ defined by $m(A) = cardA$, then the inequality (4.3) becomes equality. Indeed, taking $A_i = \{x_i\}, \forall i \in \{1, ..., n\}$, we obtain the converse inequality of (4.3)

$$a_n(\mu) \geq 2\lambda \frac{n(n-1)}{2} = \lambda \frac{n-1}{n} m^2(X) = \frac{n-1}{n}\mu(X),$$

i.e. in this case $a_n(\mu) = \frac{n-1}{n}\mu(X)$.

Theorem 4.7 *If* $\mu : \mathcal{A} \to [0, \infty)$ *is the fuzzy measure defined by the relation*

$$\mu(A \cup B) = \frac{\ln\left((\lambda+1)^{\mu(A)} + (\lambda+1)^{\mu(B)} - 1\right)}{\ln(\lambda+1)},$$

for every $A, B \in \mathcal{A}, A \cap B = \emptyset$, *then*

$$a_{n,\lambda}(\mu) \leq n\mu(X) - \frac{\ln\left(n(\lambda+1)^{\mu(X)} - n + 1\right)}{\ln(\lambda+1)},$$

for every $n \geq 2, \lambda \in (0, \infty)$.

Proof. Firstly, we observe that μ is subadditive, $\forall \lambda \in (0, \infty)$. Indeed, $\left((\lambda+1)^{\mu(A)} - 1\right)\left((\lambda+1)^{\mu(B)} - 1\right) \geq 0, \forall \lambda \in (0, \infty)$ implies

$$\begin{aligned} \mu(A \cup B) &= \frac{\ln\left((\lambda+1)^{\mu(A)} + (\lambda+1)^{\mu(B)} - 1\right)}{\ln(\lambda+1)} \\ &\leq \frac{\ln\left[(\lambda+1)^{\mu(A)}(\lambda+1)^{\mu(B)}\right]}{\ln(\lambda+1)} \\ &= \mu(A) + \mu(B), \end{aligned}$$

for every $A, B \in \mathcal{A}, A \cap B = \emptyset$. Then for every disjoint family $(A_i)_{i \in \{1,\dots,n\}} \subseteq \mathcal{A}$, by using the inequality between the geometric mean and the arithmetic mean, we have

$$\left|\mu\left(\cup_{i=1}^n A_i\right) - \sum_{i=1}^n \mu(A_i)\right| = \sum_{i=1}^n \mu(A_i) - \mu\left(\cup_{i=1}^n A_i\right)$$

$$= \sum_{i=1}^n \mu(A_i) - \frac{\ln\left(\sum_{i=1}^n (\lambda+1)^{\mu(A_i)} - (n-1)\right)}{\ln(\lambda+1)}$$

$$\leq \sum_{i=1}^{n} \mu(A_i) - \frac{\ln\left(n \sqrt[n]{(\lambda+1)^{\sum_{i=1}^{n}\mu(A_i)}} - (n-1)\right)}{\ln(\lambda+1)}$$

Because the function $h : [0, n\mu(X)] \to \mathbf{R}$ defined by

$$h(t) = t - \frac{\ln\left(n \sqrt[n]{(\lambda+1)^t} - (n-1)\right)}{\ln(\lambda+1)}$$

has the derivative

$$h'(t) = \frac{(n-1)\left((\lambda+1)^{\frac{t}{n}} - 1\right)}{n(\lambda+1)^{\frac{t}{n}} - (n-1)}$$

positive on $[0, n\mu(X)]$, we get that h is increasing and $h(n\mu(X)) = n\mu(X) - \frac{\ln\left(n(\lambda+1)^{\mu(X)} - n + 1\right)}{\ln(\lambda+1)}$. We obtain

$$\left|\mu\left(\cup_{i=1}^{n} A_i\right) - \sum_{i=1}^{n} \mu(A_i)\right| \leq n\mu(X) - \frac{\ln\left(n(\lambda+1)^{\mu(X)} - n + 1\right)}{\ln(\lambda+1)},$$

for every disjoint family $(A_i)_{i \in \{1,\ldots,n\}} \subseteq \mathcal{A}$, that is

$$a_{n,\lambda}(\mu) \leq n\mu(X) - \frac{\ln\left(n(\lambda+1)^{\mu(X)} - n + 1\right)}{\ln(\lambda+1)},$$

for all $n \geq 2, \lambda \in (0, \infty)$. □

Remark. We obtain $\lim_{\lambda \searrow 0} a_{n,\lambda}(\mu) = 0$ for all $n \geq 2$.

In the final part of this section we present three kinds of applications for the previous results.

4.1.1 *Application to calculation of fuzzy integral*

In general, the duality between fuzzy measure and its corresponding fuzzy integral takes place. This means that if $\mu : \mathcal{A} \to [0, \infty), \mathcal{A} \subseteq \mathcal{P}(X)$, is a fuzzy measure and f is a nonnegative $\mathcal{A}$-measurable function, then $\nu : \mathcal{A} \to [0, \infty)$ defined by $\nu(A) = \widetilde{\int}_A f d\mu$, where $\widetilde{\int}$ denotes a fuzzy integral, is a fuzzy measure. In general, these two fuzzy measures have not the same defect of additivity. For example, if μ is a fuzzy measure with $\mu(X) = 1$ and $\widetilde{\int}$ is the Choquet integral (see *e.g.* Murofushi-Sugeno [157]), that is

$$\nu(A) = \widetilde{\int}_A f\mathrm{d}\mu = \int_0^\infty \mu(\{x \in A : f(x) \geq \alpha\})\, \mathrm{d}\alpha,$$

where f is a bounded nonnegative $\mathcal{A}$-measurable function, then

$$a_n(\nu) \leq M a_n(\mu), \tag{4.4}$$

where $M = \sup\{f(x) : x \in X\}$. Indeed, if $A_1, ..., A_n$ are measurable, $A_i \cap A_j = \emptyset \ (i \neq j)$, then

$$\left\{x \in \bigcup_{i=1}^n A_i : f(x) \geq \alpha\right\} = \bigcup_{i=1}^n \{x \in A_i : f(x) \geq \alpha\}$$

with $\{x \in A_i : f(x) \geq \alpha\} \cap \{x \in A_j : f(x) \geq \alpha\} = \emptyset \ (i \neq j)$ and

$$\left|\int_{\cup_{i=1}^n A_i} f\mathrm{d}\mu - \sum_{i=1}^n \int_{A_i} f\mathrm{d}\mu\right| \leq \int_0^M \left|\mu\left(\left\{x \in \bigcup_{i=1}^n A_i : f(x) \geq \alpha\right\}\right)\right.$$

$$\left. - \sum_{i=1}^n \mu(\{x \in A_i : f(x) \geq \alpha\})\right| \mathrm{d}\alpha \leq M a_n(\mu).$$

For $M < 1$ we obtain $a_n(\nu) < a_n(\mu)$.

The inequality (4.4) is useful for the approximative calculation of Choquet integrals. We omit the details because the same ideas are presented below.

Some fuzzy measures and fuzzy integrals have a property of the following kind (see Suarez Diaz-Suarez Garcia [202], Sugeno-Murofushi [205]):

$$\text{if } \mu(A \cup B) = \mu(A) * \mu(B), \forall A, B \in \mathcal{A}, A \cap B = \emptyset \text{ then}$$

$$\nu(A \cup B) = \nu(A) * \nu(B), \forall A, B \in \mathcal{A}, A \cap B = \emptyset, \tag{4.5}$$

where μ and ν are as above, that is ν comes from a fuzzy integral and $*$ is a binary operation with suitable properties.

Taking into account this last fact, we can apply the estimates for the defect of additivity, to approximative calculation of the fuzzy integrals, in the following way.

Let $(X, \mathcal{A})$ be a measurable space, f be a nonnegative $\mathcal{A}$-measurable function on X and $\mu : \mathcal{A} \to [0, \infty)$ be a fuzzy measure with the defect of additivity of order n equal to $a_n(\mu)$ and which verifies the condition

$$\mu(A \cup B) = \mu(A) * \mu(B), \forall A, B \in \mathcal{A}, A \cap B = \emptyset.$$

We assume that the fuzzy measure ν constructed with the help of fuzzy integral $\widetilde{\int}$ by $\nu(A) = \widetilde{\int}_A f d\mu, \forall A \in \mathcal{A}$ satisfies the property (4.5). If A has the representation $A = A_1 \cup ... \cup A_n$, where $A_i \in \mathcal{A}, \forall i \in \{1, ..., n\}, n \geq 2$ and $A_i \cap A_j = \emptyset$ if $i \neq j$, then

$$\left| \widetilde{\int}_A f d\mu - \sum_{i=1}^{n} \widetilde{\int}_{A_i} f d\mu \right| = \left| \nu\left(\bigcup_{i=1}^{n} A_i \right) - \sum_{i=1}^{n} \nu(A_i) \right| \leq a_n(\nu).$$

Therefore, if we know the values of the fuzzy integrals on each A_i, $i \in \{1, ..., n\}$, then we can approximate the value of the fuzzy integral on A.

For example, if m is a classical measure, $\mu : \mathcal{A} \to [0, \infty)$ given by $\mu(A) = \frac{c^{m(A)} - 1}{\lambda}, \forall A \in \mathcal{A}$, where $c^{m(X)} = \lambda + 1, c > 0, \lambda \in (0, \infty)$ and $\mu(A) = \frac{\ln(\lambda m(A) + 1)}{\ln(\lambda + 1)}, \forall A \in \mathcal{A}, \lambda \in (0, \infty)$, and the integral is that given by Sugeno-Murofushi [205], then denoting $\gamma(A) = \frac{\nu(A)}{\nu(X)}$ we see that γ is a $\lambda\nu(X)$-additive fuzzy measure and applying Theorem 4.3, we get

$$\left| \widetilde{\int}_A f d\mu - \sum_{i=1}^{n} \widetilde{\int}_{A_i} f d\mu \right| \leq \nu(X) a_n(\gamma) \leq \frac{\lambda\nu(X) + n - n \sqrt[n]{\lambda\nu(X) + 1}}{\lambda},$$

and (applying Theorem 4.7)

$$\left| \widetilde{\int}_A f d\mu - \sum_{i=1}^{n} \widetilde{\int}_{A_i} f d\mu \right| \leq n\nu(X) - \frac{\ln\left(n(\lambda + 1)^{\nu(X)} - n + 1 \right)}{\ln(\lambda + 1)},$$

respectively (here $\nu(X) = \widetilde{\int}_X f d\mu$).

Remark. The ideas in this subsection can be framed into the general scheme described in Section 1.1. Indeed, let $f : X \to \mathbf{R}_+$ be fixed, bounded by $M \in \mathbf{R}_+$, *i.e.* $M = \sup\{f(x); x \in X\}$. If μ is an additive fuzzy measure then the Choquet fuzzy integral denoted by $\widetilde{\int}$, generates an additive fuzzy measure by $\nu(G) = \widetilde{\int}_G f\mathrm{d}\mu, \forall G \in \mathcal{A}$, that is we have the formula

$$\left|\widetilde{\int}_{G_1} f\mathrm{d}\mu + \widetilde{\int}_{G_2} f\mathrm{d}\mu - \widetilde{\int}_{G_1 \cup G_2} f\mathrm{d}\mu\right| \le Ma_2(\mu), \forall G_1, G_2 \in \mathcal{A}, G_1 \cap G_2 = \emptyset,$$

valid for all fuzzy measures (additive or non-additive).

Obviously, for μ additive, the last inequality becomes the additivity property of $\widetilde{\int}$ and can be framed into the scheme in Section 1.1, by defining

$$\begin{aligned} A_{G_1,G_2} &= \widetilde{\int}_{G_1} f\mathrm{d}\mu + \widetilde{\int}_{G_2} f\mathrm{d}\mu - \widetilde{\int}_{G_1 \cup G_2} f\mathrm{d}\mu, \forall G_1, G_2 \in \mathcal{A}, \\ B_{G_1,G_2} &= Mx, \forall x \in \mathbf{R}, \forall G_1, G_2 \in \mathcal{A}, \\ U &= \{\textit{the set of all fuzzy measures on } \mathcal{A}\} \end{aligned}$$

and

$$P = \text{"}\textit{the property of additivity on } \mathcal{A}\text{"}.$$

4.1.2 *Application to best approximation of a fuzzy measure*

Let us define

$$\mathcal{F} = \{\mu : \mathcal{A} \to [0,1]; \mu(\emptyset) = 0, \mu \text{ nondecreasing on the } \sigma\text{-algebra } \mathcal{A}\}$$

and

$$\mathcal{M} = \{m : \mathcal{A} \to [0,1]; m \text{ additive on the } \sigma\text{-algebra } \mathcal{A}\}.$$

It is obvious that $\mathcal{M} \subset \mathcal{F}$. Given $\mu \in \mathcal{F}$, it is natural to consider the problem of approximation of μ by elements in $\mathcal{M}$. In this sense, let us introduce $d : \mathcal{F} \times \mathcal{F} \to [0,1]$ by $d(\mu,\nu) = \sup\{|\mu(A) - \nu(A)|; A \in \mathcal{A}\} \le 1, \forall \mu, \nu \in \mathcal{F}$. Obviously d is a metric on $\mathcal{F}$. Now, given $\mu \in \mathcal{F}$, let us introduce the best approximation of μ by elements in $\mathcal{M}$, by

$$E(\mu) = \inf\{d(\mu, m); m \in \mathcal{M}\} \le 1.$$

We will use the defect to give a lower estimate for $E(\mu)$. Let $A_i \in \mathcal{A}$, $i \in \{1, ..., n\}, A_i \cap A_j = \emptyset, i \ne j$. We have

$$\left|\mu\left(\bigcup_{i=1}^{n} A_i\right) - \sum_{i=1}^{n} \mu(A_i)\right| = \left|\mu\left(\bigcup_{i=1}^{n} A_i\right) - m\left(\bigcup_{i=1}^{n} A_i\right)\right.$$

$$\left. - \sum_{i=1}^{n} (\mu(A_i) - m(A_i)) \right| \leq (n+1)\, d(\mu, m),$$

for every $\mu \in \mathcal{F}, m \in \mathcal{M}$. It follows

$$\frac{a_n(\mu)}{n+1} \leq E(\mu) \leq 1, \forall n \geq 2. \tag{4.6}$$

As a consequence, if μ is such that $\lim_{n\to\infty} \frac{a_n(\mu)}{n+1} = 1$ (for example if μ is the fuzzy measure given by Example 4.3), then we obtain $E(\mu) = 1$.

Now, we give a non-trivial estimation of the best approximation for a concrete fuzzy measure. If X is an infinite set, $\mathcal{A} = \mathcal{P}(X)$ and $m : \mathcal{A} \to [0,1]$ is defined by

$$m(A) = \begin{cases} a^{\operatorname{card} A}, & \text{if } A \text{ is finite} \\ 0, & \text{if } A \text{ is infinite,} \end{cases}$$

where $a \in (0,1)$, then m is an exponential-type measure. By using Example 4.4, the mapping $\mu : \mathcal{A} \to [0,1]$ defined by $\mu(A) = 1 - m(A)$ is a subadditive fuzzy measure. Let $A_i = \{x_i\}, x_i \in X, \forall i \in \{1, ..., n\}, x_i \neq x_j$ if $i \neq j$. Because

$$\mu(A_1) + ... + \mu(A_n) - \mu(A_1 \cup ... \cup A_n) = n - 1 - na + a^n,$$

for every $n \in \mathbf{N}, n \geq 2$, passing to limit in inequalities (4.6), with $a_n(\mu) = n - 1 - na + a^n$, we obtain $1 - a \leq E(\mu) \leq 1$.

4.1.3 *A metric on the family of fuzzy measures*

Let us define

$$\mathbf{F} = \{\mu : \mathcal{A} \to \mathbf{R}; \mu(\emptyset) = 0\}$$

and

$$\mathbf{M} = \{m : \mathcal{A} \to \mathbf{R}; m \text{ additive on the } \sigma\text{-algebra } \mathcal{A}\}.$$

By using the concept of defect of additivity in Definition 4.2 (which obviously can be considered for non-monotonic fuzzy measures too), we will introduce a metric on $\mathbf{F}$ (and so implicitly on $\mathcal{F}$-see the previous application) in the following way. Firstly, it is easy to prove that the defect of additivity of order $n \geq 2, a_n : \mathbf{F} \to \mathbf{R}_+$, satisfies the properties

(i) $a_n(\mu) = 0$,if $\mu \in \mathbf{M}$;
(ii) $a_n(\lambda\mu) = |\lambda| a_n(\mu)$, for all $\mu \in \mathbf{F}, \lambda \in \mathbf{R}$;
(iii) $a_n(\mu + \nu) \leq a_n(\mu) + a_n(\nu)$, for all $\mu, \nu \in \mathbf{F}$,
i. e. $(a_n)_{n \geq 2}$ is a nondecreasing sequence of seminorms on $(\mathbf{F}, +, \cdot)$, where $+$ and $\cdot$ are the usual sum and product with real scalars.

Let us introduce on $\mathbf{F}$ the relation $\sim$ by $\mu \sim \nu$ iff there exists $m : \mathcal{A} \to \mathbf{R}$, additive on $\mathcal{A}$, such that $\mu - \nu = m$. It is easy to see that $\sim$ is a relation of equivalence on $\mathbf{F}$, therefore let us denote $\widehat{\mathbf{F}} = \mathbf{F}/\sim$. Define $a_n : \widehat{\mathbf{F}} \to \mathbf{R}_+$ by $a_n(\widehat{\mu}) = a_n(\mu)$, where $\mu \in \widehat{\mu}$, therefore $(a_n)_{n \geq 2}$ remains a decreasing sequence of seminorms on $\widehat{\mathbf{F}}$, but moreover has the property: $a_n(\widehat{\mu}) = 0$, for all $n \geq 2$, implies $\widehat{\mu} = \mathbf{M} = \widehat{0}_{\widehat{\mathbf{F}}}$ (obviously $\widehat{\mathbf{F}}$ endowed with the usual sum and product with real scalars induced from $\mathbf{F}$, is a linear space). Then

$$d(\widehat{\mu}, \widehat{\nu}) = d(\mu, \nu) = \sum_{n=2}^{\infty} \frac{1}{2^n} \frac{a_n(\mu - \nu)}{1 + a_n(\mu - \nu)},$$

where $\mu \in \widehat{\mu}, \nu \in \widehat{\nu}$, is an invariant to translations metric on $\widehat{\mathbf{F}} \times \widehat{\mathbf{F}}$.

In connection with fuzzy measures, an abstract interpretation is given (see Murofushi-Sugeno [156]): the non-additivity of the fuzzy measure expresses interaction among subsets; so, if μ is a fuzzy measure and $A_1, ..., A_n$ are disjoint subsets, then the inequality

$$\mu(A_1 \cup ... \cup A_n) > \mu(A_1) + ... + \mu(A_n)$$

exhibits the synergy between $A_1, ..., A_n$ and the inequality

$$\mu(A_1 \cup ... \cup A_n) < \mu(A_1) + ... + \mu(A_n)$$

exhibits the incompatibility between $A_1, ..., A_n$. In this sense, the defect of additivity, concept introduced and studied by this section, gives a global measure of synergy or incompatibility for a fuzzy measure. In other fields of mathematics, a basic concept is that of real nonadditive set function. For example, the theory of capacity (Choquet [55]), the evidence theory (Shafer [196]), the surprise theory (Shackle [195]), the possibility theory (Dubois-Prade [67], Zadeh [229]). These theories are susceptible to benefit the concepts and properties in this section. Also, there exist some practical fields where the nonadditivity is important and the defect introduced by the present section could be useful, as for example, in decision problems (see *e.g.*

Yoneda-Fukami-Grabisch [228]), subjective evaluations (see *e.g.* Onisawa-Sugeno-Nishiwaki-Kawai-Harima [161], Tanaka-Sugeno [211]), information processing (see *e.g.* Prade [166], Tahani-Keller [210]), or in classification (see *e.g.* Grabisch-Nicolas [93]).

4.2 Defect of Complementarity

Because the fuzzy measures are non-additive, the sum between the measure of a set and the measure of its complement is not equal to the measure of space. As a consequence, in this section we introduce and study the concept of defect of complementarity from global and pointwise viewpoint. Also, for various classes of fuzzy measures, this defect of complementarity is calculated or estimated. Finally, some applications and interpretations are given.

We recall the definitions of fuzzy measure and of defect of additivity given in the previous section.

Definition 4.4 (see *e.g.* Ralescu-Adams [168]) Let X be a set and $\mathcal{A}$ be a σ-algebra of subsets of X. A set function $\mu : \mathcal{A} \to [0, \infty)$ is said to be a fuzzy measure if the followings hold:

(*i*) $\mu(\emptyset) = 0$;

(*ii*) $A, B \in \mathcal{A}$ and $A \subseteq B$ implies $\mu(A) \leq \mu(B)$.

Let $\mu : \mathcal{A} \to [0, \infty)$ be a fuzzy measure, where $\mathcal{A} \subseteq \mathcal{P}(X)$ is a σ-algebra.

Definition 4.5 The defect of additivity of order $n, n \in \mathbf{N}, n \geq 2$, of the fuzzy measure μ, is given by

$$\begin{aligned} a_n(\mu) &= \sup \left\{ \left| \mu\left(\bigcup_{i=1}^{n} A_i \right) - \sum_{i=1}^{n} \mu(A_i) \right| : A_i \in \mathcal{A}, \right. \\ &\qquad \left. \forall i \in \{1, ..., n\}, A_i \cap A_j = \emptyset, i \neq j \right\}. \end{aligned}$$

Now, we can introduce

Definition 4.6 The value

$$c(\mu) = \frac{\sup\{|\mu(X) - \mu(A) - \mu(A^c)| : A \in \mathcal{A}\}}{\mu(X)}$$

where $A^c = X \setminus A$, is called defect of complementarity of the fuzzy measure μ.

Similarly, the value

$$e(\mu) = 1 - \frac{\sup\{|\mu(X) - \mu(A) - \mu(A^c)| : A \in \mathcal{A}\}}{\mu(X)}$$

can be called order of complementarity. Obviously, $c(\mu) = 1 - e(\mu)$.

Remark. A non-monotonic fuzzy measure is a set function which verifies only the condition (i) in Definition 4.4 (see *e.g.* Murofushi-Sugeno-Machida [158], Murofushi-Sugeno [159]). Definition 4.5 and Definition 4.6 can be considered for non-monotonic fuzzy measures too.

The following properties hold.

Theorem 4.8 *Let $\mu, \mu' : \mathcal{A} \to [0, \infty)$ be fuzzy measures. We have:*
(i) $0 \le c(\mu) \le 1$;
(ii) $c(\mu) \le \frac{a_2(\mu)}{\mu(X)}$;
(iii) $c(\mu) = c(\mu^c)$, where μ^c is the dual of μ, that is $\mu^c(A) = \mu(X) - \mu(A^c)$, A^c being the complementary of A;
(iv) $c(\alpha\mu) = c(\mu), \forall \alpha > 0$;
(v) $c(\mu + \mu') \le \frac{\mu(X)}{(\mu+\mu')(X)} c(\mu) + \frac{\mu'(X)}{(\mu+\mu')(X)} c(\mu')$.

Proof. (i) We have

$$|\mu(X) - \mu(A) - \mu(A^c)| \le \mu(X), \forall A \in \mathcal{A}$$

which proves (i).
(ii) It is immediate by

$$\begin{aligned} &\sup\{|\mu(A \cup B) - \mu(A) - \mu(B)| : A, B \in \mathcal{A}, A \cap B = \emptyset\} \\ \ge\ &\sup\{|\mu(X) - \mu(A) - \mu(A^c)| : A \in \mathcal{A}\}. \end{aligned}$$

(iii) We observe that $\mu^c(X) = \mu(X)$. This implies

$$|\mu^c(X) - \mu^c(A) - \mu^c(A^c)| = |\mu(X) - \mu(A) - \mu(A^c)|, \forall A \in \mathcal{A}$$

which proves the equality.
(iv) It is obvious that $\alpha\mu$ is a fuzzy measure for every $\alpha > 0$, therefore

$$c(\alpha\mu) = \frac{\sup\{|\alpha\mu(X) - \alpha\mu(A) - \alpha\mu(A^c)| : A \in \mathcal{A}\}}{\alpha\mu(X)} = c(\mu).$$

(v) We have

$$\begin{aligned} & c(\mu+\mu') \\ = & \frac{\sup\{|(\mu+\mu')(X)-(\mu+\mu')(A)-(\mu+\mu')(A^c)|: A\in\mathcal{A}\}}{(\mu+\mu')(X)} \\ \leq & \frac{\sup\{|\mu(X)-\mu(A)-\mu(A^c)|: A\in\mathcal{A}\}}{(\mu+\mu')(X)} \\ & +\frac{\sup\{|\mu'(X)-\mu'(A)-\mu'(A^c)|: A\in\mathcal{A}\}}{(\mu+\mu')(X)} \\ = & \frac{\mu(X)}{(\mu+\mu')(X)}c(\mu)+\frac{\mu'(X)}{(\mu+\mu')(X)}c(\mu'). \end{aligned}$$

□

Remark. The inequality (ii) in the previous theorem can be strict. For example, if $X\neq\emptyset, A,B\in\mathcal{P}(X), \emptyset\neq A\subset B\neq X$ then

$$\mathcal{A}=\{\emptyset,A,A^c,B,B^c,A\cup B^c,A^c\cap B,X\}$$

is a σ-algebra and the function $\mu:\mathcal{A}\to[0,1]$ defined by

$$\mu(M)=\begin{cases} 0, & \text{if } M=\emptyset; \\ \alpha, & \text{if } M\in\mathcal{A}\setminus\{\emptyset,X\}; \\ 1, & \text{if } M=X, \end{cases}$$

is a fuzzy measure, $\forall\alpha\in[0,1]$. If $\alpha\in\left(\frac{1}{2},1\right)$ then the defect of complementarity of μ is

$$c(\mu)=\sup\{|1-\mu(A)-\mu(A^c)|: A\in\mathcal{A}\}=|2\alpha-1|=2\alpha-1$$

and the defect of additivity of order 2 is

$$\begin{aligned} a_2(\mu) &= \sup\{|\mu(A\cup B)-\mu(A)-\mu(B)|: A,B\in\mathcal{A}, A\cap B=\emptyset\} \\ &\geq |\mu(A\cup B^c)-\mu(A)-\mu(B^c)|=\alpha. \end{aligned}$$

We obtain $c(\mu)=2\alpha-1<\alpha\leq\frac{a_2(\mu)}{\mu(X)}$.

Example 4.10 If μ is an additive measure then $c(\mu)=0$.

Example 4.11 If μ is the fuzzy measure which models the total ignorance (see Dubois-Prade [64], p.128), that is $\mu_0:\mathcal{P}(X)\to[0,1], \mu_0(A)=0$, if $A\neq X$ and $\mu_0(A)=1$, if $A=X$, then $c(\mu_0)=1$. (Here X is supposed to have at least two distinct elements).

Example 4.12 Let $\mathcal{A}$ be a σ-algebra of subsets of X and $m : \mathcal{A} \to [0,1]$ be of exponential type, *i.e.* $m(A \cup B) = m(A)\, m(B)$ for all $A, B \in \mathcal{A}, A \cap B \neq \emptyset$. The mapping $\mu : \mathcal{A} \to [0,1]$ defined by $\mu(A) = 1 - m(A)$ is a subadditive fuzzy measure (see Example 4.4). Let us assume that there is a set $A \in \mathcal{A}$ such that $m(A) > 0$. Because

$$\mu(A^c) = 1 - m(A^c) = 1 - \frac{m(X)}{m(A)}$$

and

$$\begin{aligned}
&\sup\{\mu(A) + \mu(A^c) - \mu(X) : A \in \mathcal{A}\} \\
=\ &\sup\left\{1 - m(A) - \frac{m(X)}{m(A)} + m(X) : A \in \mathcal{A}\right\} \\
\leq\ &\sup\left\{1 - t - \frac{m(X)}{t} + m(X) : t \in [0,1]\right\} \\
=\ &\left(1 - \sqrt{m(X)}\right)^2,
\end{aligned}$$

we obtain $c(\mu) \leq \frac{1-\sqrt{m(X)}}{1+\sqrt{m(X)}}$.

Example 4.13 Two remarkable fuzzy measures are the outer measure m^* and the inner measure m_* induced by the Lebesgue measure on the interval $[a,b] \subset \mathbf{R}$ (see Example 4.5). We will prove that $c(m^*) = c(m_*) = 1$. Because m^* is the dual of m_* by using Theorem 4.8, *(iii)* we get the first equality. We know that there exists $A \subset [a,b]$ such that $m^*(A) = b - a$ and $m_*(A) = 0$ (see Gelbaum-Olmsted [88], p. 147). Because $m^*(A) + m_*([a,b] \setminus A) = b - a$ (see Halmos [99], p. 55), we obtain $m_*([a,b] \setminus A) = 0$. Therefore

$$c(m_*) \geq \frac{|m_*([a,b]) - m_*(A) - m_*([a,b] \setminus A)|}{m_*([a,b])} = 1,$$

that is $c(m_*) = 1$.

It is well-known (see Congxin-Minghu [57]) that for a null-additive fuzzy measure, that is a fuzzy measure $\mu : \mathcal{A} \to [0, +\infty)$ such that $\mu(A \cup B) = \mu(A), \forall A, B \in \mathcal{A}$ with $\mu(B) = 0$, we can repeat the classical construction of completion. We obtain the complete fuzzy measure $\widetilde{\mu} : \widetilde{\mathcal{A}} \to [0, +\infty)$, where

$$\widetilde{\mathcal{A}} = \{A \cup N : A \in \mathcal{A}, \exists B \in \mathcal{A} \text{ such that } N \subseteq B, \mu(B) = 0\}$$

and

$$\widetilde{\mu}(A \cup N) = \mu(A).$$

By completion, the defect of complementarity is preserved. In this sense we have

Theorem 4.9 *If $\mu : \mathcal{A} \to [0, +\infty)$ is a null-additive fuzzy measure then $c(\widetilde{\mu}) = c(\mu)$, where $\widetilde{\mu}$ is the completion of μ.*

Proof. Let $\widetilde{A} = A \cup N \in \widetilde{\mathcal{A}}$, where $A \in \mathcal{A}$ and there exists $B \in \mathcal{A}$ such that $N \subseteq B, \mu(B) = 0$. We get

$$\widetilde{\mu}\left(\widetilde{A}^c\right) = \widetilde{\mu}(A^c \cap N^c) \leq \widetilde{\mu}(A^c) = \mu(A^c).$$

Because the completion of a null-additive fuzzy measure is also null-additive (see Congxin-Minghu [57]), we obtain

$$\begin{aligned} \widetilde{\mu}\left(\widetilde{A}^c\right) &= \widetilde{\mu}(A^c \cap N^c) \geq \widetilde{\mu}(A^c \cap B^c) = \widetilde{\mu}((A^c \cap B^c) \cup (A^c \cap B)) \\ &= \mu(A^c \cap (B^c \cup B)) = \mu(A^c). \end{aligned}$$

As a conclusion, $\widetilde{\mu}\left(\widetilde{A}^c\right) = \mu(A^c)$. Then

$$\begin{aligned} c(\widetilde{\mu}) &= \frac{\sup\left\{\left|\widetilde{\mu}(X) - \widetilde{\mu}\left(\widetilde{A}\right) - \widetilde{\mu}\left(\widetilde{A}^c\right)\right| : \widetilde{A} \in \widetilde{\mathcal{A}}\right\}}{\widetilde{\mu}(X)} \\ &= \frac{\sup\{|\mu(X) - \mu(A) - \mu(A^c)| : A \in \mathcal{A}\}}{\mu(X)} = c(\mu). \end{aligned}$$

□

In what follows, we will prove that for an important class of fuzzy measures, the inequality (*ii*) in Theorem 4.8 becomes equality.

Definition 4.7 (see *e.g.* Bertoluzza-Cariolaro [38] or Dubois-Prade [65]) A fuzzy measure $\mu : \mathcal{A} \to [0, \infty)$ with $\mu(X) = 1$ is said to be S-decomposable if there exists a composition law $S : [0,1] \times [0,1] \to [0,1]$ such that

$$\mu(A \cup B) = \mu(A) S \mu(B), \forall A, B \in \mathcal{A}, A \cap B = \emptyset.$$

Theorem 4.10 *Let S be a triangular conorm fulfilling*

$$T(x,y) + S(x,y) = x + y, \forall x, y \in [0,1],$$

where T is the triangular norm associated with S. If the fuzzy measure $\mu : \mathcal{A} \to [0, \infty)$ is S-decomposable then $c(\mu) = a_2(\mu)$.

Proof. We will prove only the inequality $c(\mu) \geq a_2(\mu)$ because the converse inequality is given by Theorem 4.8, (ii).

Let $A, B \in \mathcal{A}, A \cap B = \emptyset$. By using the hypothesis and the inclusion $B \subseteq A^c$, we get

$$|\mu(A \cup B) - \mu(A) - \mu(B)| = |\mu(A) S \mu(B) - \mu(A) - \mu(B)|$$

$$\begin{aligned} &= \mu(A) T \mu(B) \leq \mu(A) T \mu(A^c) = \mu(A) + \mu(A^c) - \mu(A) S \mu(A^c) \\ &= |\mu(A) + \mu(A^c) - 1|. \end{aligned}$$

This implies

$$\begin{aligned} & \sup \{|\mu(A \cup B) - \mu(A) - \mu(B)| : A, B \in \mathcal{A}, A \cap B = \emptyset\} \\ \leq\ & \sup \{|\mu(A) + \mu(A^c) - 1| : A \in \mathcal{A}\} \end{aligned}$$

therefore $a_2(\mu) \leq c(\mu)$. □

Remark. We mention that in Frank [78] was proved that the family $\{(T_s, S_s) : s \in [0, \infty]\}$, where the triangular conorms S_s are given by

$$S_s(x, y) = \begin{cases} \max(x, y), & \text{if } s = 0; \\ x + y - xy, & \text{if } s = 1; \\ \min(x + y, 1), & \text{if } s = \infty; \\ 1 - \log_s \left(1 + \frac{(s^{1-x} - 1)(s^{1-y} - 1)}{s - 1}\right), & \text{if } s \in (0,1) \cup (1, \infty) \end{cases}$$

and their so-called ordinal sums are the unique pairs of triangular conorms and associated triangular norms fulfilling $T(x,y) + S(x,y) = x + y, \forall x, y \in [0,1]$. In fact, the above mentioned family contains the most important triangular conorms (and implicitly the triangular norms), namely S_0 and S_∞.

In what follows, we deal with the calculation or the estimation of defect of complementarity for some classes of fuzzy measures. The first class is that of λ-additive fuzzy measures. We recall

Definition 4.8 (see *e.g.* Grabisch [94] or Kruse [126]) Let $\lambda \in (-1, \infty)$, $\lambda \neq 0$, be a real number and $\mathcal{A}$ be a σ-algebra on the set X. A fuzzy measure μ with $\mu(X) = 1$ is called λ-additive if, whenever $A, B \in \mathcal{A}, A \cap B = \emptyset$,

$$\mu(A \cup B) = \mu(A) + \mu(B) + \lambda \mu(A) \mu(B).$$

We denote by $c_\lambda(\mu)$ the defect of complementarity of a λ-additive fuzzy measure μ.

Theorem 4.11 *If μ is a λ-additive fuzzy measure, then*

$$c_\lambda(\mu) \le \frac{\left(1-\sqrt{\lambda+1}\right)^2}{|\lambda|}.$$

Proof. The λ-additivity of μ implies $\mu(A^c) = \frac{1-\mu(A)}{1+\lambda\mu(A)}, \forall A \in \mathcal{A}$.
Then

$$|\mu(X) - \mu(A) - \mu(A^c)| = |\lambda|\,\mu(A)\,\mu(A^c) = |\lambda|\,\mu(A)\,\frac{1-\mu(A)}{1+\lambda\mu(A)},$$

for every $A \in \mathcal{A}$. Because the function $h : [0, \mu(X)] \to \mathbf{R}$ defined by $h(t) = t\frac{1-t}{1+\lambda t}$ is increasing on $\left[0, \frac{1-\sqrt{1+\lambda}}{-\lambda}\right]$, decreasing on $\left[\frac{1-\sqrt{1+\lambda}}{-\lambda}, 1\right]$ and $h\left(\frac{1-\sqrt{1+\lambda}}{-\lambda}\right) = \frac{\left(1-\sqrt{1+\lambda}\right)^2}{\lambda^2}$, we get

$$\sup\{|\mu(X) - \mu(A) - \mu(A^c)| : A \in \mathcal{A}\} \le |\lambda|\,\frac{\left(1-\sqrt{1+\lambda}\right)^2}{\lambda^2},$$

that is $c_\lambda(\mu) \le \frac{\left(1-\sqrt{1+\lambda}\right)^2}{|\lambda|}$. □

Remarks. 1) If $\lambda \to 0$ then $c_\lambda(\mu) \to 0$, in concordance with the fact that for $\lambda = 0$, the λ-measures are additive.

2) The inequality given by the previous theorem becomes equality in some situations. Indeed, we know (see Theorem 4.2 or Wierzchon [222]) that $\mu : \mathcal{A} \to [0,1]$ defined by $\mu(A) = \frac{(\lambda+1)^{\frac{m(A)}{m(X)}}-1}{\lambda}$, where $\lambda \in (-1,0) \cup (0,\infty)$ is a λ-additive fuzzy measure if m is a classical finite measure on $\mathcal{A}$. If we consider m the induced Lebesgue measure on the interval $X = [a,b], a < b$ and $A = \left[a, \frac{a+b}{2}\right]$ then

$$c_\lambda(\mu) \ge \frac{|\mu(X) - \mu(A) - \mu(A^c)|}{\mu(X)}$$

$$= \left|1 - 2\frac{(\lambda+1)^{\frac{b-a}{2(b-a)}} - 1}{\lambda}\right| = \frac{\left(1-\sqrt{1+\lambda}\right)^2}{|\lambda|},$$

that is $c_\lambda(\mu) = \frac{\left(1-\sqrt{1+\lambda}\right)^2}{|\lambda|}$.

3) The inequality proved by Theorem 4.11 gives only an estimation of the defect of complementarity. The effective values of the λ-additive fuzzy measure affect the value of the indicator. So, if $X = \{x_1, x_2, x_3\}$ and the λ-additive fuzzy measures $\mu, \mu' : \mathcal{P}(X) \to [0,1]$ are given as in the below table, then $\lambda = 1 < 2 = \lambda'$ and nevertheless $c_\lambda(\mu) > c_{\lambda'}(\mu')$.

	λ	x_1	x_2	x_3	x_1, x_2	x_2, x_3	x_1, x_3	$c(\mu)$
μ	1	0	1/2	1/3	1/2	1	1/3	1/6
μ'	2	0	1/10	3/4	1/10	1	3/4	3/20

In Theorem 4.5, starting from a functional equation classes of fuzzy measures are obtained. In what follows, for some of them we give estimations of their defect of complementarity.

Theorem 4.12 *If $\mu : \mathcal{A} \to [0,\infty)$ is the fuzzy measure defined by $\mu(A) = \lambda m^2(A), \forall A \in \mathcal{A}$, where m is a finite measure, then $c(\mu) \le \frac{1}{2}, \forall \lambda \in (0,\infty)$.*

Proof. Let $A \in \mathcal{A}$. Because μ is superadditive, we have

$$|\mu(A) + \mu(A^c) - \mu(X)| = \mu(X) - \mu(A) - \mu(A^c)$$

$$\begin{aligned} &= \lambda m^2(X) - \lambda m^2(A) - \lambda (m(X) - m(A))^2 \\ &= \lambda\left(-2m^2(A) + 2m(X)m(A)\right) = 2\lambda m(A)(m(X) - m(A)) \\ &\le 2\lambda \frac{m^2(X)}{4} = \frac{\mu(X)}{2}. \end{aligned}$$

Therefore

$$c(\mu) = \frac{\sup\{|\mu(X) - \mu(A) - \mu(A^c)| : A \in \mathcal{A}\}}{\mu(X)} \le \frac{1}{2}.$$

□

Remark. Reasoning as in Remark 2) after Theorem 4.11, if m is the induced Lebesgue measure on $X = [a,b], a < b$ then $c(\mu) = \frac{1}{2}$, where μ is given by Theorem 4.12.

Theorem 4.13 *If $\mu : \mathcal{A} \to [0,\infty)$ is the fuzzy measure defined by $\mu(A) = \frac{\ln(\lambda m(A)+1)}{\ln(\lambda+1)}, \forall A \in \mathcal{A}$, where m is a finite measure and $\lambda > 0$, then*

$$c(\mu) \le 2\frac{\ln\left(\lambda \frac{m(X)}{2} + 1\right)}{\ln(\lambda m(X) + 1)} - 1.$$

Proof. Firstly, we observe that μ is subadditive, $\forall\lambda \in (0,\infty)$. Indeed, $\left((\lambda+1)^{\mu(A)}-1\right)\left((\lambda+1)^{\mu(B)}-1\right) \geq 0, \forall\lambda \in (0,\infty)$ implies

$$\begin{aligned}\mu(A \cup B) &= \frac{\ln\left((\lambda+1)^{\mu(A)}+(\lambda+1)^{\mu(B)}-1\right)}{\ln(\lambda+1)} \\ &\leq \frac{\ln\left[(\lambda+1)^{\mu(A)}(\lambda+1)^{\mu(B)}\right]}{\ln(\lambda+1)} = \mu(A)+\mu(B),\end{aligned}$$

for every $A, B \in \mathcal{A}, A \cap B = \emptyset$. We have

$$\begin{aligned}|\mu(X)-\mu(A)-\mu(A^c)| &= \mu(A)+\mu(A^c)-\mu(X) = \\ &= \frac{\ln\frac{(\lambda m(A)+1)(\lambda m(X)-\lambda m(A)+1)}{\lambda m(X)+1}}{\ln(\lambda+1)},\end{aligned}$$

for every $A \in \mathcal{A}$. Because the function $h : [0, m(X)] \to \mathbf{R}$ defined by

$$h(t) = (\lambda t+1)(\lambda m(X)-\lambda t+1)$$

has the maximum equal to $\left(\lambda\frac{m(X)}{2}+1\right)^2$ obtained for $t = \frac{m(X)}{2}$, we get

$$\begin{aligned}c(\mu) &= \frac{\sup\{|\mu(X)-\mu(A)-\mu(A^c)| : A \in \mathcal{A}\}}{\mu(X)} \\ &\leq 2\frac{\ln\left(\lambda\frac{m(X)}{2}+1\right)}{\ln(\lambda m(X)+1)} - 1, \forall\lambda > 0.\end{aligned}$$

□

Remark. If $\lambda \to 0$ then $c(\mu) \to 0$, where μ is given by Theorem 4.13.

The defect of complementarity (see Definition 4.6) and the order of complementarity (after Definition 4.6) are global indicators. Because they do not take into account the values of measure on every set, it is possible to obtain the same value of indicators for two different fuzzy measures, in contradiction with our intuition. In this sense let us consider the following example.

Example 4.14 Let $\mathcal{A} \subset \mathcal{P}(X)$ be a σ-algebra of subsets of X, $M \in \mathcal{A}, M \neq \emptyset, M \neq X$ and $\mu : \mathcal{A} \to [0,1]$ a (additive) measure with $\mu(X) = 1$. The set function $\overline{\mu} : \mathcal{A} \to [0,1]$ defined by

$$\overline{\mu}(A) = \begin{cases} 0, & \text{if } A \subseteq M \text{ or } A \subseteq M^c \\ \mu(A), & \text{if } A \not\subseteq M \text{ and } A \not\subseteq M^c \end{cases}$$

is a fuzzy measure. Indeed, $\mu(\emptyset) = 0$, and if $A, B \in \mathcal{A}, A \subseteq B$, then the following situations are possible:

(1) ($A \not\subseteq M$ and $A \not\subseteq M^c$) and ($B \not\subseteq M$ and $B \not\subseteq M^c$) which implies $\overline{\mu}(A) = \mu(A) \leq \mu(B) = \overline{\mu}(B)$;

(2) ($A \subseteq M$ or $A \subseteq M^c$) and ($B \not\subseteq M$ and $B \not\subseteq M^c$) implies $\overline{\mu}(A) = 0 \leq \mu(B) = \overline{\mu}(B)$;

(3) ($A \subseteq M$ or $A \subseteq M^c$) and ($B \subseteq M$ or $B \subseteq M^c$) implies $\overline{\mu}(A) = 0 = \overline{\mu}(B)$;

The defect of complementarity of fuzzy measure $\overline{\mu}$ is equal to 1 because

$$c(\overline{\mu}) = \sup\{|1 - \overline{\mu}(A) - \overline{\mu}(A^c)| : A \in \mathcal{A}\} = |1 - \overline{\mu}(M) - \overline{\mu}(M^c)| = 1.$$

The fuzzy measure $\overline{\mu}$ above constructed has the same defect of complementarity with the fuzzy measure μ_0 (on $\mathcal{A}$) which models the total ignorance (see Example 4.11). But the set function $\overline{\mu}$ restricted to

$$\mathcal{B} = \{A \in \mathcal{A} : A \not\subseteq M \text{ and } A \not\subseteq M^c\}$$

is additive (that is $A, A^c \in \mathcal{B}$ implies $\overline{\mu}(A) + \overline{\mu}(A^c) = \overline{\mu}(X)$) while the fuzzy measure μ_0 can be considered as a peak of non-complementarity because $\mu_0(A) + \mu_0(A^c) \neq \mu_0(X), \forall A \in \mathcal{A}, A \neq \emptyset$ and $A \neq X$.

Motivated by the previous discussion, in what follows we introduce an indicator of complementarity that take into account every value $\mu(X) - \mu(A) - \mu(A^c), A \in \mathcal{A}$, where $\mu : \mathcal{A} \to [0, +\infty)$ is a fuzzy measure. We will prove that a cardinality of a convenient fuzzy set is a good indicator of complementarity in this sense.

Firstly, let us introduce a natural relation between fuzzy measures with respect to complementarity.

Definition 4.9 Let $\mu_1, \mu_2 : \mathcal{A} \to [0, +\infty)$ be fuzzy measures. We say that μ_2 is more defective than μ_1 (we write $\mu_1 \preceq \mu_2$) if the inequality

$$\frac{|\mu_1(X) - \mu_1(A) - \mu_1(A^c)|}{\mu_1(X)} \leq \frac{|\mu_2(X) - \mu_2(A) - \mu_2(A^c)|}{\mu_2(X)}$$

holds for every $A \in \mathcal{A}$.We say that μ_2 is strictly more defective than μ_1 (we write $\mu_1 \prec \mu_2$) if $\mu_1 \preceq \mu_2$ and there is $A \in \mathcal{A}$ such that the previous inequality is strict.

Example 4.15 With the notations in Example 4.14, $\mu \preceq \overline{\mu}$ and $\overline{\mu} \prec \mu_0$ (if $\{A \in \mathcal{A} : A \not\subseteq M \text{ and } A \not\subseteq M^c\} \neq \{X\}$).

Now, let X be a set and let us denote by $FS(X)$ the family of all fuzzy sets on X, that is $FS(X) = \left\{ \widetilde{A} \middle| \widetilde{A} : X \to [0,1] \right\}$.

Suggested by the relations $\preceq$ and $\prec$, for every fuzzy measure $\mu : \mathcal{A} \to [0, +\infty)$ we consider the fuzzy set $\widetilde{C}_\mu : \mathcal{A} \to [0,1]$ defined as

$$\widetilde{C}_\mu (A) = \frac{|\mu(X) - \mu(A) - \mu(A^c)|}{\mu(X)}.$$

Various definitions of the cardinalities of fuzzy sets have been proposed by several authors (Dubois-Prade [66], De Luca-Termini [138], Ralescu [169], Wygralak [225], [224], Zadeh [230], *etc.*) in the finite case, especially. The sigma-count Ralescu [169] (or the power, De Luca-Termini [138])

$$\sigma - count\widetilde{A} = \sum_{i=1}^{n} \widetilde{A} (x_i),$$

the $FGCount$ Blanchard [40], $\left|\widetilde{A}\right| : \mathbf{N} \to [0,1]$ defined as

$$\left|\widetilde{A}\right| (n) = \sup \left\{ \alpha : card\widetilde{A}_\alpha \geq n \right\},$$

and the fuzzy cardinality (Zadeh [230]) $\left\|\widetilde{A}\right\| : \mathbf{N} \to [0,1]$ defined as

$$\left\|\widetilde{A}\right\| (n) = \sup \left\{ \alpha : card\widetilde{A}_\alpha = n \right\},$$

where $\widetilde{A}_\alpha = \left\{ x \in X : \widetilde{A}(x) \geq \alpha \right\}, \forall \alpha \in (0,1]$, are cardinalities of the fuzzy set $\widetilde{A} \in FS(X), X = \{x_1, ..., x_n\}$ that take into account every value $\widetilde{A}(x_i), i \in \{1, ..., n\}$. Among these, the sigma-count is a scalar cardinality which is convenient for our purpose.

In the sequel, X is considered finite, $X = \{x_1, ..., x_n\}$.

Definition 4.10 Let $\mu : \mathcal{A} \to [0, \infty)$ be a fuzzy measure, where $\mathcal{A} \subseteq \mathcal{P}(X)$ is a σ-algebra. The value

$$\Sigma(\mu) = \sigma - count\widetilde{C}_\mu$$

is called the Σ-defect of complementarity of fuzzy measure μ.

Remark. Between the defect of complementarity c and the pointwise defect of complementarity Σ does not exist effective connections. So, if $X = \{x_1, x_2, x_3\}$ and μ_1, μ_2 are given as in the below table, then $c(\mu_1) <$

$c(\mu_2)$ and $\Sigma(\mu_1) < \Sigma(\mu_2)$. On the other hand, $c(\mu_3) < c(\mu_2)$ and $\Sigma(\mu_3) > \Sigma(\mu_2)$ for the λ-additive fuzzy measures μ_3 and μ_2 defined as in the below table (for $\lambda = 1$, everywhere).

	x_1	x_2	x_3	x_1, x_2	x_2, x_3	x_1, x_3	$c(\mu)$	$\Sigma(\mu)$
μ_1	0	1/2	1/3	1/2	1	1/3	1/6	2/3
μ_2	2/5	2/5	1/49	24/25	21/49	21/49	6/35	888/1225
μ_3	1/5	1/4	1/3	1/2	2/3	3/5	1/6	9/10

According to the above definition we obtain the following properties.

Theorem 4.14 *Let $\mu, \mu' : \mathcal{A} \to [0, \infty)$ be fuzzy measures. Then*
(*i*) $0 \leq \Sigma(\mu) \leq (card\mathcal{A} - 2)\, c(\mu)$;
(*ii*) *If μ is additive then* $\Sigma(\mu) = 0$;
(*iii*) $\Sigma(\mu) = \Sigma(\mu^c)$, *where μ^c is the dual of* μ;
(*iv*) $\Sigma(\alpha\mu) = \Sigma(\mu), \forall\alpha > 0$;
(*v*) $\Sigma(\mu + \mu') \leq \frac{\mu(X)}{(\mu+\mu')(X)}\Sigma(\mu) + \frac{\mu'(X)}{(\mu+\mu')(X)}\Sigma(\mu')$;
(*vi*) *If $\mu \preceq \mu'$ then* $\Sigma(\mu) \leq \Sigma(\mu')$;
(*vii*) *If $\mu \prec \mu'$ then* $\Sigma(\mu) < \Sigma(\mu')$.

Proof. (*i*) The obvious inequality $\widetilde{C}_\mu(A) \leq c(\mu), \forall A \in \mathcal{A}$ and $\widetilde{C}_\mu(\emptyset) = \widetilde{C}_\mu(X) = 0$ implies $\Sigma(\mu) = \sum_{A\in\mathcal{A}} \widetilde{C}_\mu(A) \leq (\text{card}\mathcal{A} - 2)\, c(\mu)$.
(*ii*) It is obvious.
(*iii*) Because $\mu^c(X) = \mu(X)$ and

$$|\mu^c(X) - \mu^c(A) - \mu^c(A^c)| = |\mu(X) - \mu(A) - \mu(A^c)|, \forall A \in \mathcal{A}$$

we obtain $\widetilde{C}_{\mu^c}(A) = \widetilde{C}_\mu(A), \forall A \in \mathcal{A}$.
(*iv*) It is obvious that $\alpha\mu$ is a fuzzy measure for every $\alpha > 0$.

$$\begin{aligned}\Sigma(\alpha\mu) &= \sum_{A\in\mathcal{A}} \widetilde{C}_{\alpha\mu}(A) = \sum_{A\in\mathcal{A}} \frac{|\alpha\mu(X) - \alpha\mu(A) - \alpha\mu(A^c)|}{\alpha\mu(X)} \\ &= \sum_{A\in\mathcal{A}} \widetilde{C}_\mu(A) = \Sigma(\mu).\end{aligned}$$

(*v*) We have

$$\begin{aligned}\widetilde{C}_{\mu+\mu'}(A) &= \frac{|(\mu+\mu')(X) - (\mu+\mu')(A) - (\mu+\mu')(A^c)|}{(\mu+\mu')(X)} \\ &\leq \frac{|\mu(X) - \mu(A) - \mu(A^c)|}{(\mu+\mu')(X)} + \frac{|\mu'(X) - \mu'(A) - \mu'(A^c)|}{(\mu+\mu')(X)}\end{aligned}$$

$$= \frac{\mu(X)}{(\mu+\mu')(X)}\widetilde{C}_{\mu}(A) + \frac{\mu'(X)}{(\mu+\mu')(X)}\widetilde{C}_{\mu'}(A),$$

$\forall A \in \mathcal{A}$, which implies the inequality.

(*vi*) We obtain $\widetilde{C}_{\mu}(A) \leq \widetilde{C}_{\mu'}(A), \forall A \in \mathcal{A}$. By using the monotonicity of the cardinality (see Dubois-Prade [66]) we get $\Sigma(\mu) \leq \Sigma(\mu')$.

(*vii*) It is immediate. □

Example 4.16 The maximum value of the pointwise defect of complementarity Σ (see (*i*) in the previous theorem) is attained for the fuzzy measure which models the total ignorance (see Example 4.11), that is $\Sigma(\mu_0) = 2^{\operatorname{card} X} - 2$.

Now, in what follows, we present a theoretical application of the defect of complementarity (see Definition 4.6) to the estimation of Choquet integral on the entire space X, if we know its values on the sets A and A^c.

Let $\mu : \mathcal{A} \to [0,1]$ be a λ-additive fuzzy measure (see Definition 4.8). If $f : X \to [0,+\infty)$ is bounded and $\mathcal{A}$-measurable, then the Choquet integral $\widetilde{\int}$ (see Choquet [55]) is given by

$$\begin{aligned} \nu(A) &= \widetilde{\int_A} f\mathrm{d}\mu = \int_0^{\infty} \mu(\{x \in A : f(x) \geq \alpha\})\,\mathrm{d}\alpha \\ &= \int_0^{M} \mu(\{x \in A : f(x) \geq \alpha\})\,\mathrm{d}\alpha \end{aligned}$$

where $M = \sup\{f(x) : x \in X\}$. We obtain

$$|\nu(X) - \nu(A) - \nu(A^c)|$$

$$\begin{aligned} &\leq |\lambda| \int_0^M \mu(\{x \in A : f(x) \geq \alpha\})\,\mu(\{x \in A^c : f(x) \geq \alpha\})\,\mathrm{d}\alpha \\ &\leq M|\lambda|\,\mu(A)\,\mu(A^c) = M\,|\mu(X) - \mu(A) - \mu(A^c)| \\ &\leq M \cdot c_{\lambda}(\mu), \end{aligned}$$

for every $A \in \mathcal{A}$, that is $c(\nu) \leq M \cdot c_{\lambda}(\mu)$. As a consequence, if we know the values of the Choquet's integrals $\widetilde{\int}_A f\mathrm{d}\mu$ and $\widetilde{\int}_{A^c} f\mathrm{d}\mu$, then we can approximate the value of the Choquet's integral $\widetilde{\int}_X f\mathrm{d}\mu$, by using the obvious estimation

$$\left|\widetilde{\int_X} f\mathrm{d}\mu - \left(\widetilde{\int_A} f\mathrm{d}\mu + \widetilde{\int_{A^c}} f\mathrm{d}\mu\right)\right| \leq M \cdot c_{\lambda}(\mu).$$

The same idea can be applied to other families of fuzzy measures too, like the S-decomposable fuzzy measures μ (see Definition 4.7), such that the function $F : [0,1] \to \mathbf{R}$ defined by $F(x) = S(x, y) - x - y$, is nondecreasing as function of $x, \forall y \in [0,1]$.

In Murofushi-Sugeno-Machida [158], Murofushi-Sugeno [156], a concrete interpretation of fuzzy measures was proposed as follows.

Let X be the set of workers in a workshop and suppose that they produce the same products. A group A may have various ways to work: various combinations of joint work and divided work. But let us suppose that a group A works in the most efficient way. Let $\mu(A)$ be the number of the products made by A in one hour. Then μ is a measure of the productivity of group. By the definition of μ, the following statements are natural:

$$\mu(\emptyset) = 0$$

and

$$A \subset B \text{ implies } \mu(A) \leq \mu(B),$$

that is μ is a fuzzy measure. It is obvious that μ is not necessarily additive. Let A and B be disjoint subsets of X and let us consider the productivity of the coupled group $A \cup B$. If A and B separately work, then $\mu(A \cup B) = \mu(A) + \mu(B)$. But, since in general, they interact each to other, equality may not necessarily hold. The inequality $\mu(A \cup B) > \mu(A) + \mu(B)$ shows the effective cooperation of the members of $A \cup B$ and the converse inequality $\mu(A \cup B) < \mu(A) + \mu(B)$ shows the incompatibility between the groups A and B.

Now, we will give an interpretation of the defect of complementarity in the sense of the above interpretation.

If $A \subset X$ is fixed, then the number $|\mu(X) - \mu(A) - \mu(A^c)|$ expresses the difference between the number of products made in one hour by the entire group X and the number of products made in one hour by the workers in X if they work separately in two disjoint groups A and A^c (in the situation of the effective cooperation of members of X) and viceversa (in the situation of the incompatibility of common work). The value $\sup\{|\mu(X) - \mu(A) - \mu(A^c)| : A \in \mathcal{P}(X)\}$ is the maximum variation (in number of products on one hour) if the work is organized in all possible 2-partitions of X (*i.e.* X divided into two disjoint parts) in comparison with the common work. This means that

$$[\mu(X) \cdot (1 - c(\mu)), \mu(X) \cdot (1 + c(\mu))]$$

is the interval of the possible values of number of products made in one hour if the entire group of workers X is divided into two disjoint parts.

In connection with the above interpretation, a concrete example of Choquet integral is given in Murofushi-Sugeno [159]:

Let $X = \{x_1, ..., x_n\}$. On day, each worker x_i works $f(x_i)$ hours from the opening hour. Without loss of generality, we can assume that $f(x_1) \leq ... \leq f(x_n)$. Then, we have for $i \in \{2, ..., n\}$,

$$f(x_i) - f(x_{i-1}) \geq 0$$

and

$$f(x_i) = f(x_1) + (f(x_2) - f(x_1)) + ... + (f(x_i) - f(x_{i-1})).$$

Now, let us aggregate the working hours of all the workers in the following way. First, the group X with n workers works $f(x_1)$ hours, next the group $X \setminus \{x_1\} = \{x_2, ..., x_n\}$ works $f(x_2) - f(x_1)$ hours, then the group $X \setminus \{x_1, x_2\} = \{x_3, ..., x_n\}$ works $f(x_3) - f(x_2)$ hours,..., the last worker $\{x_n\}$ works $f(x_n) - f(x_{n-1})$ hours. Because a group $A \subseteq X$ produces the amount $\mu(A)$ in one hour (see the above interpretation of fuzzy measure), the total number of products obtained by the workers is expressed by

$$\begin{aligned}
& f(x_1) \cdot \mu(X) \\
& + (f(x_2) - f(x_1)) \cdot \mu(X \setminus \{x_1\}) \\
& + (f(x_3) - f(x_2)) \cdot \mu(X \setminus \{x_1, x_2\}) \\
& + ... + \\
& + (f(x_n) - f(x_{n-1})) \cdot \mu(\{x_n\}) \\
= & \sum_{i=1}^{n} (f(x_i) - f(x_{i-1})) \cdot \mu(\{x_i, ..., x_n\}) \\
= & \widetilde{\int_X} f d\mu,
\end{aligned}$$

where $f(x_0) = 0$. Therefore, if μ describes the number of the products obtained in one hour by subgroups of X, and f describes the number of work hours of every worker, then the Choquet integral $\widetilde{\int}_X f d\mu$ represents the total number of products obtained by workers in one day.

Now, if $A \subseteq X$ then $\widetilde{\int}_A f d\mu$ and $\widetilde{\int}_{A^c} f d\mu$ represent the total number of products obtained by the workers in A and A^c on one day, respectively. Then, the value

$$P(\mu, f) = \sup \left\{ \left| \widetilde{\int}_X f d\mu - \widetilde{\int}_A f d\mu - \widetilde{\int}_{A^c} f d\mu \right| : A \in \mathcal{P}(X) \right\}$$

expresses the maximum growth or the maximum diminution (in number of products) of the production in one day, if the work is organized in all possible 2-partitions of X. If μ is a λ-additive fuzzy measure, then by using the previous results of estimation for Choquet integral, we obtain

$$P(\mu, f) \leq M \cdot c_\lambda(\mu) = f(x_n) \cdot c_\lambda(\mu).$$

At the end of this section we propose a simple numerical example.

Let us consider a group of workers, $X = \{x_1, x_2, x_3, x_4, x_5\}$, and the following table that indicates the number of products obtained in one hour by the subgroups of X.

Subgroup	Products	Subgroup	Products
$\{x_1\}$	17	$\{x_5\}$	6
$\{x_2\}$	26	$\{x_1, x_5\}$	23
$\{x_1, x_2\}$	44	$\{x_2, x_5\}$	32
$\{x_3\}$	22	$\{x_1, x_2, x_5\}$	51
$\{x_1, x_3\}$	40	$\{x_3, x_5\}$	28
$\{x_2, x_3\}$	49	$\{x_1, x_3, x_5\}$	46
$\{x_1, x_2, x_3\}$	68	$\{x_2, x_3, x_5\}$	56
$\{x_4\}$	21	$\{x_1, x_2, x_3, x_5\}$	75
$\{x_1, x_4\}$	39	$\{x_4, x_5\}$	28
$\{x_2, x_4\}$	48	$\{x_1, x_4, x_5\}$	46
$\{x_1, x_2, x_4\}$	67	$\{x_2, x_4, x_5\}$	55
$\{x_3, x_4\}$	44	$\{x_1, x_2, x_4, x_5\}$	74
$\{x_1, x_3, x_4\}$	63	$\{x_3, x_4, x_5\}$	51
$\{x_2, x_3, x_4\}$	72	$\{x_1, x_3, x_4, x_5\}$	70
$\{x_1, x_2, x_3, x_4\}$	93	$\{x_2, x_3, x_4, x_5\}$	80

In fact, the above table defines a fuzzy measure $\mu : \mathcal{P}(X) \to [0, \infty)$. If $\mu(X) = 100$ then $\mu' : \mathcal{P}(X) \to [0, 1]$ defined as $\mu'(A) = \frac{\mu(A)}{100}, \forall A \in \mathcal{P}(X)$ is a normalized fuzzy measure.

The λ-additive fuzzy measures are probably the most important fuzzy measures. As a consequence, several methods have been developed for λ-additive fuzzy measures identification (see *e.g.* Lee-Leekwang [134], Sekita [192], Wierzchon [222]). The value λ corresponding to fuzzy measure μ' is equal with 0.2359 in Wierzchon [222], (and 0.2363 or 0.2351 in Lee-Leekwang [134]). By using Theorem 4.11, we have $c_{0.2359}(\mu') \leq 0.0529$. By the interpretation of the defect of complementarity, this means that without other calculus we can say that for every division of the workers in X into two separate parts, the number of products obtained in one hour will belong to the interval $[94.71, 105.29]$. In fact, the effective value of the defect of complementarity is

$$c(\mu') = 1 - \mu'(\{x_2, x_4\}) - \mu'(\{x_1, x_3, x_5\}) = 0.06$$

and the real interval is $[94, 106]$ (see the same interpretation).

Finally, let us assume that the worker x_i works $f(x_i)$ hours from the opening hour, according to the following table

x_1	x_2	x_3	x_4	x_5
4	6	7	8	8

By using the above interpretations and the estimation given for Choquet integral (with $M = 8$ and $c_{0.2359}(\mu') = 0.06$) we get that the maximum growth or the maximum diminution of the number of products obtained in one day, is smaller than 48 if we consider all 2-partitions of X, in comparison with the common work.

4.3 Defect of Monotonicity

The main characteristics of a measure (in the classical sense) are the additivity and monotonicity. These characteristics are very effective and convenient, but often too inflexible or too rigid in applications. As a solution, the fuzzy measure introduced in Sugeno [204] (the additivity is omitted remaining the monotonicity), non-monotonic fuzzy measure in Murofushi-Sugeno-Machida [158], or equivalently, signed fuzzy measure in Murofushi-Sugeno [159] (the set function which vanishes on the empty set) and T-measure studied in Butnariu-Klement [50], [52] (the fuzzy set function is T-additive, but not necessarily monotone, T being a triangular norm) were proposed. Consequently, appears as natural the following question: how can be evaluated the absence of additivity of fuzzy measures and the absence of monotonicity of signed fuzzy measures and T-measures? An answer to the first part of this question was given by introducing and studying the concept of defect of additivity for a fuzzy measure (see Section 4.1).

In this section we give an answer to the second part of question, by introducing the concept of defect of monotonicity for set functions (particularly, for signed fuzzy measures) and for fuzzy set functions (particularly, for T-measures). For various concrete fuzzy measures and T-measures, this defect is calculated. We also use the defect of monotonicity to estimate the quantity of best approximation of a non-monotonic set function by monotonic set functions.

Definition 4.11 (see Murofushi-Sugeno [159]) Let X be a non-empty set and $\nu : \mathcal{P}(X) \to \mathbf{R}$.

(i) The set function ν is said to be additive if for every pair of disjoint subsets A and B of X, $\nu(A \cup B) = \nu(A) + \nu(B)$.

(ii) The set function ν is said to be monotone if for every pair of subsets A and B of X, $A \subseteq B$

$$\nu(A) \leq \nu(B)$$

or for every pair of subsets A and B of X, $A \subseteq B$

$$\nu(A) \geq \nu(B)$$

(iii) A measure on X is a non-negative additive set function defined on $\mathcal{P}(X)$.

(iv) A signed measure on X is an additive set function defined on $\mathcal{P}(X)$.

(v) A fuzzy measure on X is a monotone set function defined on $\mathcal{P}(X)$ which vanishes on the empty set.

(vi) A signed fuzzy measure is a set function defined on $\mathcal{P}(X)$ which vanishes on the empty set.

If ν is a set function, then we can introduce a quantity which "measures" the deviation of ν from monotonicity.

Definition 4.12 Let $\nu : \mathcal{P}(X) \to \mathbf{R}$ be a set function. The defect of monotonicity of ν is given by

$$d_{MON}(\nu) = \frac{1}{2} \sup \{ |\nu(A_1) - \nu(A)| + |\nu(A_2) - \nu(A)| - |\nu(A_1) - \nu(A_2)| ;$$

$$A_1, A, A_2 \in \mathcal{P}(X), A_1 \subseteq A \subseteq A_2\}.$$

Remarks. 1) The quantity $d_{MON}(\nu)$ was suggested by the definition of the so-called modulus of non-monotonicity of function f on $[a, b]$ defined by

$$\mu(f; \delta)_{[a,b]} = \frac{1}{2} \sup \{|f(x_1) - f(x)| + |f(x_2) - f(x)| - |f(x_1) - f(x_2)|;$$

$$x_1, x, x_2 \in [a, b], x_1 \leq x \leq x_2, |x_1 - x_2| \leq \delta\}, \delta > 0,$$

introduced and used by Sendov in approximation theory (see *e.g.* Sendov [193]).

2) If we denote

$$\begin{aligned}
d_1(\nu) &= \sup\{\nu(A) - \nu(A_1); A_1, A, A_2 \in \mathcal{P}(X), \nu(A_2) \leq \nu(A_1) \leq \nu(A)\} \\
d_2(\nu) &= \sup\{\nu(A) - \nu(A_2); A_1, A, A_2 \in \mathcal{P}(X), \nu(A_1) \leq \nu(A_2) \leq \nu(A)\} \\
d_3(\nu) &= \sup\{\nu(A_1) - \nu(A); A_1, A, A_2 \in \mathcal{P}(X), \nu(A) \leq \nu(A_1) \leq \nu(A_2)\} \\
d_4(\nu) &= \sup\{\nu(A_2) - \nu(A); A_1, A, A_2 \in \mathcal{P}(X), \nu(A) \leq \nu(A_2) \leq \nu(A_1)\}
\end{aligned}$$

then the above definition can be rewritten as

$$d_{MON}(\nu) = \max\{d_1(\nu), d_2(\nu), d_3(\nu), d_4(\nu)\}$$

which is an useful expression for calculation or estimation of the defect of monotonicity. Also, it justifies the constant $\frac{1}{2}$ in the definition of defect of monotonicity.

Theorem 4.15 *The set function $\nu : \mathcal{P}(X) \to \mathbf{R}$ is monotone if and only if $d_{MON}(\nu) = 0$.*

Proof. Suppose that ν is for example increasing, that is $A \subseteq B$ implies $\nu(A) \leq \nu(B)$. If $A_1 \subseteq A \subseteq A_2$ then we get

$$|\nu(A_1) - \nu(A)| + |\nu(A_2) - \nu(A)| - |\nu(A_1) - \nu(A_2)| = 0,$$

which obviously implies $d_{MON}(\nu) = 0$.

Conversely, let us suppose that $d_{MON}(\nu) = 0$. It follows that for all $A_1, A, A_2 \in \mathcal{P}(X), A_1 \subseteq A \subseteq A_2$ we have

$$|\nu(A_1) - \nu(A)| + |\nu(A_2) - \nu(A)| - |\nu(A_1) - \nu(A_2)| = 0.$$

Now, let us assume that ν is non-monotone on $\mathcal{P}(X)$, that is there exist $A_1 \subseteq A \subseteq A_2 \subseteq X$ such that

$$\nu(A_1) < \nu(A) \text{ and } \nu(A) > \nu(A_2)$$

or

$$\nu(A_1) > \nu(A) \text{ and } \nu(A) < \nu(A_2).$$

In both cases, we get

$$|\nu(A_1) - \nu(A)| + |\nu(A_2) - \nu(A)| - |\nu(A_1) - \nu(A_2)| > 0$$

which contradicts $d_{MON}(\nu) = 0$. The theorem is proved □

Example 4.17 If ν is a measure or a fuzzy measure, then $d_{MON}(\nu) = 0$.

Other important properties of the defect of monotonicity are given by the following

Theorem 4.16 *Let $\nu : \mathcal{P}(X) \to \mathbf{R}$ be a set function. We have:*

(i) $0 \le d_{MON}(\nu) \le 2\sup\{|\nu(A)|; A \in \mathcal{P}(X)\}$. If ν is non-negative then $0 \le d_{MON}(\nu) \le \sup\{\nu(A); A \in \mathcal{P}(X)\}$.

(ii) $d_{MON}(\alpha\nu) = |\alpha| d_{MON}(\nu), \forall \alpha \in \mathbf{R}$, where $(\alpha\nu)(A) = \alpha\nu(A)$, $\forall A \in \mathcal{P}(X)$.

(iii) $d_{MON}(\nu_B) \le d_{MON}(\nu)$, where $\nu_B(A) = \nu(A \cap B), \forall A \in \mathcal{P}(X)$, is the induced set function on $B \in \mathcal{P}(X)$.

(iv) If ν is an additive set function then $d_{MON}(\nu) \le a_2(|\nu|)$, where $|\nu|(A) = |\nu(A)|, A \in \mathcal{P}(X)$.

(v) If ν is a signed fuzzy measure then $d_{MON}(\nu) = d_{MON}(\nu^c)$, where ν^c is the dual of ν, that is $\nu^c(A) = \nu(X) - \nu(A^c), \forall A \in \mathcal{P}(X)$.

Proof. (*i*) By Remark 2 before Theorem 4.15, the results are immediate.

(*ii*)

$$d_{MON}(\alpha\nu) = \frac{1}{2}\sup\{|\alpha\nu(A_1) - \alpha\nu(A)| + |\alpha\nu(A_2) - \alpha\nu(A)|$$

$$- |\alpha\nu(A_1) - \alpha\nu(A_2)|; A_1, A, A_2 \in \mathcal{P}(X), A_1 \subseteq A \subseteq A_2\} = |\alpha| d_{MON}(\nu).$$

(*iii*) Because $A_1 \subseteq A \subseteq A_2 \subseteq X$ implies $A_1 \cap B \subseteq A \cap B \subseteq A_2 \cap B \subseteq X$, for every $B \in \mathcal{P}(X)$, we get

$$d_{MON}(\nu_B) = \frac{1}{2}\sup\{|\nu_B(A_1) - \nu_B(A)| + |\nu_B(A_2) - \nu_B(A)|$$

$$-\left|\nu_B(A_1)-\nu_B(A_2)\right|; A_1, A, A_2 \in \mathcal{P}(X), A_1 \subseteq A \subseteq A_2\}$$

$$\leq \frac{1}{2}\sup\{\left|\nu(A_1\cap B)-\nu(A\cap B)\right|+\left|\nu(A_2\cap B)-\nu(A\cap B)\right|$$

$$-\left|\nu(A_1\cap B)-\nu(A_2\cap B)\right|; A_1\cap B\subseteq A\cap B\subseteq A_2\cap B\subseteq X\}$$

$$\leq \frac{1}{2}\sup\{\left|\nu(B_1)-\nu(B)\right|+\left|\nu(B_2)-\nu(B)\right|-\left|\nu(B_1)-\nu(B_2)\right|;$$

$$B_1, B, B_2 \in \mathcal{P}(X), B_1 \subseteq B \subseteq B_2\} = d_{MON}(\nu).$$

(iv) Because $A \subseteq B$ implies $A \cap (B \setminus A) = \emptyset$ and $A \cup (B \setminus A) = B$, we obtain $\nu(B) - \nu(A) = \nu(B \setminus A)$. Let $A_1, A, A_2 \in \mathcal{P}(X), A_1 \subseteq A \subseteq A_2$. Because $(A \setminus A_1) \cup (A_2 \setminus A) = A_2 \setminus A_1$ and $(A \setminus A_1) \cap (A_2 \setminus A) = \emptyset$, we get

$$\begin{aligned}
& \left|\nu(A_1)-\nu(A)\right|+\left|\nu(A_2)-\nu(A)\right|-\left|\nu(A_1)-\nu(A_2)\right| \\
= & \left|\nu(A\setminus A_1)\right|+\left|\nu(A_2\setminus A)\right|-\left|\nu(A_2\setminus A_1)\right| \\
\leq & \sup\{\left|\left|\nu(A)\right|+\left|\nu(B)\right|-\left|\nu(A\cup B)\right|\right|; A, B\in\mathcal{P}(X), A\cap B=\emptyset\} \\
= & a_2(|\nu|)
\end{aligned}$$

and passing to supremum with $A_1, A, A_2 \in \mathcal{P}(X), A_1 \subseteq A \subseteq A_2$ we have $d_{MON}(\nu) \leq a_2(|\nu|)$.

(v) We notice that $A_1 \subseteq A \subseteq A_2$ is equivalent to $A_2^c \subseteq A^c \subseteq A_1^c$. We have

$$d_{MON}(\nu^c) = \frac{1}{2}\sup\{\left|\nu^c(A_1)-\nu^c(A)\right|+\left|\nu^c(A_2)-\nu^c(A)\right|$$

$$-\left|\nu^c(A_1)-\nu^c(A_2)\right|; A_1, A, A_2 \in \mathcal{P}(X), A_1 \subseteq A \subseteq A_2\}$$

$$= \frac{1}{2}\sup\{\left|\nu(A_1^c)-\nu(A^c)\right|+\left|\nu(A_2^c)-\nu(A^c)\right|-\left|\nu(A_1^c)-\nu(A_2^c)\right|;$$

$$A_1^c, A^c, A_2^c \in \mathcal{P}(X), A_2^c \subseteq A^c \subseteq A_1^c\}$$

$$= \frac{1}{2}\sup\{\left|\nu(B_2)-\nu(B)\right|+\left|\nu(B_1)-\nu(B)\right|-\left|\nu(B_2)-\nu(B_1)\right|;$$

$$B_1, B, B_2 \in \mathcal{P}(X), B_1 \subseteq B \subseteq B_2\} = d_{MON}(\nu).$$

The following result gives a method of calculation for the defect of monotonicity of signed fuzzy measures. Also, it proves that the second inequality in Theorem 4.16, (i), can become equality.

Theorem 4.17 *If $\nu : \mathcal{P}(X) \to \mathbf{R}_+$ is a non-negative signed fuzzy measure such that $\nu(X) = 0$, then*

$$d_{MON}(\nu) = \sup\{\nu(A); A \in \mathcal{P}(X)\}.$$

Proof. Taking $\emptyset = A_1 \subseteq A \subseteq A_2 = X$, we have

$$|\nu(A_1) - \nu(A)| + |\nu(A_2) - \nu(A)| - |\nu(A_1) - \nu(A_2)| = 2\nu(A),$$

which implies $d_{MON}(\nu) \geq \sup\{\nu(A); A \in \mathcal{P}(X)\}$. The converse inequality is also true (see Theorem 4.16, (i)), therefore

$$d_{MON}(\nu) = \sup\{\nu(A); A \in \mathcal{P}(X)\}.$$

□

If X is finite, then to any signed fuzzy measure $\nu : \mathcal{P}(X) \to \mathbf{R}$, another set function $m : \mathcal{P}(X) \to \mathbf{R}$ can be associated by (see *e.g.* Fujimoto-Murofushi [79], [80])

$$m(A) = \sum_{B \subseteq A} (-1)^{card(A \setminus B)} \nu(B), \forall A \in \mathcal{P}(X),$$

where $cardM$ denotes the number of elements of M. This correspondence proves to be one-to-one, since conversely

$$\nu(A) = \sum_{B \subseteq A} m(B), \forall A \in \mathcal{P}(X),$$

and it is called Möbius inversion (see also Fujimoto-Murofushi [79], [80]).

Theorem 4.18 *Let m and ν (as above) be connected by a Möbius inversion. If ν is additive then*

$$d_{MON}(m) = \max\{|\nu(\{x\})|; x \in X\}.$$

Proof. The set function corresponding to additive set function ν is given by (see Fujimoto-Murofushi [80]) $m(A) = \nu(A)$, if $cardA = 1$ and $m(A) = 0$, otherwise. The expression

$$E(A_1, A, A_2) = |m(A_1) - m(A)| + |m(A_2) - m(A)| - |m(A_1) - m(A_2)|,$$

where $A_1 \subseteq A \subseteq A_2 \subseteq X$, is non-null if and only if $A_1 = \emptyset, cardA = 1$ and $cardA_2 > 1$. Indeed, if $cardA_1 > 1$ then $m(A_1) = m(A) = m(A_2) = 0$ which implies $E(A_1, A, A_2) = 0$. If $cardA_1 = 1$ then $cardA = 1$ or $cardA > 1$. The first variant implies $A = A_1$ therefore $E(A_1, A, A_2) = 0$ and the second variant implies $cardA_2 > 1$, therefore $m(A) = m(A_2) = 0$ and the same conclusion. The last case is $A_1 = \emptyset$. If $A = \emptyset$ then $E(A_1, A, A_2) = 0$. If $cardA > 1$ then $cardA_2 > 1$, which implies $m(A) = m(A_2) = 0$ and therefore $E(A_1, A, A_2) = 0$. Finally, if $cardA = 1$ then we have two possibilities: $cardA_2 = 1$ which implies $m(A) = m(A_2)$ and $E(A_1, A, A_2) = 0$ or $cardA_2 > 1$ which implies $E(A_1, A, A_2) = 2|\nu(A)|$. As a conclusion, we get

$$d_{MON}(m) = \frac{1}{2} \sup_{cardA=1} 2|\nu(A)| = \max\{|\nu(\{x\})|; x \in X\}.$$

□

The concept of k-order additivity is important in fuzzy measure theory. A signed fuzzy measure ν is said to be k-order additive if its Möbius transform $m(A) = 0$, for any A such that $cardA > k$ and there exists at least one subset A of X of exactly k elements such that $m(A) \neq 0$ (see *e.g.* Grabisch-Roubens [95]). Because $m(X) = 0$, the calculus of the defect of monotonicity for positive Möbius transforms of k-order additive fuzzy measures, it is a very easy task by using Theorem 4.17. In fact, $d_{MON}(m) = \max\{m(A); A \in \mathcal{P}(X), cardA \leq k\}$.

Example 4.18 If $X = \{a, b, c, d\}$ and $\nu : \mathcal{P}(X) \to \mathbf{R}$ is defined by $\nu(\emptyset) = 0$; $\nu(\{a\}) = \nu(\{b\}) = 2$; $\nu(\{c\}) = \nu(\{d\}) = 1$; $\nu(\{b, d\}) = 3$; $\nu(\{a, b\}) = \nu(\{a, d\}) = 4$; $\nu(\{b, c\}) = \nu(\{c, d\}) = 5$; $\nu(\{a, c\}) = 6$; $\nu(\{a, b, d\}) = 7$; $\nu(\{a, b, c\}) = \nu(\{a, c, d\}) = \nu(\{b, c, d\}) = 10$; $\nu(\{a, b, c, d\}) = 15$, then we obtain its Möbius transform $m : \mathcal{P}(X) \to \mathbf{R}$, $m(\emptyset) = 0$; $m(\{a\}) = m(\{c\}) = m(\{a, c\}) = m(\{b, c\}) = m(\{c, d\}) = 2$; $m(\{b\}) = m(\{d\}) = m(\{a, b\}) = m(\{a, d\}) = m(\{b, d\}) = 1$; $m(\{a, b, c\}) = m(\{a, b, d\}) = m(\{a, c, d\}) = m(\{b, c, d\}) = m(\{a, b, c, d\}) = 0$, and therefore ν is a 2-order additive fuzzy measure. By using Theorem 4.17 and the above remark, we have $d_{MON}(m) = \max\{1, 2\} = 2$.

Now, we consider the problem of approximation of a non-monotonic set function by monotonic set functions. In this sense, let us define

$$\mathcal{S} = \{\nu : \mathcal{P}(X) \to \mathbf{R}; \nu \text{ is bounded}\}$$

and

$$\mathcal{M} = \{m : \mathcal{P}(X) \to \mathbf{R}; m \text{ is monotone and bounded}\}.$$

In addition, let us introduce $d : \mathcal{S} \times \mathcal{S} \to \mathbf{R}$ by

$$d(\nu_1, \nu_2) = \sup\{|\nu_1(A) - \nu_2(A)|; A \in \mathcal{P}(X)\}$$

which is a metric on $\mathcal{S}$.

A natural question is to estimate the quantity of best approximation

$$E_{MON}(\nu) = \inf\{d(\nu, m); m \in \mathcal{M}\}$$

by using the defect of monotonicity.

Theorem 4.19 *For any $\nu \in \mathcal{S}$ we have*

$$E_{MON}(\nu) \geq \frac{d_{MON}(\nu)}{6}.$$

Proof. For $A_1 \subseteq A \subseteq A_2 \subseteq X$ and $m \in \mathcal{M}$ we obtain

$$\begin{aligned}
& |\nu(A_1) - \nu(A)| + |\nu(A_2) - \nu(A)| - |\nu(A_1) - \nu(A_2)| \\
= \; & |\nu(A_1) - m(A_1) - (\nu(A) - m(A)) + m(A_1) - m(A)| \\
& + |\nu(A_2) - m(A_2) - (\nu(A) - m(A)) \\
& + m(A_2) - m(A)| - |\nu(A_1) - \nu(A_2)| \\
\leq \; & |(\nu - m)(A_1) - (\nu - m)(A)| + |(\nu - m)(A_2) - (\nu - m)(A)| \\
& + |m(A_1) - m(A)| + |m(A_2) - m(A)| - |m(A_1) - m(A_2)| \\
& + |m(A_1) - m(A_2)| - |\nu(A_1) - \nu(A_2)|.
\end{aligned}$$

But

$$|m(A_1) - m(A)| + |m(A_2) - m(A)| - |m(A_1) - m(A_2)| = 0$$

and

$$|m(A_1) - m(A_2)| - |\nu(A_1) - \nu(A_2)| \leq \big|\, |m(A_1) - m(A_2)|$$

$$- |\nu(A_1) - \nu(A_2)| \,\big| \leq |(m - \nu)(A_1) - (m - \nu)(A_2)|,$$

which immediately implies

$$|\nu(A_1) - \nu(A)| + |\nu(A_2) - \nu(A)| - |\nu(A_1) - \nu(A_2)| \leq 6d(\nu, m).$$

Passing to supremum with $A_1 \subseteq A \subseteq A_2 \subseteq X$, we get

$$d_{MON}(\nu) \leq 6d(\nu, m), \forall m \in \mathcal{M},$$

and passing to infimum with $m \in \mathcal{M}$, we get the theorem. □

Let us denote by $FS(X) = \left\{\widetilde{A}; \widetilde{A} : X \to [0,1]\right\}$ the family of fuzzy set on a nonempty set X. A fuzzy set function $\widetilde{\nu} : FS(X) \to \mathbf{R}$ (or $\widetilde{\nu} : \widetilde{\mathcal{A}} \to \mathbf{R}$, where $\widetilde{\mathcal{A}}$ is a family of fuzzy sets on X) is called increasing if $\widetilde{A} \subseteq \widetilde{B}$ implies $\widetilde{\nu}\left(\widetilde{A}\right) \leq \widetilde{\nu}\left(\widetilde{B}\right)$ and decreasing if $\widetilde{A} \subseteq \widetilde{B}$ implies $\widetilde{\nu}\left(\widetilde{A}\right) \geq \widetilde{\nu}\left(\widetilde{B}\right)$, where $\subseteq$ denotes the usual inclusion between fuzzy sets:

$$\widetilde{A} \subseteq \widetilde{B} \text{ if and only if } \widetilde{A}(x) \leq \widetilde{B}(x), \forall x \in X.$$

If $\widetilde{\nu}$ is increasing or decreasing then $\widetilde{\nu}$ is called monotone.

For fuzzy set functions, the concept of defect of monotonicity can be introduced as in Definition 4.12 with respect to the above defined inclusion. Also, the basic properties given by Theorem 4.15, Theorem 4.16, $(i), (ii), (iii)$ (with $\left(\widetilde{A} \cap \widetilde{B}\right)(x) = \min\left(\widetilde{A}(x), \widetilde{B}(x)\right), \forall x \in X$), (v) (with $\widetilde{A}^c \in FS(X)$ the usual complementation of $\widetilde{A}$ defined by $\widetilde{A}^c(x) = 1 - \widetilde{A}(x), \forall x \in X$, and the set X replaced by the fuzzy set $\widetilde{X}, \widetilde{X}(x) = 1, \forall x \in X$) remain true.

The most important fuzzy set functions are the so-called T-measures (see *e.g.* Butnariu-Klement [50], [52]).

Definition 4.13 A mapping $\widetilde{\nu} : FS(X) \to \mathbf{R}$ is called a T-measure if the following properties are satisfied:

(i) $\widetilde{\nu}\left(\widetilde{\emptyset}\right) = 0$, where $\widetilde{\emptyset}(x) = 0, \forall x \in X$.

(ii) For all $\widetilde{A}, \widetilde{B} \in FS(X)$ we have $\widetilde{\nu}\left(\widetilde{A} \cap_T \widetilde{B}\right) + \widetilde{\nu}\left(\widetilde{A} \cup_S \widetilde{B}\right) = \widetilde{\nu}\left(\widetilde{A}\right) + \widetilde{\nu}\left(\widetilde{B}\right)$.

(iii) For each non-decreasing sequence $\left(\widetilde{A}_n\right)_{n \in \mathbf{N}}$ in $FS(X)$ with $\left(\widetilde{A}_n\right)_{n \in \mathbf{N}} \nearrow \widetilde{A}$ we have $\lim_{n\to\infty} \widetilde{\nu}\left(\widetilde{A}_n\right) = \widetilde{\nu}\left(\widetilde{A}\right)$,

where T is a triangular norm (that is a function $T : [0,1] \times [0,1] \to [0,1]$, commutative, associative, increasing in each component and $T(x, 1) = x, \forall x \in [0,1]$), $S : [0,1] \times [0,1] \to [0,1]$ given by $S(x, y) = 1 - T(1 - x, 1 - y)$,

$\forall x, y \in [0,1]$ is the triangular conorm associated with T and

$$\left(\widetilde{A} \cap_T \widetilde{B}\right)(x) = T\left(\widetilde{A}(x), \widetilde{B}(x)\right), \forall x \in [0,1],$$
$$\left(\widetilde{A} \cup_S \widetilde{B}\right)(x) = S\left(\widetilde{A}(x), \widetilde{B}(x)\right), \forall x \in [0,1].$$

Example 4.19 If $T = T_L$, that is $T(x,y) = \max(x+y-1, 0), \forall x, y \in [0,1]$, then every non-negative T_L-measure $\widetilde{\nu}$ is increasing (see *e.g.* Butnariu [49]), therefore $d_{MON}(\widetilde{\nu}) = 0$.

Example 4.20 If T is a measurable triangular norm, S its associated conorm such that $T(x,y) + S(x,y) = x + y, \forall x, y \in [0,1], P$ a finite measure on $\mathcal{P}(X)$ such that $P(X) > 0$, then $\widetilde{\nu} : FS(X) \to \mathbf{R}$ defined by $\widetilde{\nu}\left(\widetilde{A}\right) = \int_X \widetilde{A} \mathrm{d}P$ is a finite T -measure (see Klement [118]) which is obvious monotone, therefore $d_{MON}(\widetilde{\nu}) = 0$.

Example 4.21 Let $m : \mathcal{P}(X) \to [0,1]$ be a probability measure and let us consider $\widetilde{\nu} : FS(X) \to \mathbf{R}_+$ defined by $\widetilde{\nu}\left(\widetilde{A}\right) = \int_X \left(\widetilde{A} - \widetilde{A}^2\right) \mathrm{d}m$, where $\int_X$ is the Lebesgue integral. Then $\widetilde{\nu}$ is a finite T_M-measure ($T_M(x,y) = \min(x,y), \forall x, y \in [0,1]$) with the defect of monotonicity equal to $\frac{1}{4}$. Indeed, taking $\widetilde{A} \in FS(X), \widetilde{\emptyset} = \widetilde{A}_1 \subset \widetilde{A} \subset \widetilde{A}_2 = \widetilde{X}$ we obtain (as in Theorem 4.17)

$$d_{MON}(\widetilde{\nu}) = \sup\left\{\int_X \left(\widetilde{A} - \widetilde{A}^2\right) \mathrm{d}m; \widetilde{A} \in FS(X)\right\} = \frac{1}{4}.$$

Let us note that there exist some practical fields where the monotonicity of set functions is not necessary, cases when the defect introduced in the present section could be useful, as for example, in decision problem (see Fujimoto-Murofushi [79]) or subjective evaluations (see Kwon-Sugeno [131], Onisawa-Sugeno-Nishiwaki-Kawai-Harima [161], Tanaka-Sugeno [211]).

4.4 Defect of Subadditivity and of Superadditivity

In general, the defect of a property characterized by an equality is easy to be introduced. In the previous section the defect of a property given by an inequality (the monotonicity) was introduced and studied, but in the theory of set functions there exist other concepts too characterized by inequalities. Let us recall some of them.

Definition 4.14 (see *e.g.* Dennenberg [62], Grabisch [96]) Let X be a nonempty set and $\mathcal{A} \subseteq \mathcal{P}(X)$ a σ-algebra. A set function $\nu : \mathcal{A} \to \mathbf{R}$ is called

(i) subadditive, if

$$\nu(A \cup B) \leq \nu(A) + \nu(B), \forall A, B \in \mathcal{A}, A \cap B = \emptyset;$$

(ii) superadditive, if

$$\nu(A \cup B) \geq \nu(A) + \nu(B), \forall A, B \in \mathcal{A}, A \cap B = \emptyset;$$

(iii) submodular, if

$$\nu(A \cup B) + \nu(A \cap B) \leq \nu(A) + \nu(B), \forall A, B \in \mathcal{A};$$

(iv) supermodular, if

$$\nu(A \cup B) + \nu(A \cap B) \geq \nu(A) + \nu(B), \forall A, B \in \mathcal{A};$$

(v) k-monotone ($k \geq 2$), if

$$\nu\left(\bigcup_{i=1}^{k} A_i\right) + \sum_{\emptyset \neq I \subseteq \{1,...,k\}} (-1)^{|I|} \nu\left(\bigcap_{i \in I} A_i\right) \geq 0, \forall A_1, ..., A_k \in \mathcal{A};$$

(vi) k-alternating ($k \geq 2$), if

$$\nu\left(\bigcap_{i=1}^{k} A_i\right) + \sum_{\emptyset \neq I \subseteq \{1,...,k\}} (-1)^{|I|} \nu\left(\bigcup_{i \in I} A_i\right) \leq 0, \forall A_1, ..., A_k \in \mathcal{A}.$$

Remark. In fact, 2-monotone set functions are supermodular, while 2-alternating set functions are submodular. Also, submodularity implies subadditivity and supermodularity implies superadditivity.

In this section we consider set function ν such that $\nu(\emptyset) = 0$, that is ν is a signed fuzzy measure (see Murofushi-Sugeno [159]) or a non-monotonic fuzzy measure (see Murofushi-Sugeno-Machida [158]).

Definition 4.15 Let $\nu : \mathcal{A} \to \mathbf{R}, \nu(\emptyset) = 0$, where $\mathcal{A} \subseteq \mathcal{P}(X)$ is a σ-algebra. The defect of subadditivity of order $n, n \geq 2$, of ν is given by

$$d_{SUB}^{(n)}(\nu) = \sup\left\{\nu\left(\bigcup_{i=1}^{n} A_i\right) - \sum_{i=1}^{n} \nu(A_i); A_i \in \mathcal{A}, A_i \cap A_j = \emptyset, i \neq j, i, j = \overline{1, n}\right\}$$

and the defect of superadditivity of order $n, n \geq 2$, of ν is given by

$$d_{SUP}^{(n)}(\nu) = \sup\left\{\sum_{i=1}^{n}\nu(A_i) - \nu\left(\bigcup_{i=1}^{n}A_i\right); A_i \in \mathcal{A}, A_i \cap A_j = \emptyset, i \neq j, i,j = \overline{1,n}\right\}.$$

Similarly, the defect of countable subadditivity of ν is given by

$$d_{SUB}^{\infty}(\nu) = \sup\left\{\nu\left(\bigcup_{i=1}^{\infty}A_i\right) - \sum_{i=1}^{\infty}\nu(A_i); A_i \in \mathcal{A}, A_i \cap A_j = \emptyset, i \neq j, i,j \in \mathbf{N}\right\}$$

and the defect of countable superadditivity of ν is given by

$$d_{SUP}^{\infty}(\nu) = \sup\left\{\sum_{i=1}^{\infty}\nu(A_i) - \nu\left(\bigcup_{i=1}^{\infty}A_i\right); A_i \in \mathcal{A}, A_i \cap A_j = \emptyset, i \neq j, i,j \in \mathbf{N}\right\}.$$

The main properties of the above defined defects are the following.

Theorem 4.20 *(i)* $0 \leq d_{SUB}^{(n)}(\nu) \leq a_n(\nu), 0 \leq d_{SUP}^{(n)}(\nu) \leq a_n(\nu)$, $\forall n \in \mathbf{N}, n \geq 2$, *where* $a_n(\nu)$ *is the defect of additivity of order* n *of* ν*;*

(ii) $0 \leq d_{SUB}^{\infty}(\nu) \leq a_{\infty}(\nu), 0 \leq d_{SUP}^{\infty}(\nu) \leq a_{\infty}(\nu)$, *where* $a_{\infty}(\nu)$ *is the countable defect of additivity of* ν*;*

(iii) If ν *is subadditive or countable subadditive then* $d_{SUP}^{(n)}(\nu) = a_n(\nu)$ *or* $d_{SUB}^{\infty}(\nu) = a_{\infty}(\nu)$, *respectively;*

(iv) If ν *is superadditive or countable superadditive then* $d_{SUB}^{(n)}(\nu) = a_n(\nu)$ *or* $d_{SUP}^{\infty}(\nu) = a_{\infty}(\nu)$, *respectively;*

(v) $d_{SUB}^{(n)}(\nu) \leq d_{SUB}^{(n+1)}(\nu), d_{SUP}^{(n)}(\nu) \leq d_{SUP}^{(n+1)}(\nu), \forall n \in \mathbf{N}, n \geq 2$*;*

(vi) $d_{SUB}^{(n)}(\nu) = 0$ *if and only if* ν *is subadditive;* $d_{SUP}^{(n)}(\nu) = 0$ *if and only if* ν *is superadditive;* $d_{SUB}^{\infty}(\nu) = 0$ *if and only if* ν *is countable subadditive;* $d_{SUP}^{\infty}(\nu) = 0$ *if and only if* ν *is countable superadditive;*

(vii) $d_{SUB}^{(n)}(\nu) \leq d_{SUB}^{(n-1)}(\nu) + d_{SUB}^{(2)}(\nu)$ *and* $d_{SUP}^{(n)}(\nu) \leq d_{SUP}^{(n-1)}(\nu) + d_{SUP}^{(2)}(\nu), \forall n \in \mathbf{N}, n \geq 3$.

(viii) $d_{SUB}^{(n)}(\nu) \leq (n-1)\, d_{SUB}^{(2)}(\nu)$ *and* $d_{SUP}^{(n)}(\nu) \leq (n-1)\, d_{SUP}^{(2)}(\nu)$, $\forall n \in \mathbf{N}, n \geq 3$.

Proof. $(i),(ii)$ Taking $A_i = \emptyset, \forall i = \overline{1,n}$ or $\forall i \in \mathbf{N}$, respectively, we obtain

$$\nu\left(\bigcup_{i=1}^{n} A_i\right) = \sum_{i=1}^{n} \nu(A_i) = \nu\left(\bigcup_{i=1}^{\infty} A_i\right) = \sum_{i=1}^{\infty} \nu(A_i) = 0$$

which implies the non-negativity of all defects in Definition 4.15. Because

$$\nu\left(\bigcup_{i=1}^{n} A_i\right) - \sum_{i=1}^{n} \nu(A_i) \leq \left|\nu\left(\bigcup_{i=1}^{n} A_i\right) - \sum_{i=1}^{n} \nu(A_i)\right| \tag{4.7}$$

and

$$\sum_{i=1}^{n} \nu(A_i) - \nu\left(\bigcup_{i=1}^{n} A_i\right) \leq \left|\nu\left(\bigcup_{i=1}^{n} A_i\right) - \sum_{i=1}^{n} \nu(A_i)\right|, \tag{4.8}$$

for all $A_i \in \mathcal{A}, i = \overline{1,n}$ (the inequalities being also true in the countable case) we obtain the upper estimations of defects.

(iii) If ν is subadditive then the inequality (4.8) becomes equality for all $A_i \in \mathcal{A}, i = \overline{1,n}$ and passing to supremum we have $d_{SUP}^{(n)}(\nu) = a_n(\nu)$. The countable case is similar.

(iv) Analogously with (iii) by using (4.7).

(v) It is immediate by taking $A_{n+1} = \emptyset$ in the definition of $d_{SUB}^{(n+1)}(\nu)$ and $d_{SUP}^{(n+1)}(\nu)$, respectively.

(vi) By (v), $d_{SUB}^{(n)}(\nu) = 0$ and $d_{SUP}^{(n)}(\nu) = 0$ if and only if $d_{SUB}^{(2)}(\nu) = 0$ and $d_{SUP}^{(2)}(\nu) = 0$, respectively. But the first equality is true if and only if $\nu(A \cup B) \leq \nu(A) + \nu(B), \forall A, B \in \mathcal{A}, A \cap B = \emptyset$, that is if ν is subadditive. Also, the second equality is true if and only if $\nu(A \cup B) \geq \nu(A) + \nu(B), \forall A, B \in \mathcal{A}, A \cap B = \emptyset$, that is if ν is superadditive. The countable case is immediate.

(vii) If $\{A_1, ..., A_n\} \subseteq \mathcal{A}$ is a disjoint family, then $\bigcup_{i=1}^{n-1} A_i$ and A_n are disjoint sets, such that

$$\begin{aligned} \nu\left(\bigcup_{i=1}^{n} A_i\right) - \sum_{i=1}^{n} \nu(A_i) &= \nu\left(\bigcup_{i=1}^{n} A_i\right) - \nu\left(\bigcup_{i=1}^{n-1} A_i\right) - \nu(A_n) \\ +\nu\left(\bigcup_{i=1}^{n-1} A_i\right) - \sum_{i=1}^{n-1} \nu(A_i) &\leq d_{SUB}^{(2)}(\nu) + d_{SUB}^{(n-1)}(\nu). \end{aligned}$$

Passing to supremum we obtain the desired inequality. For the defect of superadditivity the proof is similar.

(*viii*) From (*vii*), by recurrence, we obtain $d_{SUB}^{(3)}(\nu) \leq d_{SUB}^{(2)}(\nu) + d_{SUB}^{(2)}(\nu)$, afterwards $d_{SUB}^{(4)}(\nu) \leq d_{SUB}^{(3)}(\nu) + d_{SUB}^{(2)}(\nu) \leq 3d_{SUB}^{(2)}(\nu)$, $d_{SUB}^{(5)}(\nu) \leq d_{SUB}^{(4)}(\nu) + d_{SUB}^{(2)}(\nu) \leq 4d_{SUB}^{(2)}(\nu)$ and finally $d_{SUB}^{(n)}(\nu) \leq (n-1)\, d_{SUB}^{(2)}(\nu)$. For the defect of superadditivity the proof is similar. □

Example 4.22 Because the outer measure μ^* induced by a measure μ is countable subadditive and the inner measure μ_* induced by a measure μ is countable superadditive (see Halmos [99], p.60), we have

$$d_{SUB}^{(n)}(\mu^*) = d_{SUP}^{(n)}(\mu_*) = d_{SUB}^{\infty}(\mu^*) = d_{SUP}^{\infty}(\mu_*) = 0$$

and

$$\begin{aligned} d_{SUP}^{(n)}(\mu^*) &= a_n(\mu^*), d_{SUB}^{(n)}(\mu_*) = a_n(\mu_*), \\ d_{SUP}^{\infty}(\mu^*) &= a_{\infty}(\mu^*), d_{SUB}^{\infty}(\mu_*) = a_{\infty}(\mu_*). \end{aligned}$$

Remark. If the set function is subadditive or superadditive, then the calculus of defects of subadditivity and superadditivity is not interesting: they are 0 or equal to the value of the defect of additivity. Of course, when we calculate the defect of subadditivity of a set function ν (assumed non-subadditive, non-superadditive) are important the situations $\nu\left(\bigcup_{i=1}^{n} A_i\right) \geq \sum_{i=1}^{n} \nu(A_i)$. Also, the situations $\nu\left(\bigcup_{i=1}^{n} A_i\right) \leq \sum_{i=1}^{n} \nu(A_i)$ are important when we calculate the defect of superadditivity.

We recall (see Theorem 4.5) that if B is a function such that $B(0) = 0$ and $f : [0,\infty) \to [0,\infty), f(x) = B(x) + ax$ is strictly increasing and φ is given by $\varphi(x,y) = B(x+y) - B(x) - B(y)$, then $\nu_{\varphi}(A) = (f \circ m)(A)$ (where m is a classical finite measure) is a fuzzy measure which satisfies the functional equation

$$\nu_{\varphi}(A \cup B) = \nu_{\varphi}(A) + \nu_{\varphi}(A) + \varphi(m(A), m(B)), \tag{4.9}$$

for every $A, B \in \mathcal{A}, A \cap B = \emptyset$.

For set functions given by relation (4.9) we can prove the following result.

Theorem 4.21 *We have*

$$d_{SUB}^{(n)}(\nu_{\varphi}) \leq (n-1) \sup\{\varphi(x,y); x, y \in m(\mathcal{A})\}$$

and

$$d_{SUP}^{(n)}(\nu_{\varphi}) \leq -(n-1) \inf\{\varphi(x,y); x, y \in m(\mathcal{A})\},$$

where $m(\mathcal{A}) = \{y \in \mathbf{R}; \exists A \in \mathcal{A}$ *such that* $m(A) = y\}$.

Proof. We obtain

$$\nu_\varphi(A \cup B) - \nu_\varphi(A) - \nu_\varphi(A) = \varphi(m(A), m(B))$$

$$\leq \sup\{\varphi(x,y); x, y \in m(\mathcal{A})\},$$

for every $A, B \in \mathcal{A}$, $A \cap B = \emptyset$, therefore

$$d_{SUB}^{(2)}(\nu_\varphi) \leq \sup\{\varphi(x,y); x, y \in m(\mathcal{A})\}.$$

By using property $(viii)$, Theorem 4.20, we get the desired inequality. Also,

$$\nu_\varphi(A) - \nu_\varphi(A) - \nu_\varphi(A \cup B) = -\varphi(m(A), m(B))$$

$$\leq \sup\{-\varphi(x,y); x, y \in m(\mathcal{A})\} = -\inf\{\varphi(x,y); x, y \in m(\mathcal{A})\},$$

for every $A, B \in \mathcal{A}$, $A \cap B = \emptyset$. As above, we get the second inequality. □

Example 4.23 If the function φ is non-negative (or non-positive), then the calculus of the defect of subadditivity or superadditivity is not interesting (see the above remark) for the set function given by relation (4.9). Choosing, for example,

$$f(x) = B(x) = \begin{cases} \sqrt{x}, & \text{if } x \leq 4 \\ x^2 - 14, & \text{if } x \geq 4, \end{cases}$$

we get

$$\begin{aligned} \varphi(x,y) &= B(x+y) - B(x) - B(y) \\ &= \begin{cases} \sqrt{x+y} - \sqrt{x} - \sqrt{y}, & \text{if } x+y \leq 4 \\ (x+y)^2 - \sqrt{x} - \sqrt{y}, & \text{if } x+y \geq 4, x \leq 4, y \leq 4 \\ x^2 + 2xy - \sqrt{x}, & \text{if } x+y \geq 4, x \leq 4, y \geq 4 \\ y^2 + 2xy - \sqrt{y}, & \text{if } x+y \geq 4, x \geq 4, y \leq 4 \\ 2xy, & \text{if } x \geq 4, y \geq 4 \end{cases} \end{aligned}$$

which has positive and negative values. Considering the classical measure $m : \mathcal{P}(X) \to \mathbf{R}_+$ defined by $m(A) = |A|$, where $|\cdot|$ denotes the cardinality of A and $X = \{x_1, ..., x_n\}$, $n \geq 4$, we obtain $\nu_\varphi : \mathcal{P}(X) \to \mathbf{R}_+$ by

$$\nu_\varphi(A) = f(m(A)) = \begin{cases} \sqrt{|A|}, & \text{if } |A| \leq 4 \\ |A|^2 - 14, & \text{if } |A| \geq 4. \end{cases}$$

Because

$$\inf\{\varphi(m(A), m(B)); \varphi(m(A), m(B)) \leq 0, A \cap B = \emptyset\} = 2 - 2\sqrt{2}$$

we have

$$d_{SUP}^{(2)}(\nu_\varphi) = 2\sqrt{2} - 2.$$

Also,

$$\begin{aligned} d_{SUB}^{(2)}(\nu_\varphi) &= \sup\{\varphi(m(A), m(B)); \varphi(m(A), m(B)) \geq 0, A \cap B = \emptyset\} \\ &= \max\left\{60, 8n - 14, \left[\frac{n^2}{2}\right]\right\}. \end{aligned}$$

In what follows, we introduce defects for other properties given by Definition 4.14.

Definition 4.16 Let $\nu : \mathcal{A} \to \mathbf{R}, \nu(\emptyset) = 0$, where $\mathcal{A} \subseteq \mathcal{P}(X)$ is a σ-algebra. The defect of k-monotonicity of ν is given by

$$\begin{aligned} d_{k-MON}(\nu) &= \sup\left\{\sum_{\emptyset \neq I \subseteq \{1,\ldots,k\}} (-1)^{|I|+1} \nu\left(\bigcap_{i \in I} A_i\right)\right. \\ &\qquad \left. -\nu\left(\bigcup_{i=1}^{k} A_i\right); A_1, \ldots, A_k \in \mathcal{A}\right\} \end{aligned}$$

and the defect of k-alternation of ν is given by

$$\begin{aligned} d_{k-ALT}(\nu) &= \sup\left\{\sum_{\emptyset \neq I \subseteq \{1,\ldots,k\}} (-1)^{|I|} \nu\left(\bigcup_{i \in I} A_i\right)\right. \\ &\qquad \left. +\nu\left(\bigcap_{i=1}^{k} A_i\right); A_1, \ldots, A_k \in \mathcal{A}\right\}. \end{aligned}$$

If $k = 2$, $d_{2-MON}(\nu) = d_{SUPERMOD}(\nu)$ is called defect of supermodularity and $d_{2-ALT}(\nu) = d_{SUBMOD}(\nu)$ is called defect of submodularity.

Immediate properties of the above defects are the followings.

Theorem 4.22 *(i) $d_{k-MON}(\nu) \geq 0$ and $d_{k-MON}(\nu) = 0$ if and only if ν is k-monotone;*

(ii) $d_{k-ALT}(\nu) \geq 0$ and $d_{k-ALT}(\nu) = 0$ if and only if ν is k-alternating;

(iii) $d_{k-MON}(\nu) = d_{k-ALT}(\nu^c)$, *where* ν^c *is the dual of* ν, *that is* $\nu^c(A) = \nu(X) - \nu(A^c), \forall A \in \mathcal{A}$;

Proof. $(i), (ii)$ Taking $A_i = \emptyset, \forall i = \overline{1,k}$ we obtain $\nu\left(\bigcap_{i\in I} A_i\right) = 0$ and $\nu\left(\bigcup_{i\in I} A_i\right) = 0, \forall I \subseteq \{1,...,k\}$ which implies $d_{k-MON}(\nu) \geq 0$ and $d_{k-ALT}(\nu) \geq 0$.
We have $d_{k-MON}(\nu) = 0$ if and only if

$$\sum_{\emptyset \neq I \subseteq \{1,\dots,k\}} (-1)^{|I|+1} \nu\left(\bigcap_{i\in I} A_i\right) - \nu\left(\bigcup_{i=1}^{k} A_i\right) \leq 0,$$

for every $A_1,...,A_k \in \mathcal{A}$, if and only if

$$\sum_{\emptyset \neq I \subseteq \{1,\dots,k\}} (-1)^{|I|} \nu\left(\bigcap_{i\in I} A_i\right) + \nu\left(\bigcup_{i=1}^{k} A_i\right) \geq 0,$$

for every $A_1,...,A_k \in \mathcal{A}$, that is if and only if ν is k-monotone. The proof is similar for k-alternation.

(iii) Because

$$\nu^c\left(\bigcap_{i\in I} A_i\right) = \nu(X) - \nu\left(\bigcup_{i\in I} A_i^c\right)$$

and

$$\nu^c\left(\bigcup_{i\in I} A_i\right) = \nu(X) - \nu\left(\bigcap_{i\in I} A_i^c\right),$$

for every $A_1,...,A_k \in \mathcal{A}$ and $I \subseteq \{1,...,k\}$, we get

$$\begin{aligned}
&\sum_{\emptyset \neq I \subseteq \{1,\dots,k\}} (-1)^{|I|} \nu^c\left(\bigcup_{i\in I} A_i\right) + \nu^c\left(\bigcap_{i=1}^{k} A_i\right) \\
= &\sum_{\emptyset \neq I \subseteq \{1,\dots,k\}} (-1)^{|I|} \left(\nu(X) - \nu\left(\bigcap_{i\in I} A_i^c\right)\right) + \nu(X) - \nu\left(\bigcup_{i=1}^{k} A_i^c\right) \\
= &\ \nu(X) \sum_{\emptyset \neq I \subseteq \{1,\dots,k\}} (-1)^{|I|} + \sum_{\emptyset \neq I \subseteq \{1,\dots,k\}} (-1)^{|I|+1} \nu\left(\bigcap_{i\in I} A_i^c\right) \\
&+\nu(X) - \nu\left(\bigcup_{i=1}^{k} A_i^c\right),
\end{aligned}$$

for every $A_1, ..., A_k \in \mathcal{A}$. Because $\sum_{\emptyset \neq I \subseteq \{1,...,k\}} (-1)^{|I|} = -1, \forall k \in \mathbf{N}$, we obtain

$$\sum_{\emptyset \neq I \subseteq \{1,...,k\}} (-1)^{|I|} \nu^c \left(\bigcup_{i \in I} A_i \right) + \nu^c \left(\bigcap_{i=1}^{k} A_i \right)$$

$$= \sum_{\emptyset \neq I \subseteq \{1,...,k\}} (-1)^{|I|+1} \nu \left(\bigcap_{i \in I} A_i^c \right) - \nu \left(\bigcup_{i=1}^{k} A_i^c \right),$$

for every $A_1, ..., A_k \in \mathcal{A}$ which implies $d_{k-MON}(\nu) = d_{k-ALT}(\nu^c)$. □

4.5 Defect of Measurability

Let (X, ρ) be a bounded metric space and let us denote

$$CL(X) = \{A \subset X; X \setminus A \in \mathcal{T}_\rho\},$$

where $\mathcal{T}_\rho$ is the topology generated by the metric ρ. Let $\varphi : \mathcal{T}_\rho \to \mathbf{R}_+$ be a measure on $\mathcal{T}_\rho$, that is

(*i*) $\varphi(\emptyset) = 0$;

(*ii*) $A, B \in \mathcal{T}_\rho, A \subseteq B$ implies $\varphi(A) \leq \varphi(B)$;

(*iii*) $\varphi(A \cup B) = \varphi(A) + \varphi(B), \forall A, B \in \mathcal{T}_\rho, A \cap B = \emptyset$ and $\varphi(A \cup B) \leq \varphi(A) + \varphi(B), \forall A, B \in \mathcal{T}_\rho$.

Let us define $\overline{\varphi} : CL(X) \to \mathbf{R}_+$ by

$$\overline{\varphi}(A) = diam(X) - \varphi(X \setminus A), \forall A \in CL(X),$$

where $diam(X) = \sup\{\rho(x, y); x, y \in X\}$, and $\varphi^*, \varphi_* : \mathcal{P}(X) \to \mathbf{R}_+$ by

$$\varphi_*(A) = \sup\{\overline{\varphi}(F); F \in CL(X), F \subseteq A\},$$

which is called inner φ-measure of A, and

$$\varphi^*(A) = \inf\{\varphi(G); G \in \mathcal{T}_\rho, A \subseteq G\},$$

which is called exterior φ-measure of A.

Definition 4.17 Keeping the above notations and assumptions, we say that $A \in \mathcal{P}(X)$ is φ-measurable if $\varphi^*(A) = \varphi_*(A)$. The quantity

$$d_{MAS}(A) = \varphi^*(A) - \varphi_*(A)$$

is called defect of φ-measurability of A.

We present.

Theorem 4.23 *(i)* $0 \leq d_{MAS}(A) \leq \varphi(X), \forall A \in \mathcal{P}(X)$ *and* $d_{MAS}(A) = 0$ *if and only if* A *is* φ*-measurable;*

(ii) $d_{MAS}(A) = d_{MAS}(A^c), \forall A \in \mathcal{P}(X)$;

(iii) $d_{MAS}(A \cup B) \leq d_{MAS}(A) + d_{MAS}(B) + \varphi_*(A \cup B)$, $\forall A, B \in \mathcal{P}(X)$;

(iv) Let T *be the one to one transformation of the entire real line into itself, defined by* $T(x) = \alpha x + \beta$, *where* α *and* β *are real numbers and* $\alpha \neq 0$. *If for every bounded* $A \subset \mathbf{R}$ *we denote* $T(A) = \{\alpha x + \beta; x \in A\}$, *then*

$$d_{MAS}(T(A)) = |\alpha| d_{MAS}(A).$$

(Here ρ *is the usual metric on the real line, induced on* A *and* $T(A)$ *and* $\varphi^*(A), \varphi_*(A)$, *denote the outer and inner Lebesgue measures, respectively).*

Proof. *(i)* Because $\varphi^*(A) \geq \varphi_*(A) \geq 0, \forall A \in \mathcal{P}(X)$ and $\varphi^*(X) = \varphi(X)$, the first part is immediate. The second part is obvious.

(ii) We have

$$\begin{aligned}
\varphi_*(X \setminus A) &= \sup\{\overline{\varphi}(F); F \in CL(X), F \subseteq X \setminus A\} \\
&= \sup\{diam(X) - \varphi(X \setminus F); X \setminus F \in \mathcal{T}_\rho, A \subseteq X \setminus F\} \\
&= diam(X) - \inf\{\varphi(X \setminus F); X \setminus F \in \mathcal{T}_\rho, A \subseteq X \setminus F\} \\
&= diam(X) - \varphi^*(A), \forall A \in \mathcal{P}(X),
\end{aligned}$$

for every $A \in \mathcal{P}(X)$, which also implies

$$\varphi^*(X \setminus A) = diam(X) - \varphi_*(A), \forall A \in \mathcal{P}(X).$$

We obtain

$$d_{MAS}(A^c) = \varphi^*(X \setminus A) - \varphi_*(X \setminus A) = \varphi^*(A) - \varphi_*(A) = d_{MAS}(A),$$

for every $A \in \mathcal{P}(X)$.

(iii) Because φ^* is subadditive and φ_* satisfies $2\varphi_*(A \cup B) \geq \varphi_*(A) + \varphi_*(B)$ we obtain the inequality.

(iv) It follows by $\varphi^*(T(A)) = |\alpha| \varphi^*(A)$ and $\varphi_*(T(A)) = |\alpha| \varphi_*(A)$ (see *e.g.* Halmos [99], p.64). □

Example 4.24 If $X = [a, b], \rho$ is the usual metric on $[a, b]$ and φ is the Lebesgue measure induced on $[a, b]$, then there exists $A \subset \mathbf{R}$ such that

$\varphi_*([a,b]\cap A) = 0$ and $\varphi^*([a,b]\cap A) = \varphi(X) = b-a$ (see Gelbaum-Olmsted [88], p.147), which can be considered as a peak of non-additivity.

We recall that an outer measure is a mapping $\mu^o : \mathcal{P}(X) \to \overline{\mathbf{R}}_+$ which verifies:

(*i*) $\mu^o(\emptyset) = 0$;

(*ii*) $A \subseteq B$ implies $\mu^o(A) \leq \mu^o(B)$;

(*iii*) $\mu^o\left(\bigcup_{n=1}^{\infty} A_n\right) \leq \sum_{n=1}^{\infty} \mu^o(A_n), \forall (A_n)_{n\in\mathbf{N}} \subseteq \mathcal{P}(X)$.

In Measure Theory (see e.g Halmos [99]) an important concept, used in the Charathéodory's construction, is that of μ^o-measurable set, where μ^o is an outer measure.

Definition 4.18 We say that a set $A \in \mathcal{P}(X)$ is μ^o-measurable if, for every set T in $\mathcal{P}(X)$,

$$\mu^o(T) = \mu^o(T\cap A) + \mu^o(T\cap A^c).$$

In fact, the subadditivity of μ^o implies $\mu^o(T) \leq \mu^o(T\cap A) + \mu^o(T\cap A^c)$. Also, if $\mu^o(T) = +\infty$ then the converse inequality is also verified, such that we can define the defect of μ^o-measurability in the following way.

Definition 4.19 Let $\mu^o : \mathcal{P}(X) \to \overline{\mathbf{R}}_+$ be an outer measure and $A \in \mathcal{P}(X)$. The quantity

$$\begin{aligned} d_{\mu^o-MAS}(A) &= \sup\{\mu^o(T\cap A) + \mu^o(T\cap A^c) - \mu^o(T); \\ &\qquad T\in\mathcal{P}(X), \mu^o(T) < +\infty\} \end{aligned}$$

is called defect of μ^o-measurability of A.

It is obvious that $d_{\mu^o-MAS}(A) = 0$ if and only if A is μ^o-measurable. Other basic properties of the defect of μ^o-measurability are given by the below theorem.

Theorem 4.24 (*i*) $0 \leq d_{\mu^o-MAS}(A) \leq d_{SUP}^{(2)}(\mu^o), \forall A \in \mathcal{P}(X)$;

(*ii*) $d_{\mu^o-MAS}(A^c) = d_{\mu^o-MAS}(A), \forall A \in \mathcal{P}(X)$;

(*iii*) $d_{\mu^o-MAS}(A\cup B) \leq d_{\mu^o-MAS}(A) + d_{\mu^o-MAS}(B)$, $\forall A, B \in \mathcal{P}(X)$ *with* $A\cap B = \emptyset$.

Proof. (*i*) The first inequality is obvious. Because $(T\cap A)\cup(T\cap A^c) = T$ and $(T\cap A)\cap(T\cap A^c) = \emptyset, \forall T \in \mathcal{P}(X)$, we obtain

$$\begin{aligned} &\mu^o(T\cap A) + \mu^o(T\cap A^c) - \mu^o(T) \\ \leq\ &\sup\{\mu^o(A) + \mu^o(B) - \mu^o(A\cup B); A, B \in \mathcal{P}(X), A\cap B = \emptyset\}, \end{aligned}$$

for every $T \in \mathcal{P}(X)$, which implies the desired inequality.

(*ii*) It is immediate.

(*iii*) Let $A, B, T \in \mathcal{P}(X)$. We have

$$\begin{aligned}
&\mu^o(T \cap (A \cup B)) + \mu^o(T \cap (A \cup B)^c) - \mu^o(T) \\
&\leq \mu^o(T \cap (A \cup B) \cap A) + \mu^o(T \cap (A \cup B) \cap A^c) \\
&+\mu^o(T \cap (A \cup B)^c) - \mu^o(T) \leq \mu^o(T \cap A) + \mu^o(T \cap B \cap A^c) \\
&+\mu^o(T \cap A^c \cap B^c) - \mu^o(T) = \mu^o(T \cap A) + \mu^o(T \cap A^c) \\
&-\mu^o(T) + \mu^o(T \cap A^c \cap B) + \mu^o(T \cap A^c \cap B^c) - \mu^o(T \cap A^c) \\
&\leq d_{\mu^o - MAS}(A) + d_{\mu^o - MAS}(B),
\end{aligned}$$

and passing to supremum with $T \in \mathcal{P}(X), \mu^o(T) < +\infty$, we obtain

$$d_{\mu^o - MAS}(A \cup B) \leq d_{\mu^o - MAS}(A) + d_{\mu^o - MAS}(B).$$

□

Example 4.25 If $\mu^o(B) = 0$ then $d_{\mu^o - MAS}(B) = 0$.

4.6 Bibliographical Remarks

Definition 4.2, Theorems 4.1, 4.3, 4.5-4.7, Examples 4.1-4.9 and applications in Section 4.1 are from Ban-Gal [24]. Definitions 4.6, 4.9, 4.10, Theorems 4.8-4.14, Examples 4.10-4.14 and applications in Section 4.2 are from Ban-Gal [25]. Definition 4.12, Theorems 4.15-4.19, Examples 4.17, 4.18, 4.20 and 4.21 are in Ban-Gal [26]. Completely new are Definitions 4.15, 4.16, 4.17, 4.19 and the results proved in Section 4.4 and Section 4.5.

Chapter 5

Defect of Property in Real Function Theory

This chapter discusses various defects of properties in Real Function Theory.

5.1 Defect of Continuity, of Differentiability and of Integrability

In this section we deal with real functions of real variable, $f : E \to \mathbf{R}, E \subseteq \mathbf{R}$. In order to measure the continuity of f, several well-known quantities were introduced, as follows.

Definition 5.1 (see *e.g.* Sireţchi [199], p.151, p.165, [200], p.239). Let $f : E \to \mathbf{R}$. For $x_0 \in E$, the quantity defined by

$$\omega(x_0; f) = \inf \{\delta[f(V \cap E)]; V \in \mathcal{V}(x_0)\},$$

is called oscillation of f at x_0, where $\mathcal{V}(x_0)$ denotes the class of all neighborhoods of x_0 and $\delta[A] = \sup\{|a_1 - a_2|; a_1, a_2 \in A\}$ represents the diameter of the set $A \subset \mathbf{R}$.
For x_0 limit point of E, the quantity defined by

$$\widehat{\omega}(x_0; f) = \inf \{\delta[f(V \cap E \setminus \{x_0\})]; V \in \mathcal{V}(x_0)\}$$

is called pointed oscillation of f at x_0.
For $\varepsilon > 0$, the quantity defined by

$$\omega(f; \varepsilon)_E = \sup\{|f(x_1) - f(x_2)|; |x_1 - x_2| \le \varepsilon, x_1, x_2 \in E\}$$

is called modulus of continuity of f on E with step $\varepsilon > 0$.

The following results also are well-known.

Theorem 5.1 *(see e.g. Sireţchi [199], p.166, p.154, [200], p.211, p.239).*
(i) f is continuous at $x_0 \in E$ if and only if $\omega(x_0; f) = 0$;
(ii) f has finite limit at $x_0 \in E'$ if and only if $\widehat{\omega}(x_0; f) = 0$. Also

$$\overline{\lim_{x \to x_0}} f(x) - \underline{\lim_{x \to x_0}} f(x) \leq \widehat{\omega}(x_0; f)$$

and if, in addition, f is bounded on E, then

$$\widehat{\omega}(x_0; f) = \overline{\lim_{x \to x_0}} f(x) - \underline{\lim_{x \to x_0}} f(x),$$

where

$$\underline{\lim_{x \to x_0}} f(x) = \sup \{\inf f(V \cap E \setminus \{x_0\}); V \in \mathcal{V}(x_0)\}$$

and

$$\overline{\lim_{x \to x_0}} f(x) = \inf \{\sup f(V \cap E \setminus \{x_0\}); V \in \mathcal{V}(x_0)\}.$$

(iii) f is uniformly continuous on E if and only if $\inf \{\omega(f; \varepsilon); \varepsilon > 0\} = 0$.

Theorem 5.1 suggests us to introduce the following

Definition 5.2 The quantities

$$d_C(f)(x_0) = \omega(x_0; f),$$

$$d_{\lim}(f)(x_0) = \widehat{\omega}(x_0; f)$$

and

$$d_{uc}(f)(E) = \inf \{\omega(f; \varepsilon); \varepsilon > 0\}$$

can be called defect of continuity of f on x_0, defect of limit of f on x_0 and defect of uniform continuity of f on E, respectively.

Example 5.1 Let $f : \mathbf{R} \to \mathbf{R}$ be given by $f(0) = 1, f(x) = 0, \forall x \neq 0$. Then $d_C(f)(0) = \omega(0; f) = 1, d_{\lim}(f)(0) = \widehat{\omega}(0; f) = 0$ and $d_{uc}(f)(\mathbf{R}) = 1$.

Remark. The concepts in Definition 5.1 can be extended to more general classes of functions. Thus, if $(X, \mathcal{T})$ is a topological space and (Y, ρ) is a metric space, then for $f : X \to Y, E \subset X$ and $x_0 \in E$ (or $x_0 \in E'$ in the topology $\mathcal{T}$), we can define

$$\omega(x_0; f) = \inf\{\delta_\rho[f(V \cap E)]; V \in \mathcal{V}_{\mathcal{T}}(x_0)\},$$

$$\widehat{\omega}(x_0; f) = \inf\{\delta_\rho[f(V \cap E \setminus \{x_0\})]; V \in \mathcal{V}_{\mathcal{T}}(x_0)\},$$

where $\delta_\rho[A] = \sup\{\rho(y_1, y_2); y_1, y_2 \in A\}, A \subset Y$, and $\mathcal{V}_{\mathcal{T}}(x_0)$ denotes the class of all the neighborhoods of $x_0 \in X$ in the topology $\mathcal{T}$.
Also, if (X, β) is a metric space and $f : X \to Y$, then we can define the quantities

$$\omega(f; \varepsilon)_X = \sup\{\rho(f(x_1), f(x_2)); \beta(x_1, x_2) \le \varepsilon, x_1, x_2 \in X\}$$

and

$$\inf\{\omega(f; \varepsilon)_X; \varepsilon > 0\}.$$

Of course that can be defined other quantities too that measure the deviation from a property. For example, in the case of property of continuity, can be introduced the following

Definition 5.3 (see Burgin-Ŝostak [44], [45]). Let $f : X \to Y$ be with $X, Y \subset \mathbf{R}$ and $x_0 \in X$. The defect of continuity of f at x_0 can be defined by

$$\mu(x_0; f) = \sup\{|y - f(x_0)|; y \text{ is limit point of } f(x_n) \text{ when } x_n \to x_0\}.$$

The defect of continuity of f on X is defined by

$$\mu(f)_X = \sup\{\mu(x_0; f); x_0 \in X\}.$$

The next theorem presents some properties of the quantities in Definition 5.3.

Theorem 5.2 *(see Burgin-Ŝostak [44], [45]).*
(i) f is continuous on x_0 if and only if $\mu(x_0; f) = 0$;
(ii) f is continuous on X if and only if $\mu(f)_X = 0$;

(iii)

$$\begin{aligned} |\mu(x_0; f) - \mu(x_0; g)| &\leq \mu(x_0; f+g) \leq \mu(x_0; f) + \mu(x_0; g), \\ |\mu(x_0; f) - \mu(x_0; g)| &\leq \mu(x_0; f-g), \\ \mu(x_0; -f) &= \mu(x_0; f) \end{aligned}$$

and similar inequalities for $\mu(f)_X$ *hold.*

(iv)

$$\mu(x_0; f \cdot g) \leq \mu(x_0; f) \cdot \|g\|_{x_0} + \mu(x_0; g) \cdot \|f\|_{x_0},$$

where $\|g\|_{x_0} = \sup\{|g(t)|; t \in V_{x_0}\}, \|f\|_{x_0} = \sup\{|f(t)|; t \in V_{x_0}\}, V_{x_0}$ *is a neighborhood of* x_0 *and* $f, g : X \to Y$*, with* $X, Y \subset \mathbf{R}$.

(v) If we define $\mu_H(x_0; f) = \frac{\mu(x_0;f)}{M}, \mu_H(f)_X = \frac{\mu(f)_X}{M}$*, where* $M = \sup_{x \in X} f(x) - \inf_{x \in X} f(x), f : X \to Y$*, then* $\mu_H(f)_X = \mu(t \circ f)_X$*, where* $t(x) = kx, k > 0$ *fixed.*

Remarks. 1) It is obvious that $\mu(x_0; f)$ can be defined if f is a mapping between two metric spaces, i. e. $f : (X, \rho_1) \to (Y, \rho_2), x_0 \in X$,

$$\begin{aligned} \mu(x_0; f) &= \sup\{\rho_2(y, f(x_0)); y \text{ is limit point of } f(x_n) \\ &\quad \text{with respect to } \rho_2 \text{ when } \rho_1(x_n, x_0) \overset{n\to\infty}{\to} 0\Big\}. \end{aligned}$$

For example, if $h : \mathbf{R} \to \mathbf{R}$ is monotonous on $\mathbf{R}$ and satisfies $h(x+y) \leq h(x) + h(y), \forall x, y \in \mathbf{R}$, then h induces on $\mathbf{R}$ the metric $\rho_2(x, y) = h(|x-y|)$ and consequently for $f : X \to Y, X, Y \subset \mathbf{R}$, we can consider a different defect of continuity according to the above definition (see Burgin-Ŝostak [44], [45]).

2) A natural question is to find the relationships between $\omega(x_0; f)$ and $\mu(x_0; f)$, and between $d_{uc}(f)_X$ and $\mu(f)_X$.

In general, they are not equal as the following example shows.

Let $f : [-1, 1] \to \mathbf{R}$ be defined by $f(x) = -1, x \in [-1, 0), f(0) = 0, f(x) = 1, x \in (0, 1]$. We have

$$\omega(0; f) = 2, \omega(f; \varepsilon)_{[-1,1]} = 2, d_{uc}(f)([-1, 1]) = 2$$

and

$$\mu(0; f) = 1, \mu(f)_{[-1,1]} = 1.$$

Now, concerning the differentiability we introduce the following

Definition 5.4 Let $f : [a, b] \to \mathbf{R}$ and $x_0 \in [a, b]$. The quantity

$$d_{dif}(f)(x_0) = \widehat{\omega}(x_0; F),$$

where $F : [a, b] \setminus \{x_0\} \to \mathbf{R}$ is given by $F(x) = \frac{f(x)-f(x_0)}{x-x_0}$, is called defect of differentiability of f on x_0.

An immediate consequence of Theorem 5.1 is

Corollary 5.1 *(i) f is differentiable on x_0 if and only if $d_{dif}(f)(x_0) = 0$.*
(ii) If f is locally Lipschitz on x_0 (i.e. Lipschitz in a neighborhood of x_0) then

$$d_{dif}(f)(x_0) = \overline{\lim_{x \to x_0}} F(x) - \underline{\lim_{x \to x_0}} F(x).$$

Example 5.2 For $f(x) = |x|, x \in [-1, 1]$, we have $d_{dif}(f)(0) = 2$.

Theorem 5.3 *Let $f : [a, b] \to \mathbf{R}$ and $x_0 \in (a, b)$. If f is locally Lipschitz on x_0, then*

$$\widehat{\omega}(x_0; f) \leq (b-a)\, d_{dif}(f)(x_0).$$

Proof. Let $V_0 \in \mathcal{V}(x_0)$ be such that $|f(x) - f(y)| \leq M|x - y|, \forall x, y \in V_0$. By

$$f(x) - f(x_0) = \frac{f(x) - f(x_0)}{x - x_0}(x - x_0), \forall x \in [a, b] \setminus \{x_0\},$$

for any $V \in \mathcal{V}(x_0), V \subset V_0$, denoting $g(x) = x - x_0$ and F as in Definition 5.4, we get

$$\begin{aligned}
&\delta[f(V \cap [a, b] \setminus \{x_0\})] = \delta[(F \cdot g)(V \cap [a, b] \setminus \{x_0\})] \\
&= \sup_{x_1, x_2 \in V \cap [a,b] \setminus \{x_0\}} |F(x_1) \cdot g(x_1) - F(x_2) \cdot g(x_2)| \\
&\leq \sup_{x_1 \in V \cap [a,b] \setminus \{x_0\}} |F(x_1)| \cdot \delta[g(V \cap [a, b] \setminus \{x_0\})] \\
&\quad + \sup_{x_2 \in V \cap [a,b] \setminus \{x_0\}} |g(x_2)| \cdot \delta[F(V \cap [a, b] \setminus \{x_0\})] \\
&\quad \leq M \cdot \delta[V \cap [a, b]] + \delta[V] \cdot \delta[F(V \cap [a, b] \setminus \{x_0\})] \\
&\leq M \cdot \delta[V \cap [a, b]] + (b - a) \cdot \delta[F(V \cap [a, b] \setminus \{x_0\})] \\
&=: L(V),
\end{aligned}$$

where $L : \mathcal{V}(x_0) \to \mathbf{R}_+$. Obviously, if $V_1, V_2 \in \mathcal{V}(x_0), V_1 \subseteq V_2$ then $L(V_1) \leq L(V_2)$, therefore

$$\begin{aligned} \inf_{V \in \mathcal{V}(x_0)} L(V) &= \inf\left\{ \lim_{n\to\infty} L(V_n); V_n \in \mathcal{V}(x_0), V_{n+1} \subset V_n, \right. \\ &\qquad \left. \forall n \in \mathbf{N}, \cap_{n=1}^{\infty} V_n = \{x_0\} \right\}, \end{aligned}$$

which immediately implies

$$\begin{aligned} \widehat{\omega}(x_0; f) &= \inf\{\delta[f(V \cap [a,b] \setminus \{x_0\})]; V \in \mathcal{V}(x_0)\} \\ &\leq (b-a) \cdot \inf\{\delta[F(V \cap [a,b] \setminus \{x_0\})]; V \in \mathcal{V}(x_0)\} \\ &= (b-a)\, d_{dif}(f)(x_0). \end{aligned}$$

□

In what follows we deal with integrability. Firstly, let us consider the well-known

Definition 5.5 (see *e.g.* Sireţchi [199], p.311). Let $f : [a,b] \to \mathbf{R}$ be bounded. The real number

$$\sup\left\{ s_\Delta(f) = \sum_{i=1}^{n} m_i(f)(x_i - x_{i-1}); m_i(f) = \inf_{x \in (x_{i-1}, x_i)} f(x), \right. \\ \left. \Delta = \{a = x_0 < ... < x_n = b\}, n \in \mathbf{N} \right\}$$

is called the lower Darboux integral of f on $[a,b]$ and it is denoted by $\underline{\int_a^b} f(x)\mathrm{d}x$. Similarly, the real number

$$\inf\left\{ S_\Delta(f) = \sum_{i=1}^{n} M_i(f)(x_i - x_{i-1}); M_i(f) = \sup_{x \in (x_{i-1}, x_i)} f(x), \right. \\ \left. \Delta = \{a = x_0 < ... < x_n = b\}, n \in \mathbf{N} \right\}$$

is called the upper Darboux integral of f on $[a,b]$ and it is denoted by $\overline{\int_a^b} f(x)\mathrm{d}x$.

By Definition 5.5 it is natural to introduce

Definition 5.6 Let $f : [a,b] \to \mathbf{R}$ be bounded. The real number

$$d_{int}(f)([a,b]) = \overline{\int_a^b} f(x)\,\mathrm{d}x - \underline{\int_a^b} f(x)\,\mathrm{d}x$$

is called defect of integrability of f on $[a,b]$.

Remark. According to *e.g.* Sireţchi [199], p.314, f is Riemann integrable on $[a,b]$ if and only if $d_{int}(f)([a,b]) = 0$.

Example 5.3 Let $f : [0,1] \to \mathbf{R}$ be defined by $f(x) = 1$, if x is irrational, $f(x) = 0$ if x is rational. We have $\overline{\int_0^1} f(x)\mathrm{d}x = 0$, $\underline{\int_0^1} f(x)\mathrm{d}x = 1$, therefore $d_{int}(f)([0,1]) = 1$.

At the end of this section we present some applications.

Let us denote

$$\begin{aligned} B[a,b] &= \{f : [a,b] \to \mathbf{R}; f \text{ bounded on } [a,b]\}, \\ C[a,b] &= \{g : [a,b] \to \mathbf{R}; g \text{ continuous on } [a,b]\} \end{aligned}$$

and for $f \in B[a,b]$ the quantity of best approximation

$$E_c(f) = \inf\{||f - g||; g \in C[a,b]\},$$

where $||f|| = \sup\{|f(t)|; t \in [a,b]\}$.

We present

Theorem 5.4 *Let $f \in B[a,b]$ be with*

$$C_f = \{x \in [a,b]; x \textit{ is point of discontinuity of } f\}.$$

Then

$$E_c(f) \geq \frac{\sup\{\omega(x;f); x \in C_f\}}{2}.$$

Proof. Let $f \in B[a,b], x \in C_f$ and $V \in \mathcal{V}(x)$ be fixed. For any $x_1, x_2 \in V \cap [a,b]$ and $g \in C[a,b]$ we have

$$\begin{aligned} |f(x_1) - f(x_2)| &\leq |f(x_1) - g(x_1)| + |g(x_1) - g(x_2)| + |g(x_2) - f(x_2)| \\ &\leq 2||f - g|| + |g(x_1) - g(x_2)|, \end{aligned}$$

which implies

$$\delta[f(V \cap [a,b])] \leq 2||f - g|| + \delta[g(V \cap [a,b])].$$

Passing to infimum with $V \in \mathcal{V}(x)$, it follows

$$\omega(x;f) \leq 2||f - g|| + \omega(x;g) = 2||f - g||.$$

Passing now to infimum with $g \in C[a,b]$ and then to supremum with $x \in C_f$, we get the conclusion of theorem. □

Let us denote

$$L[a,b] = \{f : [a,b] \to \mathbf{R}; \|f\|_L < +\infty\},$$
$$C_1[a,b] = \{g : [a,b] \to \mathbf{R}; g \text{ is differentiable on } [a,b]\},$$

where $\|f\|_L = \sup\left\{\left|\frac{f(x)-f(y)}{x-y}\right|; x, y \in [a,b], x \neq y\right\}$ is the so-called Lipschitz norm. If we denote $E_L(f) = \inf\{\|f-g\|_L; g \in C_1[a,b]\}$, we can present

Theorem 5.5 *Let $f \in L[a,b]$ be with*

$$D_f = \{x \in [a,b]; x \text{ is point of non-differentiability of } f\}.$$

Then

$$E_L(f) \geq \sup\{d_{dif}(f)(x); x \in D_f\}.$$

Proof. Let $f \in L[a,b], x \in D_f$ and $V \in \mathcal{V}(x)$ be fixed. For any $y \in V \cap [a,b] \setminus \{x\}$ and $g \in C_1[a,b]$ we have

$$\begin{aligned}\frac{f(y)-f(x)}{y-x} &= \frac{(f-g)(y)-(f-g)(x)}{y-x} + \frac{g(y)-g(x)}{y-x} \\ &\leq \|f-g\|_L + \frac{g(y)-g(x)}{y-x},\end{aligned}$$

which implies (with the notations $F(y) = \frac{f(y)-f(x)}{y-x}, G(y) = \frac{g(y)-g(x)}{y-x}$)

$$\delta[F(V \cap [a,b] \setminus \{x\})] \leq \|f-g\|_L + \delta[G(V \cap [a,b] \setminus \{x\})].$$

Passing to infimum with $V \in \mathcal{V}(x)$, it follows

$$d_{dif}(f)(x) \leq \|f-g\|_L + d_{dif}(g)(x) = \|f-g\|_L.$$

Passing here to infimum with $g \in C_1[a,b]$ and then to supremum after $x \in D_f$, we get the desired conclusion. □

Now, let us denote by

$$R[a,b] = \{g : [a,b] \to \mathbf{R}; g \text{ is Riemann integrable on } [a,b]\}$$

and for $f \in B[a,b]$, we define the quantity of best approximation

$$E_I(f) = \inf\{\|f-g\|_I; g \in R[a,b]\},$$

where $\|f\|_I = \inf\{S_\Delta(f) - s_\Delta(f); \Delta \in \mathcal{D}[a,b]\}, \mathcal{D}[a,b]$ being the set of divisions of $[a,b]$. We present

Theorem 5.6 *For each $f \in B[a,b]$ we have*

$$E_I(f) \geq d_{int}(f)([a,b]).$$

Proof. For any division $\Delta \in \mathcal{D}[a,b]$ and for any $g \in R[a,b]$ we have

$$S_\Delta(f) = S_\Delta(f-g+g) \leq S_\Delta(f-g) + S_\Delta(g)$$

and

$$s_\Delta(f-g) + s_\Delta(g) \leq s_\Delta(f),$$

which implies

$$S_\Delta(f) - s_\Delta(f) \leq S_\Delta(f-g) - s_\Delta(f-g) + S_\Delta(g) - s_\Delta(g).$$

Passing to infimum with $\Delta \in \mathcal{D}[a,b]$ and denoting by $\nu(\Delta_n)$ the norm of division Δ_n, we get

$$\overline{\int_a^b} f(x)\,\mathrm{d}x - \underline{\int_a^b} f(x)\,\mathrm{d}x \leq \inf\{S_\Delta(f) - s_\Delta(f); \Delta \in \mathcal{D}[a,b]\}$$

$$\leq \inf_{\Delta\in\mathcal{D}[a,b]}\{S_\Delta(f-g) - s_\Delta(f-g) + S_\Delta(g) - s_\Delta(g)\}$$

$$= \inf_{\Delta_n\in\mathcal{D}[a,b],\Delta_{n+1}\subseteq\Delta_n,n\in\mathbf{N},\nu(\Delta_n)\overset{n\to\infty}{\to}0}\ \lim_{n\to\infty}(S_{\Delta_n}(f-g) - s_{\Delta_n}(f-g)$$

$$+S_{\Delta_n}(g) - s_{\Delta_n}(g)) = \inf_{\Delta\in\mathcal{D}[a,b]}\{S_\Delta(f-g) - s_\Delta(f-g)\},$$

because $g \in R[a,b]$ implies $\lim_{n\to\infty}(S_{\Delta_n}(g) - s_{\Delta_n}(g)) = 0$. Passing to infimum with $g \in R[a,b]$ we obtain the desired conclusion. □

5.2 Defect of Monotonicity, of Convexity and of Linearity

If $f : E \to \mathbf{R}$, $E \subset \mathbf{R}$, is not, for example, monotone (decreasing or increasing) on E, then a natural question is to introduce a quantity that measures the "deviation" of f from monotonicity. Similar questions can be considered with respect to the property of convexity (concavity) or linearity of f. In this section we give some answers to these questions.

Definition 5.7 Let $f : E \to \mathbf{R}$. The quantity

$$\begin{aligned} d_M(f)(E) &= \sup\{|f(x_1) - f(x)| + |f(x_2) - f(x)| \\ &\quad - |f(x_1) - f(x_2)|; x_1, x, x_2 \in E, x_1 \leq x \leq x_2\} \end{aligned}$$

is called defect of monotonicity of f on E.

Remark. The quantity $d_M(f)(E)$ was suggested by the definition of the so-called modulus of non-monotonicity defined for $\delta > 0$ by

$$\begin{aligned} \mu(f;\delta)[a,b] &= \frac{1}{2}\sup\{|f(x_1) - f(x)| + |f(x_2) - f(x)| \\ &\quad - |f(x_1) - f(x_2)|; x_1 \leq x \leq x_2, |x_1 - x_2| \leq \delta\}, \end{aligned}$$

introduced and used by Sendov in approximation theory (see *e.g.* Sendov [194]).

We have

Theorem 5.7 *The function $f : E \to \mathbf{R}$ is monotone on E if and only if $d_M(f)(E) = 0$.*

Proof. If f is monotone (nondecreasing or nonincreasing) on E then it is immediate $d_M(f)(E) = 0$.

Conversely, let us suppose that $d_M(f)(E) = 0$ but f is not monotone on E. In this case, there exists $x_1 \leq x \leq x_2$ in E,

$$f(x_1) < f(x) \text{ and } f(x) > f(x_2)$$

or

$$f(x_1) > f(x) \text{ and } f(x_2) > f(x).$$

In both cases it is to see that

$$|f(x_1) - f(x)| + |f(x_2) - f(x)| - |f(x_1) - f(x_2)| > 0,$$

which produces the contradiction $d_M(f)(E) > 0$. □

Example 5.4 Let $f : [-1,1] \to \mathbf{R}$ be given by $f(x) = |x|$. We easy get $d_M(f)([-1,1]) = 2$.

A natural question is that, given $f \in B[a,b]$, where

$$B[a,b] = \{g : [a,b] \to \mathbf{R}; g \text{ is bounded on } [a,b]\},$$

to estimate the quantity of best approximation

$$E_M(f)([a,b]) = \inf\{\|f-g\|; g \in M[a,b]\},$$

where $M[a,b] = \{g : [a,b] \to \mathbf{R}; g \text{ is monotone on } [a,b]\}$ and $\|\cdot\|$ is the uniform norm. The following result shows that $E_M(f)([a,b])$ cannot be arbitrarily small.

Theorem 5.8 *For any $f \in B[a,b]$ we have*

$$E_M(f)([a,b]) \geq \frac{d_M(f)([a,b])}{6}.$$

Proof. For $x_1 \leq x \leq x_2$ in $[a,b]$ and $g \in M[a,b]$ we obtain

$$\begin{aligned} & |f(x_1) - f(x)| + |f(x_2) - f(x)| - |f(x_1) - f(x_2)| \\ = \; & |f(x_1) - g(x_1) - (f(x) - g(x)) + g(x_1) - g(x)| \\ & + |f(x_2) - g(x_2) - (f(x) - g(x)) + g(x_2) - g(x)| - |f(x_1) - f(x_2)| \\ \leq \; & |(f-g)(x_1) - (f-g)(x)| + |(f-g)(x_2) - (f-g)(x)| \\ & + |g(x_1) - g(x)| + |g(x_2) - g(x)| - |g(x_1) - g(x_2)| \\ & + |g(x_1) - g(x_2)| - |f(x_1) - f(x_2)|. \end{aligned}$$

But

$$|g(x_1) - g(x)| + |g(x_2) - g(x)| - |g(x_1) - g(x_2)| = 0$$

and

$$\begin{aligned} |g(x_1) - g(x_2)| - |f(x_1) - f(x_2)| & \leq \; ||g(x_1) - g(x_2)| - |f(x_1) - f(x_2)|| \\ & \leq \; |(f-g)(x_1) - (f-g)(x_2)|, \end{aligned}$$

which immediately implies

$$|f(x_1) - f(x)| + |f(x_2) - f(x)| - |f(x_1) - f(x_2)| \leq 6\|f-g\|.$$

Passing to supremum with $x_1 \leq x \leq x_2$ we get

$$d_M(f)([a,b]) \leq 6\|f-g\|, \forall g \in M[a,b],$$

and passing to infimum with $g \in M[a,b]$ we get the theorem. □

Passing to convexity, it is well-known that $f : [a,b] \to \mathbf{R}$ is called convex (concave) on $[a,b]$ if

$$f(\alpha x + (1-\alpha) y) \le (\ge) \alpha f(x) + (1-\alpha) f(y), \forall \alpha \in [0,1], x, y \in [a,b].$$

We can introduce the following

Definition 5.8 Let $f : [a,b] \to \mathbf{R}$ be bounded on $[a,b]$. The quantities

$$d_{CONC}(f)([a,b]) = \sup\{K(f)(x) - f(x); x \in [a,b]\}$$

and

$$d_{CONV}(f)([a,b]) = \sup\{f(x) - k(f)(x); x \in [a,b]\}$$

are called defect of concavity and of convexity of f on $[a,b]$, respectively, where $K(f) : [a,b] \to \mathbf{R}$ is the least concave majorant of f and $k(f) : [a,b] \to \mathbf{R}$ is the greatest convex minorant of f on $[a,b]$.

Remark. We have the formulas

$$K(f)(x) = \inf\{g(x); g \text{ concave on } [a,b] \text{ and } f(t) \le g(t), \forall t \in [a,b]\},$$

$$k(f)(x) = \sup\{h(x); h \text{ convex on } [a,b] \text{ and } h(t) \le f(t), \forall t \in [a,b]\},$$

$$K(f)(x) = \sup\left\{\frac{(x-z) f(y) + (y-x) f(z)}{y-z}; a \le z \le x \le y \le b, z \ne y\right\}$$

$$= \sup\{(1-\alpha) f(y) + \alpha f(z); \alpha \in [0,1], a \le z \le x \le y \le b, z \ne y\}$$

(see *e.g.* Mitjagin-Semenov [151]),

$k(f)(x) = -K(-f)(x), x \in [a,b].$

Remark. Obviously $K(f)$ is concave on $[a,b]$ and $k(f)$ is convex on $[a,b]$.

Theorem 5.9 *Let $f \in B[a,b]$. Then:*
(i) $d_{CONC}(f)([a,b]) = 0$ if and only if f is concave on $[a,b]$;
(ii) $d_{CONV}(f)([a,b]) = 0$ if and only if f is convex on $[a,b]$.

Proof. (*i*) If $d_{CONC}(f)([a,b]) = 0$ then it follows $f(x) = K(f)(x), \forall x \in [a,b]$, so f is concave on $[a,b]$. The converse of (*i*) is obvious.

(*ii*) The proof is similar. □

Example 5.5 Let $f : [0, 2\pi] \to \mathbf{R}$ be given by $f(x) = \sin x, x \in [0, 2\pi]$. Then simple geometrical considerations show that

$$K(f)(x) = \begin{cases} \sin x, & x \in [0, x_0] \\ \sin x_0 + (x - x_0)\cos x_0, & x \in [x_0, 2\pi], \end{cases}$$

where x_0 is the solution in $\left(\frac{\pi}{2}, \pi\right)$ of the equation $x = \text{tg}x + 2\pi$ (actually $y = \sin x_0 + (x - x_0)\cos x_0$ is the equation of the tangent to the graphic of $f(x)$, that passes through the point $(2\pi, 0)$).
Denoting now $F(x) = K(f)(x) - f(x), x \in [0, 2\pi]$, we have $F(x) = 0, \forall x \in [0, x_0]$ and $F'(x) = 0, x \in (x_0, 2\pi)$ becomes $\cos x = \cos x_0, x \in (x_0, 2\pi)$, that is $x^* = 2\pi - x_0$ is the point of maximum for F and consequently $d_{CONC}(f)([0, 2\pi]) = -2\pi \cos x_0$.
Similarly we get $d_{CONV}(f)([0, 2\pi]) = -2\pi \cos x_0$.

Remarks. 1) Let $f \in B[a,b]$. If we introduce the quantities

$$E_{CONC}(f)([a,b]) = \inf\{\|f - g\|; g \text{ is concave on } [a,b]\}$$

and

$$E_{CONV}(f)([a,b]) = \inf\{\|f - g\|; g \text{ is convex on } [a,b]\},$$

then obviously

$$E_{CONC}(f)([a,b]) \le d_{CONC}(f)([a,b])$$

and

$$E_{CONV}(f)([a,b]) \le d_{CONV}(f)([a,b]).$$

(Here $\|\cdot\|$ denotes again the uniform norm).

2) Because the property of concavity is dual to that of convexity, the defect of concavity (convexity) of f on $[a,b]$ can also be considered (in a sense) as degree of convexity (concavity, respectively) of f.

Different concepts of defects of concavity and convexity in the spirit of above Remark 2 might be the following.

Definition 5.9 Let $p \in \mathbf{N}$ and $f : [a,b] \to \mathbf{R}$ be such that the derivative of order p, $f^{(p)}$, exists, Lebesgue measurable and bounded on $[a,b]$ and $\int_a^b \left|f^{(p)}(x)\right| \mathrm{d}x \neq 0$. The quantities

$$K_+^{(p)}(f)([a,b]) = \frac{\int_a^b f_+^{(p)}(x)\,\mathrm{d}x}{\int_a^b \left|f^{(p)}(x)\right| \mathrm{d}x}$$

and

$$K_-^{(p)}(f)([a,b]) = \frac{\int_a^b f_-^{(p)}(x)\,\mathrm{d}x}{\int_a^b \left|f^{(p)}(x)\right| \mathrm{d}x}$$

are called degree of convexity (or defect of concavity) of order p of f on $[a,b]$, respectively. Here $f_+^{(p)}(x) = f^{(p)}(x)$ if $f^{(p)}(x) \geq 0, f_+^{(p)}(x) = 0$ if $f^{(p)}(x) < 0$ and $f_-^{(p)}(x) = -f^{(p)}(x)$ if $f^{(p)}(x) \leq 0, f_-^{(p)}(x) = 0$ if $f^{(p)}(x) > 0$.

Remark. It is well-known that a function $f : [a,b] \to \mathbf{R}$ satisfying $f^{(p)}(x) \geq 0$ (or ≤ 0) on $[a,b]$, is called convex (concave) of order p.

Theorem 5.10 *Let $f : [a,b] \to \mathbf{R}$ be satisfying the conditions in Definition 5.9. We have:*

(i) $K_+^{(p)}(f)([a,b]), K_-^{(p)}(f)([a,b]) \geq 0$ *and* $K_+^{(p)}(f)([a,b]) + K_-^{(p)}(f)([a,b]) = 1$.

(ii) If $f^{(p)}(x) \geq 0$, *a.e.* $x \in [a,b]$ *then* $K_+^{(p)}(f)([a,b]) = 1$, *if* $f^{(p)}(x) \leq 0$, *a.e.* $x \in [a,b]$ *then* $K_-^{(p)}(f)([a,b]) = 1$.

(iii) Let us suppose, in addition, that $f^{(p)}$ is continuous on $[a,b]$. Then f is convex (concave) of order p on $[a,b]$ if and only if $K_+^{(p)}(f)([a,b]) = 1$ ($K_-^{(p)}(f)([a,b]) = 1$, respectively).

Proof. *(i)* It is immediate by definition.

(ii) It is immediate.

(iii) Let us suppose $K_+^{(p)}(f)([a,b]) = 1$. It follows $K_-^{(p)}(f)([a,b]) = 0$, $\int_a^b f_-^{(p)}(x)\mathrm{d}x = 0, f_-^{(p)}(x) = 0, \forall x \in [a,b]$, that is $f^{(p)}(x) \geq 0, \forall x \in [a,b]$. □

Remarks. 1) If $p = 1$ then $K_+^{(1)}(f)([a,b])$ represents the degree of nondecreasing monotonicity (or equivalently, the defect of nonincreasing monotonicity) and $K_-^{(1)}(f)([a,b])$ represents the degree of nonincreasing monotonicity (or equivalently, the defect of nondecreasing monotonicity). Therefore they are, in a way, more refined than those in Definition 5.7, because make difference between increasing and decreasing monotonicity. If $p=0$ then $K_{+}^{(0)}(f)([a,b])$ and $K_{-}^{(0)}(f)([a,b])$ measure the positive and negative signature of f on $[a,b]$, respectively.

2) Starting from the definition of convex functions, we can introduce another concept of defect of convexity, by the quantity

$$\begin{aligned} d^{*}_{CONV}(f)([a,b]) &= \sup\{f(\lambda x+(1-\lambda)y)-(\lambda f(x)+(1-\lambda)f(y));\\ &\qquad \lambda\in[0,1], x,y\in[a,b]\}. \end{aligned}$$

It is not difficult to show that f is convex on $[a,b]$ if and only if $d^{*}_{CONV}(f)([a,b])=0$.

If, for example, f is strongly concave on $[a,b]$, that is there exists $\alpha>0$ such that for all $x,y\in[a,b]$ and all $\lambda\in[0,1]$ we have

$$\lambda(1-\lambda)\alpha(x-y)^{2}\leq f(\lambda x+(1-\lambda)y)-(\lambda f(x)+(1-\lambda)f(y)),$$

then

$$\begin{aligned} d^{*}_{CONV}(f)([a,b]) &\geq \sup\left\{\lambda(1-\lambda)\alpha(x-y)^{2}; x,y\in[a,b], \lambda\in[0,1]\right\}\\ &= \frac{\alpha(b-a)^{2}}{4}. \end{aligned}$$

3) Somehow related but still different ideas concerning the measurement of degree of convexity of a real function (of one or of several variables), can be found in *e.g.* Roberts-Varberg [174], p. 264 and Crouzeix-Lindberg [59].

At the end of this section we introduce a concept of defect of linearity for a function $f\in C^{2}[a,b]=\{f:[a,b]\to\mathbf{R}; \exists f'' \text{ continuous on } [a,b]\}$. For fixed $x_0\in[a,b]$, let us consider the Taylor's formula

$$f(x)=f(x_0)+(x-x_0)f'(x_0)+\frac{(x-a)^{2}f''(\xi)}{2}, x\in[a,b],$$

where $\xi\in(a,b)$. Then

$$|f(x)-(f(x_0)+(x-x_0)f'(x_0))|=\frac{(x-a)^{2}}{2}|f''(\xi)|$$

and if we denote $\|f''\|=\sup\{|f''(\xi)|;\xi\in[a,b]\}$, then for the next quantity, called defect of linearity of f on $[a,b]$,

$$d_{LIN}(f)([a,b])=\sup\{|f(x)-(f(x_0)+(x-x_0)f'(x_0))|; x,x_0\in[a,b]\},$$

we have

$$d_{LIN}(f)([a,b]) \leq \frac{(b-a)^2}{2}\|f''\|, \forall f \in C^2[a,b].$$

Obviously $f \in C^2[a,b]$ is linear if and only if $d_{LIN}(f)([a,b]) = 0$.

5.3 Defect of Equality for Inequalities

In this section we introduce and study two concepts that measure the "quality" of an inequality: the absolute defect of equality and the averaged defect of equality. For some remarkable inequalities we calculate and estimate these defects.

Firstly, we introduce the following.

Definition 5.10 Let L, R be two functions, $L : D_1 \to \mathbf{R}, R : D_2 \to \mathbf{R}$, where $D_1, D_2 \subseteq \mathbf{R}^n, n \in \mathbf{N}$, such that

$$L(x) \leq R(x), \forall x \in D \subseteq D_1 \cap D_2. \tag{5.1}$$

The absolute defect of equality for (5.1) is defined by

$$d_D^{ab}(L,R) = \sup\{R(x) - L(x); x \in D\}.$$

If, in addition, D is Lebesgue measurable and $L(x), R(x)$ are continuous on D, then the averaged defect of equality on $[-r,r]^n$ for (5.1) is defined by

$$d_D^{av}(L,R)(r) = \frac{\int_{D\cap[-r,r]^n}(R(x) - L(x))\, d\mu_n}{\mu_n(D\cap[-r,r]^n)},$$

where $[-r,r]^n = \underbrace{[-r,r] \times ... \times [-r,r]}_{n}$ and μ_n is the Lebesgue measure on $\mathbf{R}^n$.

Concerning these concepts, we present

Theorem 5.11 *(i) If $L(x)$ and $R(x)$ are given by (5.1), then we have:*

$$\begin{aligned} d_D^{ab}(L+Q, R+Q) &= d_D^{ab}(L,R), \\ d_D^{av}(L+Q, R+Q)(r) &= d_D^{av}(L,R)(r), \forall r > 0 \end{aligned}$$

for any $Q : D \to \mathbf{R}$
and

$$\begin{aligned} d_D^{ab}(kL, kR) &= kd_D^{ab}(L, R), \\ d_D^{av}(kL, kR)(r) &= kd_D^{av}(L, R)(r), \forall r > 0, \end{aligned}$$

for any constant $k \geq 0$.

(ii) *If* $D \subseteq \mathbf{R}^n, 0 \leq L_1(x) \leq R_1(x), \forall x \in D$ *and* $0 \leq L_2(x) \leq R_2(x), \forall x \in D$ *then*

$$d_D^{ab}(L_1 L_2, R_1 R_2) \leq d_D^{ab}(L_1, R_1) \sup_{x \in D} L_2(x) + d_D^{ab}(L_2, R_2) \sup_{x \in D} R_1(x)$$

and

$$d_D^{av}(L_1 L_2, R_1 R_2)(r) \leq d_D^{av}(L_1, R_1)(r) \sup_{x \in D} L_2(x) + d_D^{av}(L_2, R_2)(r) \sup_{x \in D} R_1(x)$$

for every $r > 0$.

(iii) $d_D^{av}(L, R)(r) \leq d_{D \cap [-r,r]^n}^{ab}(L, R)$.

(iv) $A(x) \leq B(x) \leq C(x), \forall x \in D$ *implies* $d_D^{ab}(A, C) \geq d_D^{ab}(B, C)$ *and* $d_D^{av}(A, C)(r) \geq d_D^{av}(B, C)(r), \forall r > 0$.

Proof. (i) It is immediate.

(ii) We get

$$d_D^{ab}(L_1 L_2, R_1 R_2) = \sup_{x \in D} \{R_1(x) R_2(x) - L_1(x) L_2(x)\}$$

$$\begin{aligned} &= \sup_{x \in D} \{R_1(x)(R_2(x) - L_2(x)) + L_2(x)(R_1(x) - L_1(x))\} \\ &\leq \sup_{x \in D} R_1(x) \sup_{x \in D} \{R_2(x) - L_2(x)\} + \sup_{x \in D} L_2(x) \sup_{x \in D} \{R_1(x) - L_1(x)\} \\ &= d_D^{ab}(L_1, R_1) \sup_{x \in D} L_2(x) + d_D^{ab}(L_2, R_2) \sup_{x \in D} R_1(x). \end{aligned}$$

The proof in the case of averaged defect is similar.

(iii) We have

$$R(x) - L(x) \leq \sup\{R(x) - L(x); x \in D \cap [-r, r]^n\}, \forall x \in D \cap [-r, r]^n,$$

which implies the inequality.

(iv) It is immediate. □

Remark. While the first two relations in Theorem 5.11, (i) are obvious, the last two relations also are natural because by multiplying two real numbers with the same positive constant, the distance between them one modifies.

As a possible application of the above introduced concepts, let us consider the following inequalities:

$$L(x) \leq R_1(x), \forall x \in D, \tag{5.2}$$

$$L(x) \leq R_2(x), \forall x \in D, \tag{5.3}$$

where $D \subseteq \mathbf{R}^n$. A natural question is to see which inequality is better in certain sense. An answer might be given by comparing their defects. For example, we can say that (5.2) is better than (5.3) if $d_D^{ab}(L, R_1) \leq d_D^{ab}(L, R_2)$. More general, we can compare two inequalities $L_1(x) \leq R_1(x), x \in D_1$ and $L_2(x) \leq R_2(x), x \in D_2$, saying that the first one is better than the second one if, for example, $d_{D_1}^{ab}(L_1, R_1) \leq d_{D_2}^{ab}(L_2, R_2)$. Another variant is to say that (5.2) is better than (5.3) if $\lim_{r \to \infty} \frac{d_D^{av}(L,R_1)(r)}{d_D^{av}(L,R_2)(r)} < 1$.

For another possible application, let us consider the system of nonlinear inequalities

$$A_i(x) \leq 0, i \in \{1, ..., n\}, x \in D \subseteq \mathbf{R}^n, \tag{5.4}$$

and let us suppose that the system is difficult to be solved. We can drop this shortcomings by replacing this system with a simpler one

$$B_i(x) \leq 0, i \in \{1, ..., n\}, x \in D \subseteq \mathbf{R}^n, \tag{5.5}$$

where we have $A_i(x) \leq B_i(x), \forall x \in D$. Obviously, any solution of (5.5) is also solution of (5.4). Let us denote by M the set of all solutions of (5.4) and by M^* the set of all solutions of (5.5).

A natural question is to estimate the distance (of Hausdorff kind, for example) between M and M^*, with respect to the defects $d_D^{ab}(A_i, B_i), i \in \{1, ..., n\}$ or with respect to the defects $d_D^{av}(A_i, B_i)(r), r > 0, i \in \{1, ..., n\}$.

Example 5.6 We will use the above ideas for the inequalities between the harmonic mean, geometric mean and arithmetic mean. That is, take $L_1(x, y) = \frac{2xy}{x+y}$, $R_1(x, y) = L_2(x, y) = \sqrt{xy}$ and $R_2(x, y) = \frac{x+y}{2}$, where

$(x, y) \in D = (0, +\infty) \times (0, +\infty)$. We have

$$\begin{aligned} d^{ab}_{(0,r]^2}(L_2, R_2) &= \sup\left\{\frac{x+y}{2} - \sqrt{xy}; x, y \in (0, r]\right\} \\ &= \sup\left\{\frac{\left(\sqrt{x} - \sqrt{y}\right)^2}{2}; x, y \in (0, r]\right\} = \frac{r}{2} \end{aligned}$$

and

$$d^{av}_D(L_2, R_2)(r) = \frac{\frac{1}{18}r^3}{r^2} = \frac{1}{18}r$$

because

$$\int\int_{(0,r]^2}\left(\frac{x+y}{2} - \sqrt{xy}\right)\mathrm{d}x\mathrm{d}y = \int_0^r\left(\frac{x}{2}r + \frac{r^2}{4} - \sqrt{x}\frac{r^{\frac{3}{2}}}{\frac{3}{2}}\right)\mathrm{d}x = \frac{1}{18}r^3.$$

The calculus of $d^{ab}_{(0,r]^2}(L_1, R_1)$ is not simple. Let us consider the function $f_y : (0, r] \to \mathbf{R}$ defined by $f_y(x) = \sqrt{x}\cdot\sqrt{y} - \frac{2xy}{x+y}$, where y is a parameter, $y \in (0, r]$. The solutions of the equation $f'_y(x) = 0$ are obtained by solving $\frac{1}{2}\sqrt{\frac{y}{x}} = \frac{2y^2}{(x+y)^2}$. But this equation is equivalent to $(x+y)^4 - 16xy^3 = 0$, that is

$$(x-y)\left(x^3 + 5x^2y + 11xy^2 - y^3\right) = 0.$$

Using the Cardan's method we get the real solutions of the above equation,

$$\begin{aligned} x_1 &= y, \\ x_2 &= \frac{2y}{3}\left(\left(17 + 3\sqrt{33}\right)^{\frac{1}{3}} + \left(17 - 3\sqrt{33}\right)^{\frac{1}{3}}\right) - \frac{5y}{3} \in \left(\frac{y}{12}, \frac{y}{11}\right). \end{aligned}$$

Because $f'_y(x) \geq 0$ on $(0, x_2] \cup [x_1, r]$ and $f'_y(x) \leq 0$ on (x_2, x_1) we get

$$\sup\{f_y(x); x \in (0, r]\} = \max\{f_y(r), f_y(x_2)\}.$$

Let $g_1 : (0, r] \to \mathbf{R}$ be defined by $g_1(y) = f_y(r) = \sqrt{r}\cdot\sqrt{y} - \frac{2ry}{r+y}$. As above, the real solutions of the equation $g'_1(y) = 0$ are

$$\begin{aligned} y_1 &= r, \\ y_2 &= \frac{2r}{3}\left(\left(17 + 3\sqrt{33}\right)^{\frac{1}{3}} + \left(17 - 3\sqrt{33}\right)^{\frac{1}{3}}\right) - \frac{5r}{3} \in \left(\frac{r}{12}, \frac{r}{11}\right) \end{aligned}$$

and $g_1'(y) > 0$ on $(0, y_2)$, $g_1'(y) < 0$ on (y_2, r), therefore

$$\sup\{g_1(y); y \in (0,r]\} = g_1(y_2) = r\left(\sqrt{\frac{2\alpha-5}{3}} - \frac{2\alpha-5}{\alpha-1}\right),$$

where $\alpha = (17+3\sqrt{33})^{\frac{1}{3}} + (17-3\sqrt{33})^{\frac{1}{3}}$.
Let $g_2 : (0,r] \to \mathbf{R}$ be defined by $g_2(y) = f_y(x_2) = \sqrt{x_2}\cdot\sqrt{y} - \frac{2x_2 y}{x_2+y}$. We obtain

$$\sup\{g_2(y); y \in (0,r]\}$$

$$= \sup\left\{\sqrt{\left(\frac{2y}{3}\alpha - \frac{5y}{3}\right)y} - \frac{2\left(\frac{2y}{3}\alpha - \frac{5y}{3}\right)}{\frac{2y}{3}\alpha - \frac{5y}{3} + y}; y \in (0,r]\right\}$$

$$= r\left(\sqrt{\frac{2\alpha-5}{3}} - \frac{2\alpha-5}{\alpha-1}\right),$$

because $\sqrt{\frac{2\alpha-5}{3}} > \frac{2\alpha-5}{\alpha-1}$.
With the above notation, we have

$$d^{ab}_{(0,r]^2}(L_1, R_1) = \sup\left\{\sqrt{xy} - \frac{2xy}{x+y}; x, y \in (0,r]\right\}$$

$$\begin{aligned}
&= \sup\{\sup\{f_y(x); x \in (0,r]\}; y \in (0,r]\}\\
&= \sup\{\max\{f_y(x_2), f_y(r)\}; y \in (0,r]\}\\
&= \max\{\sup\{f_y(x_2); y \in (0,r]\}, \{f_y(r); y \in (0,r]\}\}\\
&= r\left(\sqrt{\frac{2\alpha-5}{3}} - \frac{2\alpha-5}{\alpha-1}\right),
\end{aligned}$$

where $\alpha = (17+3\sqrt{33})^{\frac{1}{3}} + (17-3\sqrt{33})^{\frac{1}{3}}$.
Now, we can calculate the averaged defect of equality for the inequality between harmonic mean and geometric mean.
If μ_2 is the Lebesgue measure on $\mathbf{R}^2$ then

$$\iint_{(0,r]^2}\left(\sqrt{xy} - \frac{2xy}{x+y}\right)\mathrm{d}\mu_2 = \int_0^r\left(\int_0^r\left(\sqrt{xy} - \frac{2xy}{x+y}\right)\mathrm{d}x\right)\mathrm{d}y$$

$$= \frac{2}{3} r^{\frac{3}{2}} \int_0^r \sqrt{y} \mathrm{d}y - 2r \int_0^r y \mathrm{d}y + 2 \int_0^r y^2 \ln(y+r) \mathrm{d}y$$

$$-2 \int_0^r y^2 \ln(y) \mathrm{d}y = \left(\frac{2}{3} \ln 4 - \frac{8}{9} \right) r^3$$

and

$$d_D^{av}(L_1, R_1)(r) = \left(\frac{2}{3} \ln 4 - \frac{8}{9} \right) r \approx 0.023r.$$

Because

$$d_{(0,r]^2}^{ab}(L_1, R_1) = r \left(\sqrt{\frac{2\alpha - 5}{3}} - \frac{2\alpha - 5}{\alpha - 1} \right) \approx 0.135r \leq \frac{r}{2} = d_{(0,r]^2}^{ab}(L_2, R_2)$$

for every $r > 0$ and $\frac{d_D^{av}(L_1,R_1)(r)}{d_D^{av}(L_2,R_2)(r)} = \frac{\frac{2}{3}\ln 4 - \frac{8}{9}}{\frac{1}{18}} \approx \frac{0.023}{0.055} < 1, \forall r > 0$, the inequality between harmonic mean and geometric mean is better (in the sense of above application) than the inequality between geometric mean and arithmetic mean, for both defects of equality (absolute or averaged).

Example 5.7 Let us consider the following particular case of Cauchy-Buniakowski-Schwarz inequality

$$(ax + by)^2 \leq (a^2 + b^2)(x^2 + y^2), \forall a, b, x, y \in \mathbf{R}.$$

We have $d_{[-r,r]^4}^{ab}(L, R) = 4r^4$ and $d_{\mathbf{R}^4}^{av}(L, R)(r) = \frac{2}{9} r^4$, where $L(a, b, x, y) = (ax + by)^2$ and $R(a, b, x, y) = (a^2 + b^2)(x^2 + y^2)$. Indeed,

$$d_{[-r,r]^4}^{ab}(L, R) = \sup \Big\{ (a^2 + b^2)(x^2 + y^2) - (ax + by)^2 ;$$

$$a, b, x, y \in [-r, r] \Big\} = 4r^4,$$

obtained for $a = b = x = -y = r$ and

$$d_{\mathbf{R}^4}^{av}(L, R)(r) = \frac{\frac{32r^8}{9}}{(2r)^4} = \frac{2}{9} r^4,$$

because

$$\int_{[-r,r]^4} \left((a^2 + b^2)(x^2 + y^2) - (ax + by)^2 \right) \mathrm{d}\mu_4$$

$$
\begin{aligned}
&= \int_{[-r,r]^4} \left(a^2y^2 + b^2x^2 - 2abxy\right) \mathrm{d}\mu_4 \\
&= 4r^2 \int_{-r}^{r} a^2 \mathrm{d}a \int_{-r}^{r} y^2 \mathrm{d}y + 4r^2 \int_{-r}^{r} b^2 \mathrm{d}b \int_{-r}^{r} x^2 \mathrm{d}a \\
&\quad -2 \int_{-r}^{r} a \mathrm{d}a \int_{-r}^{r} b \mathrm{d}b \int_{-r}^{r} x \mathrm{d}x \int_{-r}^{r} y \mathrm{d}y \\
&= 2 \cdot 4r^2 \cdot \frac{2r^3}{3} \cdot \frac{2r^3}{3} - 2r^8 = \frac{14r^8}{9}.
\end{aligned}
$$

Example 5.8 It is well-known that $f : [a,b] \to \mathbf{R}$ is called convex on $[a,b]$ if

$$f\left(\sum_{i=1}^{n} \alpha_i x_i\right) \leq \sum_{i=1}^{n} \alpha_i f(x_i), \forall \alpha_i \in [0,1], \sum_{i=1}^{n} \alpha_i = 1, \forall x_i \in [a,b].$$

By a known result (see Cîrtoaje [56]) if f is differentiable and strictly convex on $[a,b]$, then we can write

$$d_{[a,b]^n}^{ab}(L,R) = \max\left\{\sum_{i=1}^{n} \alpha_i f(x_i) - f\left(\sum_{i=1}^{n} \alpha_i x_i\right); x_i \in \{a,b\}\right\},$$

for fixed α_i. As a particular case, let us consider another special case of Cauchy-Buniakowski-Schwarz inequality

$$(x_1 + \ldots + x_n)^2 \leq n\left(x_1^2 + \ldots + x_n^2\right), \forall x_i \in [a,b], i \in \{1,\ldots,n\}, n \in \mathbf{N}.$$

We have

$$d_{[a,b]^n}^{ab}(L,R) = \left[\frac{n^2}{4}\right](b-a)^2,$$

where $L(x_1,\ldots,x_n) = (x_1 + \ldots + x_n)^2$ and $R(x_1,\ldots,x_n) = n\left(x_1^2 + \ldots + x_n^2\right)$. Indeed, taking $f : [a,b] \to \mathbf{R}, f(x) = x^2$ and $\alpha_i = \frac{1}{n}, \forall i \in \{1,\ldots,n\}$, from the above formula we get

$$
\begin{aligned}
d_{[a,b]^n}^{ab}(L,R) &= n^2 \cdot \sup\left\{\sum_{i=1}^{n} \frac{1}{n} x_i^2 - \left(\sum_{i=1}^{n} \frac{1}{n} x_i\right)^2; x_i \in [a,b]\right\} \\
&= n^2 \cdot \max\left\{\sum_{i=1}^{n} \frac{1}{n} x_i^2 - \left(\sum_{i=1}^{n} \frac{1}{n} x_i\right)^2; x_i \in \{a,b\}\right\}
\end{aligned}
$$

$$= \max\left\{ \sum_{1\leq i<j\leq n} (x_i - x_j)^2 ; x_i, x_j \in \{a,b\} \right\}$$

$$= \left[\frac{n^2}{4}\right](b-a)^2,$$

the last equality being proved by induction.

Example 5.9 The simple inequality

$$\frac{x+y}{2} \leq \sqrt{\frac{x^2+y^2}{2}}, \forall x,y \in [0,+\infty)$$

has the absolute defect of equality on $[0,r]$, $r \in \mathbf{R}_+$, equal to $\frac{\sqrt{2}-1}{2}r$ and the value of averaged defect of equality is $\frac{\sqrt{2}\ln(\sqrt{2}+1)-1}{6}r$, $\forall r > 0$. The first result is immediate because the maximum of the function $f : [0,r]\times[0,r] \to \mathbf{R}$, $f(x,y) = \sqrt{\frac{x^2+y^2}{2}} - \frac{x+y}{2}$ is attained for $x = 0$ and $y = r$ (or $y = 0$ and $x = r$). We prove the second result:

$$\int_0^r \int_0^r \left(\sqrt{\frac{x^2+y^2}{2}} - \frac{x+y}{2} \right) dxdy = \int_0^r \frac{1}{\sqrt{2}} \frac{r\sqrt{y^2+r^2}}{2} dy$$

$$+\int_0^r \frac{y^2}{2\sqrt{2}} \ln\left(r+\sqrt{y^2+r^2}\right) dy - \int_0^r \frac{y^2}{2\sqrt{2}} \ln y dy - \int_0^r \frac{r^2}{4} dy$$

$$-\int_0^r \frac{r}{2} y dy = \frac{r}{2\sqrt{2}} \left(\frac{r^2\sqrt{2}}{2} + \frac{r^2}{2} \ln r \left(1+\sqrt{2}\right) - \frac{r^2}{2} \ln r \right)$$

$$+\frac{1}{2\sqrt{2}} \left(\frac{r^3}{3} \ln\left(r\left(1+\sqrt{2}\right)\right) - \frac{r^3}{9}\left(\sqrt{2}-1\right)^2 - \frac{r^3}{18}\left(\sqrt{2}-1\right) \right)$$

$$+\frac{1}{2\sqrt{2}} \left(\frac{r^3}{6} \left(1 - \ln r\left(\sqrt{2}+1\right) + \ln r\right) \right) - \frac{1}{2\sqrt{2}} \left(\frac{r^3}{3} \ln r - \frac{r^3}{9} \right)$$

$$-\frac{r^3}{4} - \frac{r^3}{4} = \left(-\frac{1}{6} + \frac{\sqrt{2}}{6} \ln\left(\sqrt{2}+1\right) \right) r^3.$$

Finally, we deal with inequalities in Hilbert spaces and in normed spaces.

For example, let us consider the space $\mathbf{R}^2$ endowed with the usual inner product $\langle \cdot, \cdot \rangle$ and norm $\|\cdot\|$, and let use take the following two inequalities

$$|\langle x, y \rangle| \leq \|x\| \cdot \|y\|, \forall x, y \in \mathbf{R}^2, \tag{5.6}$$

$$\|x + y\| \leq \|x\| + \|y\|, \forall x, y \in \mathbf{R}^2. \tag{5.7}$$

If $f, g : \mathbf{R}^2 \to \mathbf{R}$ satisfy $f(x) \leq g(x), \forall x \in \mathbf{R}^2$, let us define the defect of equality by

$$d_{\mathbf{R}^2}(f, g) = \lim_{r \to \infty} \frac{\sup \{g(x) - f(x); x \in B(0; r)\}}{\text{area} B(0; r)}.$$

In the case of inequality (5.6) we get $d_{\mathbf{R}^2}(f, g) = \frac{1}{\pi}$ while for the inequality (5.7) we get $d_{\mathbf{R}^2}(f, g) = \frac{2\sqrt{2}}{\pi}$ (that is, the first inequality is "better" than the second one). Indeed,

$$\sup \{\|x\| \cdot \|y\| - |\langle x, y \rangle|; x, y \in \mathbf{R}^2\} = r^2$$

is obtained for $x = (r, 0)$ and $y = (0, r)$ (or $x = (0, r), y = (r, 0)$) and

$$\sup \{\|x\| + \|y\| - \|x + y\|; x, y \in \mathbf{R}^2\} = 2\sqrt{2}r^2$$

is obtained for $x = (-r, r)$ and $y = (r, -r)$ (or $x = (r, -r), y = (-r, r)$).

5.4 Bibliographical Remarks and Open Problems

The concepts and results in Section 5.3 are in Ban-Gal [30]. Theorems 5.3-5.6, Theorem 5.8, Definitions 5.8, 5.9 and Theorem 5.10 are completely new.

Open problem 5.1 Characterize the set $Y = \{f \in B[a, b]\}$ such that for $f \in Y$ and $E_M(f)([a, b])$ defined in Section 5.2, there exists (uniquely or not) an element $g^* \in M[a, b]$, such that $E_M(f)([a, b]) = \|f - g^*\|$.

Open problem 5.2 A central problem in approximation theory is that of shape preserving approximation by operators. One of the most known result in this sense is that the Bernstein polynomials preserve the convexity of order p of f, for any $p \in \mathbf{N}$. If, for example, $f : [0, 1] \to \mathbf{R}$ is monotone nondecreasing or convex, *etc.*, then the Bernstein polynomials

$$B_n(f)(x) = \sum_{k=0}^{n} C_n^k x^k (1-x)^{n-k} f\left(\frac{k}{n}\right)$$

are monotone nondecreasing, or convex, *etc.*, respectively.
But if f is not, for example, monotone on $[0, 1]$, then it is natural to ask how much of its degree of monotonicity is preserved by the Bernstein polynomials $B_n(f)(x)$. More exactly, it is an open question if there exists a constant $r \geq 1$ (independent of at least n but possibly independent of f too), such that

$$d_M(B_n(f))([0,1]) \leq r d_M(f)([0,1]), \forall n \in \mathbf{N},$$

where d_M is the defect of monotonicity in Definition 5.7.
Similarly, are open questions if there exist constants $r, s, t, u, v \geq 1$ (independent of at least n, but possibly independent of f too), such that

$$\begin{aligned} d_{CONV}(B_n(f))([0,1]) &\leq r d_{CONV}(f)([0,1]), \forall n \in \mathbf{N}, \\ K_{-}^{(0)}(B_n(f))([0,1]) &\leq s K_{-}^{(0)}(f)([0,1]), \forall n \in \mathbf{N}, \\ K_{-}^{(1)}(B_n(f))([0,1]) &\leq t K_{-}^{(1)}(f)([0,1]), \forall n \in \mathbf{N}, \\ d_{CONV}^{*}(B_n(f))([0,1]) &\leq u d_{CONV}^{*}(f)([0,1]), \forall n \in \mathbf{N}, \end{aligned}$$

and

$$d_{LIN}(B_n(f))([0,1]) \leq v d_{LIN}(f)([0,1]), \forall n \in \mathbf{N}.$$

Open problem 5.3 It is well-known that $f : [a, b] \to \mathbf{R}$ is called quasi-convex on $[a, b]$ if

$$f(\lambda x + (1-\lambda) y) \leq \max\{f(x), f(y)\}, \forall x, y \in [a,b], \lambda \in [0,1].$$

We can define the defect of quasi-convexity of f by the quantity

$$\begin{aligned} d_{QCONV}(f)([a,b]) = \sup \quad &\{f(\lambda x + (1-\lambda) y) - \max\{f(x), f(y)\}; \\ &\lambda \in [0,1], x, y \in [a,b]\}. \end{aligned}$$

An open question is the study of this defect for various classes of bounded functions on $[a, b]$.

Open problem 5.4 The concept of convexity can be generalized in an interesting way, as follows.
Let $K \subset \mathbf{R}^n$, K be nonempty and let $f : K \to \mathbf{R}$ and $\eta : K \times K \times [0,1] \to \mathbf{R}$. According to Hanson [100], Yang-Chen [227], the set K is called semi-invex with respect to η, if for any $u, v \in K$ and any $t \in [0,1]$, we have $u + t\eta(v,u,t) \in K$, and f is called semipreinvex with respect to η, if

$$f(u + t\eta(v,u,t)) \leq (1-t) f(u) + tf(v), \forall u, v \in K, t \in [0,1]$$

where $\lim_{t \to 0} t\eta(v,u,t) = 0$. Similarly, $f : K \to \mathbf{R}$ is called quasisemipreinvex with respect to η, if

$$f(u + t\eta(v,u,t)) \leq \max\{f(u), f(v)\}, \forall u, v \in K, t \in [0,1].$$

In these cases, for $f : K \to \mathbf{R}$, where K is a nonempty bounded semi-invex set with respect to η, it is natural to introduce the defects of semipreinvexity and of quasisemipreinvexity by

$$\begin{aligned} d_{INV}(f)(K) &= \sup\{f(u + t\eta(v,u,t)) - [(1-t) f(u) + tf(v)]; \\ & \quad t \in [0,1], u, v \in K\} \end{aligned}$$

and

$$\begin{aligned} d_{QINV}(f)(K) &= \sup\{f(u + t\eta(v,u,t)) - \max\{f(u), f(v)\}; \\ & \quad t \in [0,1], u, v \in K\}, \end{aligned}$$

respectively.
If $\eta(v,u,t) = v - u$ then we recapture the defects d^*_{CONV} and d^*_{QCONV} in Remark 2 after Theorem 5.10 and in Open Problem 5.3, respectively.
It would be of interest to study the quantities $d_{INV}(f)$ and $d_{QINV}(f)$ for various classes of bounded functions f and for various choices of $\eta(v,u,t)$.

Chapter 6

Defect of Property in Functional Analysis

In this chapter we introduce and study various defects of property in functional analysis as, for example: defect of orthogonality, defect of convexity, of linearity, of balancing of sets, defect of subadditivity (additivity) of functionals and so on.

6.1 Defect of Orthogonality in Real Normed Spaces

In a real normed space $(E, ||\cdot||)$, the concept of orthogonality can be stated in different ways, as follows.

Definition 6.1 Let $X, Y \subseteq E$. We say that X is orthogonal to Y in the:

(i) a-isosceles sense if

$$||x - ay|| = ||x + ay||, \forall x \in X, \forall y \in Y,$$

where $a \in \mathbf{R} \setminus \{0\}$ is fixed. We write $X \perp_I Y$ (see James [107]).

(ii) a-Pythagorean sense if

$$||x - ay||^2 = ||x||^2 + a^2 ||y||^2, \forall x \in X, \forall y \in Y,$$

where $a \in \mathbf{R} \setminus \{0\}$ is fixed. We write $X \perp_P Y$ (see James [107]).

(iii) Birkhoff sense if

$$||x|| \leq ||x + \lambda y||, \forall \lambda \in \mathbf{R}, \forall x \in X, \forall y \in Y.$$

We write $X \perp_B Y$ (see Birkhoff [39]).

(iv) Diminnie-Freese-Andalafte sense if

$$(1+a^2)\|x+y\|^2 = \|ax+y\|^2 + \|x+ay\|^2, \forall x \in X, \forall y \in Y,$$

where $a \in \mathbf{R}\setminus\{1\}$ is fixed. We write $X \perp_{DFA} Y$ (see Diminnie-Freese-Andalafte [63]).

(v) Kapoor-Prasad sense if

$$\|ax+by\|^2 + \|x+y\|^2 = \|ax+y\|^2 + \|x+by\|^2, \forall x \in X, \forall y \in Y,$$

where $a, b \in (0,1)$ are fixed. We write $X \perp_{KP} Y$ (see Kapoor-Prasad [113]).

(vi) Singer sense if $\|x\| \cdot \|y\| = 0$ or

$$\left\| \frac{x}{\|x\|} + \frac{y}{\|y\|} \right\| = \left\| \frac{x}{\|x\|} - \frac{y}{\|y\|} \right\|, \forall x \in X, \forall y \in Y.$$

We write $X \perp_S Y$ (see Singer [197]).

(vii) usual sense if

$$\langle x, y \rangle = 0, \forall x \in X, \forall y \in Y,$$

where, in addition, the norm $\|\cdot\|$ is generated by the inner product $\langle \cdot, \cdot \rangle$. We write $X \perp Y$.

Remark. There are other kinds of orthogonality too, see *e.g.* the papers of Alonso-Benitez [1], [2], Alonso-Soriano [3], where relationships between them also are proved.

If X and Y are not orthogonal (in a given sense), it is natural the question to find out a quantity that " measures" the deviation of X, Y from orthogonality.

We answer this question for each concept in Definition 6.1, introducing and studying the properties of such called defects of orthogonality. Also, some applications to Fourier series with respect to non-orthogonal systems and to approximation theory are given at the end of this section.

Corresponding to the concepts of orthogonality in Definition 6.1, we introduce the following

Definition 6.2 Let $(E, \|\cdot\|)$ be a real normed space and $X, Y \subseteq E$. We call defect of orthogonality of X with respect to Y, of:

(i) a-isosceles kind, $a \in \mathbf{R} \setminus \{0\}$ fixed, the quantity

$$d_I^{\perp}(X, Y; a) = \sup_{x \in X, y \in Y} \left| \|x - ay\|^2 - \|x + ay\|^2 \right|.$$

(ii) a-Pythagorean kind, $a \in \mathbf{R} \setminus \{0\}$ fixed, the quantity

$$d_P^{\perp}(X, Y; a) = \sup_{x \in X, y \in Y} \left| \|x\|^2 + a^2 \|y\|^2 - \|x - ay\|^2 \right|.$$

(iii) Birkhoff kind, the quantity

$$d_B^{\perp}(X, Y) = \sup_{x \in X, y \in Y} \sup_{\lambda \in \mathbf{R}} \left\{ \|x\|^2 - \|x + \lambda y\|^2 \right\}.$$

(iv) Diminnie-Freese-Andalafte kind, the quantity

$$d_{DFA}^{\perp}(X, Y; a) = \sup_{x \in X, y \in Y} \left| \left(1 + a^2\right) \|x + y\|^2 - \left(\|ax + y\|^2 + \|x + ay\|^2 \right) \right|,$$

where $a \in \mathbf{R} \setminus \{1\}$ is fixed.

(v) Kapoor-Prasad kind, the quantity

$$d_{KP}^{\perp}(X, Y; a, b) = \sup_{x \in X, y \in Y} \left| \|ax + by\|^2 + \|x + y\|^2 \right.$$

$$\left. - \left(\|ax + y\|^2 + \|x + by\|^2 \right) \right|,$$

where $a, b \in (0, 1)$ are fixed.

(vi) Singer kind, the quantity $d_S^{\perp}(X, Y)$ defined by $d_S^{\perp}(X, Y) = 0$ if $X = \{0\}$ or $Y = \{0\}$ and

$$d_S^{\perp}(X, Y) = \sup_{x \in X \setminus \{0\}, y \in Y \setminus \{0\}} \left| \left\| \frac{x}{\|x\|} + \frac{y}{\|y\|} \right\|^2 - \left\| \frac{x}{\|x\|} - \frac{y}{\|y\|} \right\|^2 \right|,$$

contrariwise.

(vii) usual kind

$$d^{\perp}(X, Y) = \sup_{x \in X, y \in Y} |\langle x, y \rangle|,$$

if, in addition, the norm $\|\cdot\|$ is generated by the inner product $\langle \cdot, \cdot \rangle$ on E.

By the Definitions 6.1 and 6.2, it is immediate the following

Theorem 6.1 *$X \perp_* Y$ if and only if $d_*^{\perp}(X, Y) = 0$, where $*$ represents any kind of orthogonality in Definition 6.1.*

Because it is well-known that in an inner product space, the orthogonalities $(i)-(vi)$ one reduce to (vii) in Definition 6.1, it is natural to see what happens with the quantities in Definition 6.2 for the case of inner product spaces. In this sense we present

Theorem 6.2 *Let $(E, \langle \cdot, \cdot \rangle)$ be a real inner product space endowed with the norm $\|x\| = \sqrt{\langle x, x \rangle}, x \in E$. We have:*

(i) $d_I^{\perp}(X, Y; a) = 4|a| \cdot d^{\perp}(X, Y)$.

(ii) $d_P^{\perp}(X, Y; a) = 2|a| \cdot d^{\perp}(X, Y)$.

(iii) $d_B^{\perp}(X, Y) = 0$ *if* $Y = \{0\}$ *and* $d_B^{\perp}(X, Y) = \sup_{x \in X, y \in Y \setminus \{0\}} \frac{\langle x, y \rangle^2}{\langle y, y \rangle}$ $\leq \sup_{x \in X} \|x\|^2 \leq (diam(X))^2$ *if* $Y \neq \{0\}$.

(iv) $d_{DFA}^{\perp}(X, Y; a) = 2(a-1)^2 \cdot d^{\perp}(X, Y)$.

(v) $d_{KP}^{\perp}(X, Y; a, b) = 2(1-a)(1-b) \cdot d^{\perp}(X, Y)$.

(vi) $d_S^{\perp}(X, Y) = 0$ *if* $X = \{0\}$ *or* $Y = \{0\}$ *and* $d_S^{\perp}(X, Y) = d^{\perp}(X', Y')$, *where* $X' = P(X), Y' = P(Y), P(x) = \frac{x}{\|x\|}$ *(i. e. X' and Y' are the projections of X and Y on the unit sphere $\{u \in E; \|u\| = 1\}$).*

Proof. (i) We have $\langle u, u \rangle = \|u\|^2, \forall u \in E$. We get

$$\left| \|x - ay\|^2 - \|x + ay\|^2 \right| = |-4a \cdot \langle x, y \rangle| = 4|a| \cdot |\langle x, y \rangle|,$$

which proves $d_I^{\perp}(X, Y; a) = 4|a| \cdot d^{\perp}(X, Y)$.

(ii) The equality

$$\left| \|x\|^2 + a^2 \|y\|^2 - \|x - ay\|^2 \right| = |2a \cdot \langle x, y \rangle| = 2|a| \cdot |\langle x, y \rangle|$$

proves $d_P^{\perp}(X, Y; a) = 2|a| \cdot d^{\perp}(X, Y)$.

(iii) We have

$$\|x\|^2 - \|x + \lambda y\|^2 = -\lambda^2 \cdot \langle y, y \rangle - 2\lambda \cdot \langle x, y \rangle,$$

for all $\lambda \in \mathbf{R}$. For fixed $x \in X, y \in Y$, let us denote $f(\lambda) = -\lambda^2 \cdot \langle y, y \rangle - 2\lambda \cdot \langle x, y \rangle, \lambda \in \mathbf{R}$. If $y = 0$ then $f(\lambda) = 0, \forall \lambda \in \mathbf{R}$. Let $y \in Y \setminus \{0\}$. The function $f(\lambda)$ is concave on $\mathbf{R}$ and consequently has its maximum value for $\lambda = -\frac{\langle x, y \rangle}{\langle y, y \rangle}$, that is $f\left(-\frac{\langle x, y \rangle}{\langle y, y \rangle}\right) = \frac{\langle x, y \rangle^2}{\langle y, y \rangle}$, which proves the desired relations (by using the Cauchy-Schwarz inequality too).

(iv) The equalities

$$\left| (1 + a^2) \|x + y\|^2 - \left(\|ax + y\|^2 + \|x + ay\|^2 \right) \right|$$

$$= \left|\langle x,y\rangle \cdot \left(2+2a^2-4a\right)\right| = 2\,(a-1)^2 \cdot |\langle x,y\rangle|$$

imply $d^{\perp}_{DFA}(X,Y;a) = 2\,(a-1)^2 \cdot d^{\perp}(X,Y)$.

(v) Because

$$\begin{aligned} & \left| \|ax+by\|^2 + \|x+y\|^2 - \left(\|ax+y\|^2 + \|x+by\|^2\right)\right| \\ = \; & |2 \cdot \langle x,y\rangle (ab+1-a-b)| = 2\,(1-a)\,(1-b) \cdot |\langle x,y\rangle| \end{aligned}$$

we obtain $d^{\perp}_{KP}(X,Y;a,b) = 2\,(1-a)\,(1-b) \cdot d^{\perp}(X,Y)$.

(vi) Simple calculations show

$$\left| \left\| \frac{x}{\|x\|} + \frac{y}{\|y\|} \right\|^2 - \left\| \frac{x}{\|x\|} - \frac{y}{\|y\|} \right\|^2 \right| = 4\left\langle \frac{x}{\|x\|}, \frac{y}{\|y\|} \right\rangle = 4\frac{\langle x,y\rangle}{\|x\| \cdot \|y\|} \leq 4,$$

which implies

$$d^{\perp}_{S}(X,Y) = 4\sup\left\{ \frac{|\langle x,y\rangle|}{\|x\| \cdot \|y\|}; x \in X\setminus\{0\}, y \in Y\setminus\{0\} \right\} = 4 \cdot d^{\perp}(X',Y'),$$

where X' and Y' are the projections of X and Y on the unit sphere $\{u \in X; \|u\| = 1\}$ (i. e. $X' = P(X), Y' = P(Y)$, where $P(x) = \frac{x}{\|x\|}$), respectively. The theorem is proved. □

Remark. Theorem 6.2 shows, among others, that in an inner product space, $d^{\perp}_{I}(X,Y;a) = 2d^{\perp}_{P}(X,Y;a)$. But in the general case when $(E, \|\cdot\|)$ is real normed space, between $d^{\perp}_{P}$ and $d^{\perp}_{I}$ there is not such of connection, as the following examples show.

Let us consider $E = C[0,1] = \{f : [0,1] \to \mathbf{R}; f \text{ continuous on } [0,1]\}$ endowed with the uniform norm $\|f\| = \max\{|f(x)|; x \in [0,1]\}$. It is well-known that this norm cannot derived from an inner product. Denoting $C^1[0,1] = \{f \in C[0,1]; f' \in C[0,1]\}$ let us take $m > 0$ a constant and

$$\begin{aligned} X &= \left\{f \in C^1[0,1]; f(0) = 0, f(1) = 2, f'(x) \geq m > 0, \forall x \in [0,1]\right\}, \\ Y &= \left\{g \in C^1[0,1]; g(0) = 1, g(1) = 0, -m \leq g'(x) \leq 0, \forall x \in [0,1]\right\}, \\ U &= \{f \in C[0,1]; f(0) = 0, f(1) = 1, f(x) \geq 0, f \text{ increasing on } [0,1]\}, \\ V &= \{g \in C[0,1]; g(0) = 1, g(1) = 0, g(x) \geq 0, g \text{ decreasing on } [0,1]\}. \end{aligned}$$

We have $d^{\perp}_{I}(X,Y;1) = 0, d^{\perp}_{P}(X,Y;1) = 1$ and $d^{\perp}_{I}(U,V;1) > 1$, $d^{\perp}_{P}(U,V;1) = 1$. Indeed, let $f \in X, g \in Y$. By hypothesis it follows that $f-g$ and $f+g$ are non-decreasing on $[0,1]$, which implies $\|f+g\|^2 =$

$(f(1)+g(1))^2 = 4, \|f-g\|^2 = (f(1)-g(1))^2 = 4$ because $(f-g)(0) = -1$ and $(f-g)(1) = 2$. Consequently,

$$\left|\|f-g\|^2 - \|f+g\|^2\right| = 0,$$

and $d_I^{\perp}(X, Y; 1) = 0$. Also, it is immediate

$$\left|\|f\|^2 + \|g\|^2 - \|f-g\|^2\right| = |4+1-4| = 1,$$

which implies $d_P^{\perp}(X, Y; 1) = 1$. On the other hand, let $f \in U, g \in V$. We have $\|f\|^2 = \|g\|^2 = 1, \|f-g\|^2 = 1$, which implies $d_P^{\perp}(U, V; 1) = 1$. Then because

$$\left|\|f-g\|^2 - \|f+g\|^2\right| = \left|1 - \|f+g\|^2\right|,$$

by choosing $f(x) = x, \forall x \in [0,1]$ and $g(x) = -\frac{1}{3}x + 1$ if $x \in \left[0, \frac{3}{4}\right], g(x) = -3x + 3$ if $x \in \left[\frac{3}{4}, 1\right]$, obviously

$$\|f+g\| = \sup\{|f(x) + g(x)|; x \in [0,1]\} \geq f\left(\frac{3}{4}\right) + g\left(\frac{3}{4}\right) = \frac{3}{2} = 1.5.$$

Therefore $\|f+g\|^2 \geq (1.5)^2 = 2.25$ and $\left|1 - \|f+g\|^2\right| \geq 1.25$, which immediately implies $d_I^{\perp}(U, V; 1) \geq 1.25 > 1$.

The properties of defects in Theorem 6.2 can be used to characterize the prehilbertian spaces. Thus, we have

Theorem 6.3 *Let $(E, \|\cdot\|)$ be a real normed space. E is an inner product space if and only if for all $X, Y \subset E$ we have*

$$d_I^{\perp}(X, Y; 1) = 2d_P^{\perp}(X, Y; 1)$$

and

$$d_I^{\perp}(X, Y; -1) = 2d_P^{\perp}(X, Y; -1).$$

Proof. Firstly, let us suppose that E is an inner product space. Then, by Theorem 6.2, $(i), (ii)$, we get

$$d_I^{\perp}(X, Y; a) = 2d_P^{\perp}(X, Y; a), \forall X, Y \subset E, a \in \mathbf{R} \setminus \{0\}.$$

Choosing $a = \pm 1$, we obtain the desired condition.

Conversely, let us suppose that for all $X, Y \subset E$, we have

$$d_I^{\perp}(X,Y;1) = 2d_P^{\perp}(X,Y;1)$$

and

$$d_I^{\perp}(X,Y;-1) = 2d_P^{\perp}(X,Y;-1).$$

Take $X = \{x\}, Y = \{y\}$ with arbitrary $x, y \in E$. We have two possibilities:

Case 1. $\|x+y\| \geq \|x-y\|$;

Case 2. $\|x+y\| < \|x-y\|$.

In the case 1, we get

$$\begin{aligned}
2\left(\|x\|^2+\|y\|^2\right) &= 2\left(\|x\|^2+\|y\|^2-\|x-y\|^2\right)+2\|x-y\|^2 \\
&\leq 2d_P^{\perp}(X,Y;1)+2\|x-y\|^2 = d_I^{\perp}(X,Y;1)+2\|x-y\|^2 \\
&= \left|\|x-y\|^2-\|x+y\|^2\right|+2\|x-y\|^2 = \|x+y\|^2 \\
&\quad -\|x-y\|^2+2\|x-y\|^2 = \|x+y\|^2+\|x-y\|^2.
\end{aligned}$$

In the case 2, we get

$$\begin{aligned}
2\left(\|x\|^2+\|y\|^2\right) &= 2\left(\|x\|^2+\|y\|^2-\|x+y\|^2\right)+2\|x+y\|^2 \\
&\leq 2d_P^{\perp}(X,Y;-1)+2\|x+y\|^2 = d_I^{\perp}(X,Y;-1)+2\|x+y\|^2 \\
&= \left|\|x+y\|^2-\|x-y\|^2\right|+2\|x+y\|^2 = \|x-y\|^2 \\
&\quad -\|x+y\|^2+2\|x+y\|^2 = \|x-y\|^2+\|x+y\|^2.
\end{aligned}$$

As a conclusion

$$2\left(\|x\|^2+\|y\|^2\right) \leq \|x+y\|^2+\|x-y\|^2, \forall x,y \in E.$$

According to Schoenberg [190], it follows that E is inner product space, which proves the theorem. □

Remark. From the proof of Theorem 6.3 it easily follows that $(E, \|\cdot\|)$ is an inner product space if and only if, for all $X, Y \subset E$, we have

$$2d_P^{\perp}(X,Y;1) \leq d_I^{\perp}(X,Y;1)$$

and

$$2d_P^{\perp}(X,Y;-1) \leq d_I^{\perp}(X,Y;-1).$$

Concerning the defect of orthogonality of Cartesian product we present

Theorem 6.4 *Let $(E_1, \|\cdot\|_1), (E_2, \|\cdot\|_2)$ be two real normed linear spaces and let us introduce on $E_1 \times E_2$ the norm $\|(u_1,u_2)\| = \sqrt{\|u_1\|_1^2 + \|u_2\|_2^2}$, $\forall (u_1,u_2) \in E_1 \times E_2$. For all $X_1, Y_1 \subseteq E_1, X_2, Y_2 \subseteq E_2$, we have*

$$d_*^{\perp}(X_1 \times X_2, Y_1 \times Y_2) \leq d_*^{\perp}(X_1, Y_1) + d_*^{\perp}(X_2, Y_2),$$

for any kind " $*$ " of orthogonality in Definition 6.1, excepting Singer orthogonality.

Proof. By using the simple properties $|x+y| \leq |x| + |y|, \forall x, y \in \mathbf{R}$ and $\sup\{a_i + b_i; i \in I\} \leq \sup\{a_i; i \in I\} + \sup\{b_i; i \in I\}$ we obtain the inequalities. For example, if $d_*^{\perp}$ is the defect of orthogonality of a-isosceles kind then we have

$$d_I^{\perp}(X_1 \times X_2, Y_1 \times Y_2; a) = \sup_{x \in X_1 \times X_2, y \in Y_1 \times Y_2} \left| \|x - ay\|^2 - \|x + ay\|^2 \right|$$

$$= \sup_{x_i \in X_i, y_i \in Y_i, i \in \{1,2\}} \left| \|(x_1 - ay_1, x_2 - ay_2)\|^2 - \|(x_1 + ay_1, x_2 + ay_2)\|^2 \right|$$

$$= \sup_{x_i \in X_i, y_i \in Y_i, i \in \{1,2\}} \left| \|x_1 - ay_1\|_1^2 + \|x_2 - ay_2\|_2^2 \right.$$

$$\left. - \|x_1 + ay_1\|_1^2 - \|x_2 + ay_2\|_2^2 \right|$$

$$\leq \sup_{x_1 \in X_1, y_1 \in Y_1} \left| \|x_1 - ay_1\|_1^2 - \|x_1 + ay_1\|_1^2 \right|$$

$$+ \sup_{x_2 \in X_2, y_2 \in Y_2} \left| \|x_2 - ay_2\|_2^2 - \|x_2 + ay_2\|_2^2 \right|$$

$$= d_I^{\perp}(X_1, Y_1; a) + d_I^{\perp}(X_2, Y_2; a).$$

The proofs for the other kinds of orthogonality are similar. □

Remark. In the above theorem, the result for usual kind of orthogonality is obtained if we consider the inner product on $E_1 \times E_2$ defined by $\langle (u_1, u_2), (v_1, v_2) \rangle = \langle u_1, v_1 \rangle_1 + \langle u_2, v_2 \rangle_2$, where $\langle \cdot, \cdot \rangle_1$ and $\langle \cdot, \cdot \rangle_2$ represent the inner products on E_1 and E_2, respectively.

In what follows, we present other properties of the defects of orthogonalities.

Theorem 6.5 *Let $(E, \|\cdot\|)$ be a real normed space and $X, Y \subseteq E$. We have:*

(*i*) $d_I^\perp (Y, X; a) = a^2 d_I^\perp \left(X, Y; \frac{1}{a}\right), \forall a \in \mathbf{R} \setminus \{0\}$. *Also, if $X = -X$ and $Y = -Y$ (i.e. X and Y are symmetric with respect to 0) then* $d_I^\perp (X, Y; a) = \frac{1}{a^2} d_I^\perp \left(X^{\perp_I}, Y^{\perp_I}; a\right), \forall a \in \mathbf{R} \setminus \{0\}$, *where* $X^{\perp_I} = \{u \in E; u \perp_I x = 0, \forall x \in X\}$.

(*ii*) $d_P^\perp (Y, X; a) = a^2 d_P^\perp \left(X, Y; \frac{1}{a}\right), \forall a \in \mathbf{R} \setminus \{0\}$.

(*iii*) $d_B^\perp (\alpha X, \beta Y) = \alpha^2 d_B^\perp (X, Y), \forall \alpha, \beta \in \mathbf{R}$.

(*iv*) $d_{DFA}^\perp (Y, X; a) = d_{DFA}^\perp (X, Y; a), \forall a \in \mathbf{R} \setminus \{1\}$.

(*v*) $d_{KP}^\perp (Y, X; a, b) = d_{KP}^\perp (X, Y; b, a), \forall a, b \in (0, 1)$.

(*vi*) $d_S^\perp (X, Y) = d_S^\perp (Y, X)$ *and* $d_S^\perp (\alpha X, \beta Y) = d_S^\perp (X, Y), \forall \alpha, \beta \in \mathbf{R} \setminus \{0\}$. *Also, if $X = -X \neq \{0\}, Y = -Y \neq \{0\}$ and $X^{\perp_S}, Y^{\perp_S} \neq \{0\}$ then we have* $d_S^\perp (X, Y) = d_S^\perp \left(X^{\perp_S}, Y^{\perp_S}\right)$.

(*vii*) $d^\perp (X, Y) = d^\perp (Y, X)$ *and* $d^\perp (\alpha X, \beta Y) = |\alpha| \cdot |\beta| \cdot d^\perp (X, Y)$.

Proof. (*i*) The equality

$$\sup_{y \in Y, x \in X} \left| \|y - ax\|^2 - \|y + ax\|^2 \right| = a^2 \sup_{x \in X, y \in Y} \left| \left\| x - \frac{1}{a} y \right\|^2 - \left\| x + \frac{1}{a} y \right\|^2 \right|$$

proves $d_I^\perp (Y, X; a) = a^2 d_I^\perp \left(X, Y; \frac{1}{a}\right)$.

Let us denote

$$\begin{aligned} V_1 &= \left| \|x - ay\|^2 - \|x + ay\|^2 \right|, x \in X, y \in Y, \\ V_2 &= \left| \|x' - ay'\|^2 - \|x' + ay'\|^2 \right|, x' \in X^{\perp_I}, y' \in Y^{\perp_I}. \end{aligned}$$

By hypothesis we have $\|x - ax'\| = \|x + ax'\|$ and $\|y - ay'\| = \|y + ay'\|$. Now let us denote $x + ax' = u, x - ax' = v, y + ay' = s, y - ay' = t$. We get $x = \frac{1}{2}(u + v), y = \frac{1}{2}(s + t), x' = \frac{1}{2a}(u - v), y' = \frac{1}{2a}(s - t)$,

$$x - ay = \frac{1}{2}(u + v - as - at),$$

$$\begin{aligned}
x + ay &= \frac{1}{2}(u + v + as + at), \\
x' - ay' &= \frac{1}{2a}(u - v) - \frac{1}{2}(s - t) = \frac{1}{2a}(u - v - as + at), \\
x' + ay' &= \frac{1}{2a}(u - v) + \frac{1}{2}(s - t) = \frac{1}{2a}(u - v + as - at), \\
V_1 &= \frac{1}{4}\left| \|u + v - as - at\|^2 - \|u + v + as + at\|^2 \right|, \\
V_2 &= \frac{1}{4a^2}\left| \|u - v - as + at\|^2 - \|u - v + as - at\|^2 \right|
\end{aligned}$$

and

$$d_I^{\perp}(X, Y; a) = \sup_{\|u\|=\|v\|,\|s\|=\|t\|} \{V_1; u \in X + aX^{\perp_I},$$

$$v \in X - aX^{\perp_I}, s \in Y + aY^{\perp_I}, t \in Y - aY^{\perp_I}\},$$

$$d_I^{\perp}(X^{\perp_I}, Y^{\perp_I}; a) = \sup_{\|u\|=\|v\|,\|s\|=\|t\|} \{V_2; u \in X + aX^{\perp_I},$$

$$v \in X - aX^{\perp_I}, s \in Y + aY^{\perp_I}, t \in Y - aY^{\perp_I}\}.$$

By hypothesis it follows that $X^{\perp_I}, Y^{\perp_I}$ and $X + aX^{\perp_I}, X - aX^{\perp_I}, Y + aY^{\perp_I}, Y - aY^{\perp_I}$ are symmetric with respect to 0, which immediately implies the equality.

(ii)

$$\begin{aligned}
d_P^{\perp}(Y, X; a) &= \sup_{y \in Y, x \in X} \left| \|y\|^2 + a^2\|x\|^2 - \|y - ax\|^2 \right| \\
&= a^2 \sup_{y \in Y, x \in X} \left| \frac{1}{a^2}\|y\|^2 + \|x\|^2 - \left\| \frac{1}{a}y - x \right\|^2 \right| \\
&= a^2 \sup_{x \in X, y \in Y} \left| \|x\|^2 + \frac{1}{a^2}\|y\|^2 - \left\| x - \frac{1}{a}y \right\|^2 \right| \\
&= a^2 d_P^{\perp}\left(X, Y; \frac{1}{a}\right).
\end{aligned}$$

(iii)

$$d_B^{\perp}(\alpha X, \beta Y) = \sup_{x' \in \alpha X, y' \in \alpha Y} \sup_{\lambda \in \mathbf{R}} \left(\|x'\|^2 - \|x' + \lambda y'\|^2 \right)$$

$$\begin{aligned}
&= \sup_{x\in X, y\in Y} \sup_{\lambda\in\mathbf{R}} \left(\|\alpha x\|^2 - \|(\alpha x) + \lambda(\alpha y)\|^2\right) \\
&= \alpha^2 \sup_{x\in X, y\in Y} \sup_{\lambda\in\mathbf{R}} \left(\|x\|^2 - \|x+\lambda y\|^2\right) \\
&= \alpha^2 d_B^{\perp}(X,Y).
\end{aligned}$$

(iv) It is a consequence of the symmetry of expression

$$\left(1+a^2\right)\|x+y\|^2 - \left(\|ax+y\|^2 + \|x+ay\|^2\right),$$

$\forall a \in \mathbf{R}\setminus\{1\}$.

(v) Because

$$\begin{aligned}
&\|bx+ay\|^2 + \|x+y\|^2 - \left(\|bx+y\|^2 + \|x+ay\|^2\right) \\
= \ &\|ay+bx\|^2 + \|y+x\|^2 - \left(\|ay+x\|^2 + \|y+bx\|^2\right), \forall x\in X, y\in Y,
\end{aligned}$$

we obtain the equality.

(vi) The symmetry of the expression $\left\|\frac{x}{\|x\|}+\frac{y}{\|y\|}\right\|^2 - \left\|\frac{x}{\|x\|}-\frac{y}{\|y\|}\right\|^2$ implies the first equality.

If $X=\{0\}$ or $Y=\{0\}$ then $\alpha X=\{0\}$ or $\beta Y=\{0\}$, therefore

$$d_S^{\perp}(\alpha X,\beta Y) = d_S^{\perp}(X,Y) = d_S^{\perp}(Y,X) = 0.$$

If $X\neq\{0\}$ and $Y\neq\{0\}$ then

$$d_S^{\perp}(\alpha X,\beta Y) = \sup_{x'\in\alpha X\setminus\{0\}, y'\in\beta Y\setminus\{0\}} \left| \left\|\frac{x'}{\|x'\|}+\frac{y'}{\|y'\|}\right\|^2 - \left\|\frac{x'}{\|x'\|}-\frac{y'}{\|y'\|}\right\|^2 \right|$$

$$\begin{aligned}
&= \sup_{x\in X\setminus\{0\}, y\in Y\setminus\{0\}} \left| \left\|\frac{\alpha}{|\alpha|}\frac{x}{\|x\|}+\frac{\beta}{|\beta|}\frac{y}{\|y\|}\right\|^2 - \left\|\frac{\alpha}{|\alpha|}\frac{x}{\|x\|}-\frac{\beta}{|\beta|}\frac{y}{\|y\|}\right\|^2 \right| \\
&= \begin{cases} \sup_{x\in X\setminus\{0\}, y\in Y\setminus\{0\}} \left| \left\|\frac{x}{\|x\|}+\frac{y}{\|y\|}\right\|^2 - \left\|\frac{x}{\|x\|}-\frac{y}{\|y\|}\right\|^2 \right|, & \text{if } \alpha\beta>0 \\ \sup_{x\in X\setminus\{0\}, y\in Y\setminus\{0\}} \left| \left\|\frac{x}{\|x\|}-\frac{y}{\|y\|}\right\|^2 - \left\|\frac{x}{\|x\|}+\frac{y}{\|y\|}\right\|^2 \right|, & \text{if } \alpha\beta<0 \end{cases} \\
&= d_S^{\perp}(X,Y).
\end{aligned}$$

Now, we prove the second assertion. Let us denote

$$V_1 = \left| \left\| \frac{x}{\|x\|} + \frac{y}{\|y\|} \right\|^2 - \left\| \frac{x}{\|x\|} - \frac{y}{\|y\|} \right\|^2 \right|, x \in X, y \in Y,$$

$$V_2 = \left| \left\| \frac{x'}{\|x'\|} + \frac{y'}{\|y'\|} \right\|^2 - \left\| \frac{x'}{\|x'\|} - \frac{y'}{\|y'\|} \right\|^2 \right|, x' \in X^{\perp_S}, y' \in Y^{\perp_S}.$$

By hypothesis we have $\left\| \frac{x}{\|x\|} + \frac{x'}{\|x'\|} \right\| = \left\| \frac{x}{\|x\|} - \frac{x'}{\|x'\|} \right\|$ and $\left\| \frac{y}{\|y\|} + \frac{y'}{\|y'\|} \right\| = \left\| \frac{y}{\|y\|} - \frac{y'}{\|y'\|} \right\|$. Denoting $\frac{x}{\|x\|} + \frac{x'}{\|x'\|} = u$, $\frac{x}{\|x\|} - \frac{x'}{\|x'\|} = v$, $\frac{y}{\|y\|} + \frac{y'}{\|y'\|} = s$, $\frac{y}{\|y\|} - \frac{y'}{\|y'\|} = t$,we easily get $\frac{x}{\|x\|} = \frac{1}{2}(u+v)$, $\frac{y}{\|y\|} = \frac{1}{2}(s+t)$, $\frac{x'}{\|x'\|} = \frac{1}{2}(u-v)$, $\frac{y'}{\|y'\|} = \frac{1}{2}(s-t)$,

$$V_1 = \frac{1}{4} \left| \|u+v+s+t\|^2 - \|u+v-s-t\|^2 \right|,$$
$$V_2 = \frac{1}{4} \left| \|u-v+s-t\|^2 - \|u-v-s+t\|^2 \right|,$$

with $\|u\| = \|v\|, \|s\| = \|t\|$. We define $P(x) = \frac{x}{\|x\|}$, $x \in X$ and we denote $P(X) = \{P(x); x \in X\}$. By hypothesis we get that $X^{\perp_S}, P(X), U(X) = P(X) + P(X^{\perp_S}), V(X) = P(X) - P(X^{\perp_S}), Y^{\perp_S}, P(Y), P(Y^{\perp_S})$, $U(Y) = P(Y) + P(Y^{\perp_S})$ and $V(Y) = P(Y) - P(Y^{\perp_S})$ are symmetric with respect to 0, which implies

$$\begin{aligned} d_S^{\perp}(X,Y) &= \sup_{\|u\|=\|v\|,\|s\|=\|t\|} \left\{ \left| \|u+v+s+t\|^2 - \|u+v-s-t\|^2 \right| ; \right. \\ &\quad \left. u \in U(X), v \in V(X), s \in U(Y), t \in V(Y) \right\} \\ &= \sup_{\|u\|=\|v\|,\|s\|=\|t\|} \left\{ \left| \|u-v+s-t\|^2 - \|u-v-s+t\|^2 \right| ; \right. \\ &\quad \left. u \in U(X), v \in V(X), s \in U(Y), t \in V(Y) \right\} \\ &= d_S^{\perp}\left(X^{\perp_S}, Y^{\perp_S}\right). \end{aligned}$$

(*vii*) The property of the inner product $\langle x, y \rangle = \langle y, x \rangle, \forall x \in X, y \in Y$ implies the first equality. For the second one, we have

$$\begin{aligned} d^{\perp}(\alpha X, \beta Y) &= \sup_{x' \in \alpha X, y' \in \beta Y} |\langle x', y' \rangle| = \sup_{x \in X, y \in Y} |\langle \alpha x, \beta y \rangle| \\ &= \sup_{x \in X, y \in Y} |\alpha| |\beta| \cdot |\langle x, y \rangle| = |\alpha| |\beta| \cdot d^{\perp}(X,Y). \end{aligned}$$

□

Remark. If $a = 1$ then $d_I^\perp(X,Y;1) = d_I^\perp(X^{\perp_I}, Y^{\perp_I}; 1)$, that is the defect of isosceles orthogonality for $a = 1$ is invariant with respect to the corresponding orthogonality.

Theorem 6.6 *If two real normed spaces $(E_1, \|\cdot\|_1)$ and $(E_2, \|\cdot\|_2)$ are isomorphic (i.e. there exists a linear bijective mapping $f : E_1 \to E_2$ such that $\|f(x)\|_2 = \|x\|_1, \forall x \in E_1$) then*

$$d_*^\perp(f(X), f(Y)) = d_*^\perp(X,Y), \forall X, Y \subseteq E_1,$$

for any kind "$*$" of orthogonality in Definition 6.1, excepting usual orthogonality. If $(E_1, \langle\cdot,\cdot\rangle_1)$ and $(E_2, \langle\cdot,\cdot\rangle_2)$ are isomorphic real inner product spaces (i.e. there exists a linear mapping $f : E_1 \to E_2$ such that $\langle f(x), f(y)\rangle_2 = \langle x, y\rangle_1, \forall x, y \in E_1$) then the above equality is also true for the usual kind of orthogonality.

Proof. We present the proof for Kapoor-Prasad kind of defect of orthogonality:

$$\begin{aligned}
& d_{KP}^\perp(f(X), f(Y); a, b) \\
&= \sup_{x' \in f(X), y' \in f(Y)} \Big| \|ax' + by'\|_2^2 + \|x' + y'\|_2^2 \\
&\qquad - \left(\|ax' + y'\|_2^2 + \|x' + by'\|_2^2 \right) \Big| \\
&= \sup_{x \in X, y \in Y} \Big| \|af(x) + bf(y)\|_2^2 + \|f(x) + f(y)\|_2^2 \\
&\qquad - \left(\|af(x) + f(y)\|_2^2 + \|f(x) + bf(y)\|_2^2 \right) \Big| \\
&= \sup_{x \in X, y \in Y} \Big| \|ax + by\|_1^2 + \|x + y\|_1^2 - \left(\|ax + y\|_1^2 + \|x + by\|_1^2 \right) \Big| \\
&= d_{KP}^\perp(X, Y; a, b).
\end{aligned}$$

For a-isosceles, a-Pythagorean, Birkhoff, Diminnie-Freese-Andalafte and Singer kind defect of orthogonality the proof is similar.

For the usual kind of orthogonality we have

$$\begin{aligned}
d^\perp(f(X), f(Y)) &= \sup\{|\langle x', y'\rangle_2| ; x' \in f(X), y' \in f(Y)\} \\
&= \sup\{|\langle x, y\rangle_1| ; x \in X, y \in Y\} = d^\perp(X,Y).
\end{aligned}$$

□

In the final part of this section we present some applications to Fourier series with respect to non-orthogonal systems and to approximation theory. An useful concept is the following.

Definition 6.3 Let $(E, \langle \cdot, \cdot \rangle)$ be a real inner product space and $X \subseteq E$. We call defect of orthogonality of X, the quantity

$$d^{\perp}(X) = \sup \{|\langle x, y \rangle| ; x, y \in X, x \neq y\}.$$

Remark. Obviously X is orthogonal system in E if and only if $d^{\perp}(X) = 0$.

We present

Theorem 6.7 *Let $(E, \langle \cdot, \cdot \rangle)$ be a real inner space and $X, Y \subset E$ with $d^{\perp}(X) = d^{\perp}(Y) = 0, \langle x, x \rangle = 1, \forall x \in X, \langle y, y \rangle = 1, \forall y \in Y$. We have*

$$d^{\perp}(X + Y) \leq 1 + 2d^{\perp}(X, Y).$$

Proof. For $x, x' \in X, y, y' \in Y, x + y \neq x' + y'$, we get

$$\begin{aligned} |\langle x + y, x' + y' \rangle| &= |\langle x, x' \rangle + \langle x, y' \rangle + \langle x', y \rangle + \langle y, y' \rangle| \\ &\leq |\langle x, x' \rangle| + |\langle x, y' \rangle| + |\langle x', y \rangle| + |\langle y, y' \rangle|. \end{aligned}$$

We have three possibilities:
Case 1. $x \neq x', y = y'$, which implies $|\langle x + y, x' + y' \rangle| \leq 1 + 2d^{\perp}(X, Y)$;
Case 2. $x = x', y \neq y'$, which implies $|\langle x + y, x' + y' \rangle| \leq 1 + 2d^{\perp}(X, Y)$;
Case 3. $x \neq x', y \neq y'$, which implies $|\langle x + y, x' + y' \rangle| \leq 2d^{\perp}(X, Y)$;

As a conclusion, in all the cases we get $|\langle x + y, x' + y' \rangle| \leq 1 + 2d^{\perp}(X, Y)$ and passing to supremum we obtain the conclusion of theorem. □

Now, let us suppose that X is at most countable, i. e. $X = \{x_1, ..., x_n, ...\}$ is finite or countable. For a given $x \in E \backslash X$, we can consider the Fourier coefficients with respect to X, that is $c_k = \langle x, x_k \rangle, k \in \{1, 2, ...\}$ and the Fourier sums with respect to X, $s_n = \sum_{k=1}^{n} c_k x_k, n \in \{1, 2, ...\}$.

If $d^{\perp}(X) = 0$ then the theory of the above Fourier sums is well-known. Then, it is natural to ask what happens if $d^{\perp}(X) > 0$. An answer is given by

Theorem 6.8 *Let $(E, \langle \cdot, \cdot \rangle)$ be real inner product space and $X = \{x_1, ..., x_n\} \subset E$ with $\langle x_k, x_k \rangle = 1, \forall k \in \{1, ..., n\}$. Then for any*

$x \in E$ the following estimate

$$\left| \|x - s_n\|^2 - \left(\|x\|^2 - \sum_{i=1}^{n} |c_i|^2 \right) \right| \le d^{\perp}(X) \cdot \sum_{i,j=1, i \neq j}^{n} |c_i| \cdot |c_j|$$

holds.

Proof. It is well-known (see *e.g.* Muntean [154], p.130) that there exists exactly one $y^* \in \text{span} X = \{\alpha_1 x_1 + ... + \alpha_n x_n; x_1, ..., x_n \in X, \alpha_1, ..., \alpha_n \in \mathbf{R}, n \in \mathbf{N}\}$ such that

$$\|x - y^*\| = \inf \{\|x - z\| ; z \in \text{span} X\},$$

and $y^* = c_1 x_1 + ... + c_n x_n$, where $c_k = \langle x, x_k \rangle, k \in \{1, ..., n\}$. We have

$$\begin{aligned} \|x - s_n\|^2 &= \|x - y^*\|^2 = \left\langle x - \sum_{i=1}^{n} c_i x_i, x - \sum_{j=1}^{n} c_j x_j \right\rangle \\ &= \|x\|^2 - \sum_{i=1}^{n} |c_i|^2 + \sum_{i,j=1,n}^{n} c_i c_j \cdot \langle x_i, x_j \rangle - \sum_{i=1}^{n} |c_i|^2 . \end{aligned}$$

By

$$\begin{aligned} \left| \sum_{i,j=1}^{n} c_i c_j \cdot \langle x_i, x_j \rangle - \sum_{i=1}^{n} |c_i^2| \right| &= \left| \sum_{i,j=1, i \neq j}^{n} c_i c_j \cdot \langle x_i, x_j \rangle \right| \\ &\le d^{\perp}(X) \cdot \sum_{i,j=1, i \neq j}^{n} |c_i| \cdot |c_j| , \end{aligned}$$

the theorem is proved. □

Corollary 6.1 *In the hypothesis of Theorem 6.8, for any $x \in E$ we have*

$$\left| \|x - s_n\|^2 - \left(\|x\|^2 - \sum_{i=1}^{n} |c_i|^2 \right) \right| \le d^{\perp}(X) \cdot \|x\|^2 \cdot n(n-1) .$$

Proof. By $|c_i| = |\langle x, x_i \rangle| \le \|x\| \cdot \|x_i\| = \|x\|$ (because $\langle x_i, x_i \rangle = \|x_i\|^2 = 1$, by hypothesis), it follows

$$\sum_{i,j=1, i \neq j}^{n} |c_i| \cdot |c_j| = \sum_{i=1}^{n} \sum_{j=1, j \neq i}^{n} |c_i| \cdot |c_j| \le \|x\|^2 \cdot \sum_{i=1}^{n} \sum_{j=1, j \neq i}^{n} 1 = \|x\|^2 n(n-1) ,$$

which together with Theorem 6.8, proves the corollary. □

Corollary 6.2 *If $X = \{x_1, ..., x_n, ...\}$ is countable, such that $\langle x_i, x_i \rangle = \frac{1}{i^4}, i \in \mathbf{N}$, then*

$$\left| \left\| x - \sum_{k=1}^{\infty} c_k x_k \right\|^2 - \left(\|x\|^2 - \sum_{i=1}^{\infty} |c_i|^2 \right) \right| \leq d^{\perp}(X) \cdot \|x\|^2 \cdot \frac{\pi^4}{36}.$$

Proof. By hypothesis, $\|x_i\| = \frac{1}{i^2}$, which implies

$$|c_i| \leq \|x\| \cdot \|x_i\| \leq \frac{\|x\|}{i^2},$$

$$\left(\sum_{i=1}^{n} |c_i| \right) \left(\sum_{j=1, j \neq i}^{n} |c_j| \right) \leq \|x\|^2 \cdot \left(\sum_{i=1}^{n} \frac{1}{i^2} \right) \left(\sum_{j=1}^{n} \frac{1}{j^2} \right).$$

By Theorem 6.8 we get

$$\left| \|x - s_n\| - \left(\|x\|^2 - \sum_{i=1}^{n} |c_i|^2 \right) \right| \leq d^{\perp}(X) \cdot \|x\|^2 \cdot \left(\sum_{i=1}^{n} \frac{1}{i^2} \right)^2, \forall n \in \mathbf{N}.$$

Passing to limit with $n \to +\infty$, we obtain the corollary. □

Remark. If X is orthonormal then $d^{\perp}(X) = 0$ and the inequality in Theorem 6.8 becomes the well-known equality $\|x - s_n\|^2 = \|x\|^2 - \sum_{i=1}^{n} |c_i|^2$. Also, Corollary 6.1 can easily be framed into the general scheme in Section 1.1. Indeed, for $x \in E, X = \{x_1, x_2, ..., x_n\}$ with $\langle x_i, x_i \rangle = 1, \forall i \in \{1, ..., n\}$, let us define

$$\begin{aligned} A_x(X) &= \|x - s_n\|^2 - \left(\|x\|^2 - \sum_{i=1}^{n} |c_i|^2 \right), \forall x \in E, \\ B_x(u) &= \left[\|x\|^2 n(n-1) \right] u, \forall x \in E, \\ U &= \mathcal{P}_n(E) = \{X = \{x_1, ..., x_n\} \subset E; \langle x_i, x_i \rangle = 1, \forall i \in \{1, ..., n\}\} \end{aligned}$$

and

$$P = \text{"the property of orthogonality"}.$$

In what follows we present some applications to the best approximation theory. Thus, using the idea of Singer in [198], p. 86, we can introduce the following.

Definition 6.4 Let $(E, \|\cdot\|)$ be a real normed space, $G \subset E$ and $x \in E$ be fixed. We say that $g_0 \in G$ is element of best approximation of x by elements by G, with respect to the orthogonality $\perp_*$, if $x - g_0 \perp_* G$, where $\perp_*$ can be any orthogonality in Definition 6.1.

Remark. If $\perp_*$ is $\perp_B$ then by Singer [198], Lemma 1.14, p.85 it follows that if G is linear subspace of $E, x \in E \backslash G$, then $x - g_0 \perp_B G$ if and only if g_0 is element of best approximation of x, in the classical sense.

Theorem 6.9 *Let us suppose that $G \subset E$ is a linear subspace of E and $x \in E \backslash G$. If there exists $g_0 \in G$, such that for a given $a \in \mathbf{R} \backslash \{0\}$, we have $x - g_0 \perp_P G$ (i.e. if g_0 is element of best approximation of x with respect to $\perp_P$), then g_0 is element of best approximation in usual sense (that is, $\|x - y_0\| = \inf\{\|x - g\|; g \in G\}$), $x - g_0 \perp_R g_0$ ($\perp_R$ is orthogonality in Roberts sense (see e.g. Singer [198], p.86)) and $\|g_0\| \leq \|x\|, \|x - 2g_0\| = \|x\|$.*

Proof. By $x - g_0 \perp_P G$, we get

$$\|x - g_0\|^2 = \|x - g_0 - ag\|^2 - a^2 \|g\|^2 \leq \|x - g_0 - ag\|^2, \forall g \in G,$$

and passing to infimum with $g \in G$, we get

$$\|x - g_0\| = \inf\{\|x - g\|; g \in G\},$$

because G is linear subspace. Moreover, by

$$\|x - g_0 - ag\|^2 = \|x - g_0\|^2 + a^2 \|g\|^2, \forall g \in G,$$

choosing firstly $g = \frac{\alpha}{a} g_0$ and secondly $g = -\frac{\alpha}{a} g_0$, we obtain

$$\|x - g_0 - \alpha g_0\|^2 = \|x - g_0\|^2 + \alpha^2 \|g_0\|^2$$

and

$$\|x - g_0 + \alpha g_0\|^2 = \|x - g_0\|^2 + \alpha^2 \|g_0\|^2,$$

which implies

$$\|x - g_0 - \alpha g_0\|^2 = \|x - g_0 + \alpha g_0\|^2, \forall \alpha \in \mathbf{R},$$

that is $x - g_0 \perp_R g_0$, where $\perp_R$ means the orthogonality in Roberts' sense (see *e.g.* Singer [198], p. 86).

The last equalities imply

$$\|g_0\| \leq \|x\| \text{ (take } \alpha = 1 \text{ in } \|x - g_0 + \alpha g_0\|^2 = \|x - g_0\|^2 + \alpha^2 \|g_0\|^2 \text{)}$$

and

$$\|x - 2g_0\| = \|x\| \text{ (take } \alpha = 1 \text{ in the last equality).}$$

□

It is natural to search for sufficient conditions involving $\perp_P$ orthogonality, which imply the existence of best approximation elements (in usual sense). Thus we present

Theorem 6.10 *Let $(E, \|\cdot\|)$ be a real normed space, $f \in E^*$, $H = \{y \in E; f(y) = 0\}$. If there exists $z \in E$ satisfying $|f(z)| = 1$ and $z \perp_P H$ for $a = 1$, (that is, if 0 is best approximation of z by elements in H, with respect to $\perp_P$ for $a = 1$), then for all $x \in E \setminus H$ with $|f(x)| < 1$, there exists element of best approximation in H, in classical sense (i.e. $\exists g_0 \in H$ with $\|x - g_0\| = \inf\{\|x - g\|; g \in H\}$).*

Proof. Let us define $g_0 = x - \frac{f(x)}{f(z)} z$. We obviously have $f(g_0) = 0$, therefore $g_0 \in H$. Also, $\frac{f(z)}{f(x)}(g - g_0) \in H$, for all $g \in H$ (here by $x \in E \setminus H$ obviously $f(x) \neq 0$). By $z \perp_P H$, it follows

$$\|z - g\|^2 = \|z\|^2 + \|g\|^2, \forall g \in H.$$

Replacing above g by $\frac{f(z)}{f(x)}(g - g_0)$, we get

$$\|x - g_0\|^2 = \frac{|f(x)|^2}{|f(z)|^2} \|z\|^2 \leq \|z\|^2 = \left\| z - \frac{f(z)}{f(x)}(g - g_0) \right\|^2$$

$$- \left\| \frac{f(z)}{f(x)}(g - g_0) \right\|^2 = \|x - g\|^2 - \frac{\|g - g_0\|^2}{|f(x)|^2} \leq \|x - g\|^2, \forall g \in H,$$

which proves the theorem. □

Definition 6.4 suggests in a natural way the following

Definition 6.5 The quantity $0 < E_*^\perp(x) = \inf\{d_*^\perp(x - g, G); g \in G\}$ will be called almost best approximation of x with respect to $\perp_*$ and an element $g_0 \in G$ with $E_*^\perp(x) = d_*^\perp(x - g_0, G)$ will be called element of almost best approximation of x (by elements in G), with respect to $\perp_*$. If, in addition, $E_*^\perp(x) = d_*^\perp(x - g_0, G) = 0$, then g_0 is element of best approximation defined by Definition 6.4.

Of course that a natural question is to find out conditions on G such that for each $x \in E$ to exist $g_0 \in G$ with $E_*^\perp(x) = d_*^\perp(x - g_0, G)$. For an answer to this question, we will need the following

Lemma 6.1 *If we denote $F(g) = \sup\{|A(x-g, g')| ; g' \in G\}, \forall g \in E$, then*

$$|F(g_1) - F(g_2)| \le \sup\{|A(x-g_1, g') - A(x-g_2, g')| ; g' \in G\},$$

where $A : E \times E \to \mathbf{R}$.

Proof. We have

$$\begin{aligned} & |A(x-g_1, g')| - |A(x-g_2, g')| \\ \le \; & \big| |A(x-g_1, g')| - |A(x-g_2, g')| \big| \\ \le \; & |A(x-g_1, g') - A(x-g_2, g')|, \end{aligned}$$

and passing to supremum after $g' \in G$ we get

$$F(g_1) \le F(g_2) + \sup\{|A(x-g_1, g') - A(x-g_2, g')| ; g' \in G\}.$$

Similarly

$$F(g_2) \le F(g_1) + \sup\{|A(x-g_1, g') - A(x-g_2, g')| ; g' \in G\},$$

which implies the lemma. □

Theorem 6.11 *Let $(E, \|\cdot\|)$ be a real normed space, $G \subset E$ be compact and $x \in E$ ($x \in E \backslash G$ for $\perp_S$). Then for each $\perp_*$ in Definition 6.1 (excepting $\perp_B$), there exists $g^* \in G$ such that $E_*^\perp(x) = d_*^\perp(x - g^*, G)$.*

Proof. (*i*) In the case of $\perp_I$, let us take

$$A(x-g, g') = \|x - g - ag'\|^2 - \|x - g + ag'\|^2, a \in \mathbf{R}\backslash\{0\}.$$

For $g_n, g, g' \in G$ with $\|g_n - g\| \overset{n\to\infty}{\to} 0$, we get

$$|A(x-g_n, g') - A(x-g, g')|$$

$$\begin{aligned}
&\leq \left| \|x - g_n - ag'\|^2 - \|x - g - ag'\|^2 \right| \\
&\quad + \left| \|x - g_n + ag'\|^2 - \|x - g + ag'\|^2 \right| \\
&\leq 2\left(2\|x\| + 2\|g_n\| + 2|a| \cdot \|g'\|\right)\|g_n - g\| \\
&\leq K\|g_n - g\| \overset{n\to\infty}{\to} 0,
\end{aligned}$$

because G is bounded too.

Consequently, with the notation in Lemma 6.1, $F(g)$ is continuous on G and there exists $g^* \in G$ with $F(g^*) = \inf\{F(g); g \in G\} = E_I^{\perp}(x)$.

(ii) For $\perp_P$ we have

$$A(x - g, g') = \|x - g\|^2 + a^2\|g'\|^2 - \|x - g - ag'\|^2$$

and

$$\begin{aligned}
&|A(x - g_n, g') - A(x - g, g')| \\
\leq\ & \left| \|x - g_n\|^2 - \|x - g\|^2 \right| + \left| \|x - g_n - ag'\|^2 - \|x - g - ag'\|^2 \right| \\
\leq\ & K\|g_n - g\|,
\end{aligned}$$

where $K > 0$, because G is bounded. The rest of the proof is similar to the above case (i).

(iii) For $\perp_{DFA}$ we have

$$A(x - g, g') = (1 + a^2)\|x - g_n + g'\|^2 - \|a(x - g) + g'\|^2 - \|x - g + ag'\|^2,$$

the proof being similar.

(iv) For $\perp_{KP}$ we take

$$\begin{aligned}
A(x - g, g') &= \|a(x - g) + bg'\|^2 + \|x - g + g'\|^2 \\
&\quad - \|a(x - g) + g'\|^2 - \|x - g + bg'\|^2
\end{aligned}$$

and we reason similarly.

(v) For $\perp_S$ we have

$$A(x - g, g') = \left\| \frac{x - g}{\|x - g\|} + \frac{g'}{\|g'\|} \right\|^2 - \left\| \frac{x - g}{\|x - g\|} - \frac{g'}{\|g'\|} \right\|^2,$$

defined for $x \in E \backslash G$ and $g' \neq 0$. We obtain (for $\|g_n - g\| \overset{n\to\infty}{\to} 0$)

$$|A(x - g_n, g') - A(x - g, g')|$$

$$\begin{aligned}
&\leq \left| \left\| \frac{x-g_n}{\|x-g_n\|} + \frac{g'}{\|g'\|} \right\|^2 - \left\| \frac{x-g}{\|x-g\|} + \frac{g'}{\|g'\|} \right\|^2 \right| \\
&\quad + \left| \left\| \frac{x-g_n}{\|x-g_n\|} - \frac{g'}{\|g'\|} \right\|^2 - \left\| \frac{x-g}{\|x-g\|} - \frac{g'}{\|g'\|} \right\|^2 \right| \\
&\leq 8 \cdot \left\| \frac{x-g_n}{\|x-g_n\|} - \frac{x-g}{\|x-g\|} \right\| \\
&= \frac{8}{\|x-g_n\| \cdot \|x-g\|} \cdot \|(x-g_n)\|x-g\| - (x-g)\|x-g_n\|\| \\
&\leq \frac{16}{\|x-g_n\|} \cdot \|g-g_n\| \to 0.
\end{aligned}$$

(vi) In the case of orthogonality with respect to an inner product $\langle \cdot, \cdot \rangle$ on $E \times E$, we have $A(x-g, g') = \langle x-g, g' \rangle$ and consequently

$$|A(x-g_n, g') - A(x-g, g')| = |\langle g-g_n, g' \rangle| \leq \|g-g_n\| \cdot \|g'\| \leq K \|g-g_n\|$$

which proves the theorem. □

Now, let us introduce the following.

Definition 6.6 Let $(E, \|\cdot\|)$ be a real normed space and $G \subset E$. We define $\pi_*^{\perp}(G)(x) = \emptyset$ or $\pi_*^{\perp}(G)(x) = \{g^* \in G; E_*^{\perp}(x) = d_*^{\perp}(x-g^*, G)\}$ and $P_*^{\perp}(G)(x) = \emptyset$ or $P_*^{\perp}(G)(x) = \{g^* \in G; 0 = E_*^{\perp}(x) = d_*^{\perp}(x-g^*, G)\}$. We say that G is $\perp_*$-Chebysev, if $\forall x \in E, P_*^{\perp}(G)(x)$ has exactly one element.

Theorem 6.12 *(i) If $G \subset E$ is closed, then for all $x \in E, \pi_*^{\perp}(G)(x)$ and $P_*^{\perp}(G)(x)$ are closed, for any $\perp_*$ in Definition 6.1.*

(ii) If G is $\perp$-Chebyshev, where $\perp$ is the usual orthogonality in an inner product space, then

$$P^{\perp}(G)(\lambda x + (1-\lambda) P^{\perp}(G)(x)) = P^{\perp}(G)(x), \forall x \in E, \lambda \in [0,1].$$

If G is $\perp_*$-Chebyshev, with $\perp_*$ different from $\perp$, then

$$P_*^{\perp}(\lambda G)(\lambda x + (1-\lambda) P_*^{\perp}(G)(x)) = P_*^{\perp}(G)(x), \forall x \in E, \lambda \in (0,1].$$

(iii) If G is such that $\pi_*^{\perp}(G)(x) \neq \emptyset, \forall x \in E$ and $\mu \in \mathbf{R} \setminus \{0\}$ is fixed, then for all $g^* \in \pi_*^{\perp}(G)(x)$ and $\perp_*$ different from $\perp_S$, we have

$$E_*^{\perp}(G)(\mu x + (1-\mu) g^*) \leq \mu^2 d_*^{\perp}\left(x - g^*, \frac{G}{\mu}\right).$$

If $\perp_*$ is $\perp_S$ then

$$E_S^{\perp}(G)(\mu x + (1-\mu) g^*) \leq d_S^{\perp}\left(x - g^*, \frac{G}{\mu}\right).$$

Proof. (*i*) The case $\pi_*^{\perp}(G)(x) = P_*^{\perp}(G)(x) = \emptyset$ is trivial. Therefore, let us suppose $\pi_*^{\perp}(G)(x) \neq \emptyset$ and $P_*^{\perp}(G)(x) \neq \emptyset$. Let $g_n^* \in \pi_*^{\perp}(G)(x), n \in \mathbf{N}$ with $\|g_n^* - g^*\| \overset{n\to\infty}{\to} 0$, *i.e.* $g^* \in G$. By the notations in Lemma 6.1, we have $E_*^{\perp}(x) = d_*^{\perp}(x - g_n^*, G) = F(g_n^*)$, which by the proof of Theorem 6.11 implies that $F(g_n^*) \overset{n\to\infty}{\to} F(g^*)$, *i.e.* $F(g^*) = E_*^{\perp}(x)$. The proof for $P_*^{\perp}(G)(x)$ is similar.

(*ii*) Let G be $\perp$-Chebyshev and $P^{\perp}(G)(x) = g^*$, which implies $E^{\perp}(x) = d^{\perp}(x - g^*, G) = 0$ and therefore $\langle x - g^*, g' \rangle = 0, \forall g' \in G$. For any $\lambda \in [0,1], g' \in G$, we get $0 = \langle \lambda(x - g^*), g' \rangle = \langle \lambda x + (1-\lambda) g^* - g^*, g' \rangle$, that is $P^{\perp}(G)(\lambda x + (1-\lambda) P^{\perp}(G)(x)) = P^{\perp}(G)(x)$.

Now, let us suppose that G is $\perp_*$-Chebyshev, where $\perp_*$ is different from $\perp$ and let us denote $g^* = P_I^{\perp}(G)(x)$, *i.e.* $\|x - g^* - ag'\| = \|x - g^* + ag'\|$, $\forall g' \in G$. Multiplying by $\lambda \in (0,1]$, we get

$$\|\lambda x + (1-\lambda) g^* - g^* - a\lambda g'\| = \|\lambda x + (1-\lambda) g^* - g^* + a\lambda g'\|,$$

and taking into account that λG also is $\perp_*$-Chebyshev, it follows $P_I^{\perp}(\lambda G)(\lambda x + (1-\lambda) P_I^{\perp}(G)(x)) = P_I^{\perp}(G)(x)$.

If $g^* = P_P^{\perp}(G)(x)$, we have $\|x - g^*\|^2 + a^2\|g'\|^2 = \|x - g^* - ag'\|^2$, $\forall g' \in G$. Multiplying both members by λ^2, it follows

$$\|\lambda x + (1-\lambda) g^* - g^*\|^2 + a^2\|\lambda g'\|^2 = \|\lambda x + (1-\lambda) g^* - g^* - a\lambda g'\|^2,$$

which means $P_P^{\perp}(\lambda G)(\lambda x + (1-\lambda) P_P^{\perp}(G)(x)) = P_P^{\perp}(G)(x)$.

Similar reasonings apply for $\perp_B, \perp_{DFA}, \perp_{KP}$ and $\perp_S$.

(*iii*) Let us suppose that $\pi_*^{\perp}(G)(x) \neq \emptyset, \forall x \in E$ and let $g^* \in \pi_*^{\perp}(G)(x)$. We have

$$E_*^{\perp}(G)(x) = \inf\{d_*^{\perp}(x - g, G); g \in G\} = d_*^{\perp}(x - g^*, G).$$

Also, denoting $g_0 \in \pi_*^{\perp}(G)(\mu x + (1-\mu) g^*)$, we get

$$\begin{aligned} &E_*^{\perp}(G)(\mu x + (1-\mu) g^*) \\ = \;& \inf\{d_*^{\perp}(\mu x + (1-\mu) g^* - g, G); g \in G\} \\ = \;& d_*^{\perp}(\mu x + (1-\mu) g^* - g_0, G) \leq d_*^{\perp}(\mu(x - g^*), G) \end{aligned}$$

$$= \sup \left\{ \left| A\left(\mu\left(x-g^{*}\right), \mu g'\right) \right| ; g' \in \frac{G}{\mu} \right\},$$

where $A(x, y)$ is given by the proof of Theorem 6.11. But it is easy to see, that for all the cases of $\perp_*$, different from $\perp_S$, we have $A(\mu x, \mu y) = \mu^2 A(x, y)$, and for $\perp_S$ we have $A(\mu x, \mu y) = A(x, y)$, which proves the theorem. □

Remark. If, for example $G \subset \mu G$, then

$$E_*^{\perp}(G)(\mu x + (1-\mu) g^*) \leq \mu^2 E_*^{\perp}(G)(x).$$

Finally, we notice that the defects of orthogonalities can have some geometrical interpretations. For example, the defect of orthogonality of Singer kind in plane, can be looked as the defect of perpendicularity of two straight lines. Let us consider $E = \mathbf{R}^2$ and the norm on $\mathbf{R}^2$ defined by $\|(x_1, x_2)\| = \sqrt{x_1^2 + x_2^2}$. If X and Y are straight lines in $\mathbf{R}^2$ such that $\{(0,0)\} \dot{\in} X$ and $\{(0,0)\} \in Y$ then $d_S^{\perp}(X, Y) = 4\left|\cos(\theta_1 - \theta_2)\right|$, where θ_1, θ_2 are the angles of X and Y with the positive axis Ox, respectively. Indeed, let

$$\begin{aligned} X &= \left\{(x, y) \in \mathbf{R}^2; y = m_1 x, x \in \mathbf{R}\right\} \\ Y &= \left\{(x, y) \in \mathbf{R}^2; y = m_2 x, x \in \mathbf{R}\right\}, \end{aligned}$$

that is $m_1 = tg\theta_1, m_2 = tg\theta_2, \theta_1, \theta_2 \in \left(-\frac{\pi}{2}, \frac{\pi}{2}\right)$. We have (see Theorem 6.2, (vi))

$$d_S^{\perp}(X, Y)$$

$$= 4\sup\left\{ \frac{\left|\left\langle (x_1, y_1), (x_2, y_2)\right\rangle\right|}{\|(x_1, y_1)\| \cdot \|(x_2, y_2)\|}; (x_1, y_1), (x_2, y_2) \in \mathbf{R}^2 \setminus \{(0,0)\} \right\}$$

$$= 4\sup\left\{ \frac{\left|x_1 x_2 + m_1 x_1 m_2 x_2\right|}{\sqrt{x_1^2 + m_1^2 x_1^2} \cdot \sqrt{x_2^2 + m_2^2 x_2^2}}; x_1, x_2 \in \mathbf{R} \setminus \{0\} \right\}$$

$$= 4\frac{\left|1 + m_1 m_2\right|}{\sqrt{1 + m_1^2}\sqrt{1 + m_2^2}}$$

$$= 4\left|\cos\left(\theta_1 - \theta_2\right)\right|.$$

The above result and Theorem 6.1 imply that X is orthogonal to Y in the Singer sense if and only if $\theta_1 - \theta_2 = \pm\frac{\pi}{2}$, that is if and only if the straight lines X and Y are orthogonal in geometrical sense.

Remark. By using the Birkhoff's orthogonality in a real normed space $(E, \|\cdot\|)$, can be introduced the so-called rectangle constant of E by

$$\mu(E) = \sup\left\{\frac{\|x\| + \|y\|}{\|x + y\|}; x, y \in X, \|x\| + \|y\| \neq 0, x \perp_B y\right\},$$

which satisfies the properties $\sqrt{2} \leq \mu(E) \leq 3$ and if the dimension of E, $\dim(E) \geq 3$, then $\mu(E) = \sqrt{2}$ if and only if the norm $\|\cdot\|$ is generated by an inner product (see Joly [108]-[109]). According to Gastinel-Joly [87], $\mu(E)$ can be thought of as a measure of the failure of Pythagora's theorem.

Denoting $d(E) = \mu(E) - \sqrt{2}$, the constant $d(E)$ can be thought as defect of Hilbert space for a Banach space E (because if $\dim(E) \geq 3$, then $d(E) = 0$ is equivalent to the fact that E is Hilbert space).

Also, suggested by $\mu(E)$, we can introduce, for example, another constant, by

$$\mu^*(E) = \sup\left\{\frac{\|x\|^2 + \|y\|^2}{\|x + y\|^2}; x, y \in X, \|x\| + \|y\| \neq 0, x \perp_B y\right\},$$

which by the inequalities

$$\|x + y\|^2 \leq (\|x\| + \|y\|)^2 \leq 2\left(\|x\|^2 + \|y\|^2\right)$$

and

$$\frac{\|x\|^2 + \|y\|^2}{\|x + y\|^2} \leq \frac{(\|x\| + \|y\|)^2}{\|x + y\|^2}$$

satisfies $\frac{1}{2} \leq \mu^*(E) \leq 9$, if $\dim(E) \geq 3$.

6.2 Defect of Property for Sets in Normed Spaces

Let $(X, \|\cdot\|)$ be a real normed space. The following concepts are well-known.

Definition 6.7 A subset $Y \subseteq X$ is called:

(i) convex, if

$$\lambda y_1 + (1-\lambda)\, y_2 \in Y, \forall y_1, y_2 \in Y, \lambda \in [0,1] .$$

(ii) linear, if

$$\alpha y_1 + \beta y_2 \in Y, \forall y_1, y_2 \in Y, \alpha, \beta \in \mathbf{R}.$$

(iii) balanced, if

$$\alpha y \in Y, \forall y \in Y, |\alpha| \leq 1.$$

(iv) absorbent, if

$$\forall x \in X, \exists \lambda > 0 \text{ such that } x \in \lambda Y.$$

Remark. The following characterizations also are well-known:
Y is convex if and only if $Y = convY$, where

$$convY = \left\{ \sum_{i=1}^{n} \alpha_i y_i ; n \in \mathbf{N}, y_i \in Y, \alpha_i \geq 0, i \in \{1, ..., n\}, \sum_{i=1}^{n} \alpha_i = 1 \right\}$$

represents the convex hull of Y;
Y is linear if and only if $Y = spanY$, where

$$spanY = \left\{ \sum_{i=1}^{n} \alpha_i y_i ; n \in \mathbf{N}, y_i \in Y, \alpha_i \in \mathbf{R}, i \in \{1, ..., n\} \right\};$$

Y is balanced if and only if $Y = balY$, where

$$balY = \{\alpha y ; y \in Y, |\alpha| \leq 1\} .$$

Now, if $Y \subseteq X$ is not convex, linear, balanced or absorbent, then it is natural to define quantities that measure the deviation (defect) of Y with respect to these properties. In this sense we introduce

Definition 6.8 Let $(X, \|\cdot\|)$ be a real normed space and $D(A, B)$ be a certain distance between the subsets $A, B \subset X$ (D will be specified later). For $Y \subseteq X$, we call:

(i) defect of convexity of Y (with respect to D), the quantity

$$d_{CONV}(D)(Y) = D(Y, convY);$$

(ii) defect of linearity of Y, the quantity

$$d_{LIN}(D)(Y) = D(Y, spanY);$$

(iii) defect of balancing of Y, the quantity

$$d_{BAL}(D)(Y) = D(Y, balY);$$

(iv) defect of absorption of bounded Y, the quantity $d_{ABS}(D)(Y) = 0$, if Y is absorbent and

$$d_{ABS}(D)(Y) = D\left(Y, \overline{B}(0, R_Y)\right)$$

if Y is not absorbent, where

$$\begin{aligned} R_Y &= \sup\{\|y\|; y \in Y\}, \\ \overline{B}(0, R_Y) &= \{x \in X; \|x\| \leq R_Y\}. \end{aligned}$$

For symmetry, let us denote $absY = \overline{B}(0, R_Y)$.

A natural candidate for D might be the Hausdorff-Pompeiu distance

$$D_H(Y_1, Y_2) = \max\{d(Y_1, Y_2), d(Y_2, Y_1)\},$$

where $d(Y_1, Y_2) = \sup\{d(y_1, Y_2); y_1 \in Y_1\}, d(y_1, Y_2) = \inf\{\|y_1 - y_2\|; y_2 \in Y_2\}$. In this case we present

Theorem 6.13 *Let $(X, \|\cdot\|)$ be a real normed space and let us suppose that $Y \subseteq X, Y \neq \emptyset$ is closed. Then:*
(i) Y is convex if and only if $d_{CONV}(D_H)(Y) = 0$.
(ii) Y is linear if and only if $d_{LIN}(D_H)(Y) = 0$.
(iii) Y is balanced if and only if $d_{BAL}(D_H)(Y) = 0$.
(iv) bounded Y is absorbent if and only if $d_{ABS}(D_H)(Y) = 0$.

Proof. If Y is convex then obviously $d_{CONV}(D_H)(Y) = D_H(Y, convY) = 0$. Conversely, let us suppose $D_H(Y, convY) = 0$. It follows $\overline{Y} = \overline{convY}$ and because $Y = \overline{Y}$, we get $Y = \overline{convY} \supset convY$, which gives $Y = convY$, *i.e.* Y is convex.

For linearity, because the necessity is obvious, let us suppose $d_{LIN}(D_H)(Y) = 0$, *i.e.* $D_H(Y, spanY) = 0$. It follows $\overline{Y} = \overline{spanY}$ and therefore $Y = \overline{spanY} \supset spanY$, which gives $Y = spanY$ and therefore Y is linear.

Now, by $d_{BAL}(D_H)(Y) = 0$, it follows $Y = \overline{balY} \supset balY$, which implies $Y = balY$, that is Y is balanced.

Finally, if Y is absorbent then by definition, $d_{ABS}(D_H)(Y) = 0$. Conversely, $d_{ABS}(D_H)(Y) = 0$ implies $\overline{Y} = \overline{B}(0, R_Y)$, that is $Y = \overline{B}(0, R_Y)$ and therefore Y is absorbent. □

Remark. For $D = D_H$, the quantity $d_{CONV}(D_H)(Y)$ is known, it is called measure of nonconvexity (see *e.g.* Eisenfeld-Lakshmikantham [70]) and it is used in fixed point theory (see *e.g.* Rus [183]-[185], Cano [53], Amoretti-Cano [4], Ewert [74]). The following properties also are known (see *e.g.* Ewert [74]):

$$\begin{aligned} d_{CONV}(D_H)(Y_1+Y_2) &\leq d_{CONV}(D_H)(Y_1)+d_{CONV}(D_H)(Y_2), \\ d_{CONV}(D_H)(\lambda Y) &= \lambda d_{CONV}(D_H)(Y), \forall \lambda > 0, \end{aligned}$$

$$|d_{CONV}(D_H)(Y_1)-d_{CONV}(D_H)(Y_2)| \leq 2D_H(Y_1,Y_2),$$

for all bounded nonempty subsets $Y, Y_1, Y_2 \subset X$.

Note that, for example, the first property is a generalization of the simple result which states that the sum of two convex sets is also a convex set.

Similar properties hold for $d_{LIN}(D_H), d_{BAL}(D_H)$ and $d_{ABS}(D_H)$, as follows.

Theorem 6.14 *Let $(X, \|\cdot\|)$ be a real normed space. For all bounded sets $Y, Y_1, Y_2 \subset X$, with $Y_1, Y_2 \neq \emptyset$, we have*

(*i*)

$$\begin{aligned} d_{LIN}(D_H)(Y) &= d_{LIN}(D_H)(\overline{Y}); \\ d_{LIN}(D_H)(Y) &= 0 \textit{ if and only if } \overline{Y} \textit{ is linear;} \\ d_{LIN}(D_H)(\lambda Y) &= \lambda d_{LIN}(D_H)(Y), \forall \lambda > 0; \\ d_{LIN}(D_H)(Y_1+Y_2) &\leq d_{LIN}(D_H)(Y_1)+d_{LIN}(D_H)(Y_2). \end{aligned}$$

(*ii*)

$$\begin{aligned} d_{BAL}(D_H)(Y) &= d_{BAL}(D_H)(\overline{Y}); \\ d_{BAL}(D_H)(Y) &= 0 \textit{ if and only if } \overline{Y} \textit{ is balanced;} \\ d_{BAL}(D_H)(\lambda Y) &= \lambda d_{BAL}(D_H)(Y), \forall \lambda > 0; \\ d_{BAL}(D_H)(Y_1+Y_2) &\leq d_{BAL}(D_H)(Y_1)+d_{BAL}(D_H)(Y_2); \\ D_H(balY_1, balY_2) &\leq D_H(Y_1,Y_2); \end{aligned}$$

$$|d_{BAL}(D_H)(Y_1)-d_{BAL}(D_H)(Y_2)| \leq 2D_H(Y_1,Y_2).$$

(*iii*)

$$d_{ABS}(D_H)(Y) = d_{ABS}(D_H)(\overline{Y});$$
$$d_{ABS}(D_H)(\lambda Y) = \lambda d_{ABS}(D_H)(Y), \forall \lambda > 0;$$

$$|d_{ABS}(D_H)(Y_1) - d_{ABS}(D_H)(Y_2)| \leq 2D_H(Y_1, Y_2)$$

$$+D_H\left(\overline{B}(0, R_{Y_1}), \overline{B}(0, R_{Y_2})\right).$$

Proof. We will use the following equivalent formula for the Hausdorff-Pompeiu distance between the nonempty bounded sets $Y_1, Y_2 \subset X$,

$$D_H(Y_1, Y_2) = \inf\{\varepsilon > 0; Y_1 \subset B(Y_2, \varepsilon), Y_2 \subset B(Y_1, \varepsilon)\},$$

where $B(A, \varepsilon) = \bigcup_{x \in A} B(x, \varepsilon), B(x, \varepsilon) = \{y \in X; \|y - x\| < \varepsilon\}$.

(*i*) Let $d_{LIN}(D_H)(\overline{Y}) = r$. For each $\varepsilon > 0$ we get

$$spanY \subset span\overline{Y} \subset B\left(\overline{Y}, r + \frac{1}{2}\varepsilon\right) \subset B(Y, r + \varepsilon),$$

which implies $d_{LIN}(D_H)(Y) \leq d_{LIN}(D_H)(\overline{Y})$. On the other hand, $span\overline{Y} \subset \overline{spanY}$. Indeeed, let $y \in span\overline{Y}$. It follows $y = \sum_{i=1}^{m} \alpha_i y_i, m \in \mathbf{N}, y_i \in \overline{Y}, i \in \{1, ..., m\}$, that is $\exists y_i^{(n)} \in Y, \left\|y_i^{(n)} - y_i\right\| \overset{n\to\infty}{\to} 0, \forall i \in \{1, ..., m\}$. Let us take $z_n = \sum_{i=1}^{m} \alpha_i y_i^{(n)} \in spanY$. We have

$$\|z_n - y\| \leq \sum_{i=1}^{m} |\alpha_i| \cdot \left\|y_i^{(n)} - y_i\right\| \overset{n\to\infty}{\to} 0,$$

which means that $y \in \overline{spanY}$. Therefore,

$$d_{LIN}(D_H)(\overline{Y}) = D_H\left(span\overline{Y}, \overline{Y}\right) \leq D_H\left(\overline{spanY}, \overline{Y}\right)$$

$$= D_H(spanY, Y) = d_{LIN}(D_H)(Y),$$

and as a conclusion, $d_{LIN}(D_H)(\overline{Y}) = d_{LIN}(D_H)(Y)$.

Now, let us suppose that $\overline{Y}$ is linear, it follows $d_{LIN}(D_H)(\overline{Y}) = 0$ and by the previous equality we get $d_{LIN}(D_H)(Y) = 0$. Conversely, the condition $d_{LIN}(D_H)(Y) = 0$, implies $spanY \subset B(Y, \varepsilon)$, for any $\varepsilon > 0$. Then

$$Y \subset spanY \subset \bigcap_{\varepsilon > 0} B(Y, \varepsilon) = \overline{Y}$$

which implies $\overline{spanY} = \overline{Y}$. But $spanY$ is linear subspace and by *e.g.* Popa [165], p.13, it follows $\overline{spanY}$ is linear subspace, which implies $\overline{Y}$ is linear.
For the next property, we get

$$\begin{aligned} d_{LIN}(D_H)(\lambda Y) &= D_H(\lambda Y, span(\lambda Y)) = D_H(\lambda Y, \lambda spanY) \\ &= \lambda D_H(Y, spanY) = \lambda d_{LIN}(D_H)(Y), \end{aligned}$$

for all $\lambda > 0$.
Finally, let us put $d_{LIN}(D_H)(Y_1) = r_1$ and $d_{LIN}(D_H)(Y_2) = r_2$. Then for each $\varepsilon > 0$ we have $spanY_1 \subset B\left(Y_1, r_1 + \frac{1}{2}\varepsilon\right)$ and $spanY_2 \subset B\left(Y_2, r_2 + \frac{1}{2}\varepsilon\right)$. Hence

$$\begin{aligned} span(Y_1 + Y_2) &\subset spanY_1 + spanY_2 \\ &\subset B\left(Y_1, r_1 + \frac{1}{2}\varepsilon\right) + B\left(Y_2, r_2 + \frac{1}{2}\varepsilon\right) \\ &\subset B(Y_1 + Y_2, r_1 + r_2 + \varepsilon), \end{aligned}$$

which gives $d_{LIN}(D_H)(Y_1 + Y_2) \leq r_1 + r_2$ and the proof is completed.

(*ii*) Let $d_{BAL}(D_H)\left(\overline{Y}\right) = r$. For each $\varepsilon > 0$ we get

$$balY \subset bal\overline{Y} \subset B\left(\overline{Y}, r + \frac{1}{2}\varepsilon\right) \subset B(Y, r + \varepsilon),$$

which implies $d_{BAL}(D_H)(Y) \leq d_{BAL}(D_H)\left(\overline{Y}\right)$. On the other hand, $bal\overline{Y} \subset \overline{balY}$. Indeed, let $y \in bal\overline{Y}$. It follows $y = \lambda y_0, |\lambda| \leq 1, y_0 \in \overline{Y}$, that is $\exists y_n \in Y, n \in \mathbf{N}$, with $\|y_n - y_0\| \overset{n\to\infty}{\to} 0$. Let us take $z_n = \lambda y_n \in balY, n \in \mathbf{N}$. We have $\|z_n - \lambda y_0\| = |\lambda| \cdot \|y_n - y_0\| \overset{n\to\infty}{\to} 0$, that is $y = \lambda y_0 \in \overline{balY}$. Therefore,

$$\begin{aligned} d_{BAL}(D_H)\left(\overline{Y}\right) &= D_H\left(bal\overline{Y}, \overline{Y}\right) \leq D_H\left(\overline{balY}, \overline{Y}\right) \\ &= D_H(balY, Y) = d_{BAL}(D_H)(Y), \end{aligned}$$

and as a conclusion, $d_{BAL}(D_H)\left(\overline{Y}\right) = d_{BAL}(D_H)(Y)$.
Now, let us suppose that $\overline{Y}$ is balanced. It follows $d_{BAL}(D_H)\left(\overline{Y}\right) = 0$ and by the previous equality, we get $d_{BAL}(D_H)(Y) = 0$. Conversely, the condition $d_{BAL}(D_H)(Y) = 0$, implies $balY \subset B(Y, \varepsilon)$, for any $\varepsilon > 0$. Then $Y \subset balY \subset \bigcap_{\varepsilon > 0} B(Y, \varepsilon) = \overline{Y}$, which implies $\overline{balY} = \overline{Y}$. But $balY$ is balanced and by *e.g.* Popa [165], p.15 it follows $\overline{balY}$ is balanced, which implies that $\overline{Y}$ is balanced.
The property $d_{BAL}(D_H)(\lambda Y) = \lambda d_{BAL}(D_H)(Y), \lambda > 0$, is immediate by the properties of D_H and by $bal\lambda Y = \lambda balY, \forall \lambda > 0$.

For the next property, let us put $d_{BAL}(D_H)(Y_1) = r_1$ and $d_{BAL}(D_H)(Y_2) = r_2$. Then for each $\varepsilon > 0$, we have $balY_1 \subset B\left(Y_1, r_1 + \frac{1}{2}\varepsilon\right)$ and $balY_2 \subset B\left(Y_2, r_2 + \frac{1}{2}\varepsilon\right)$. Hence

$$\begin{aligned} bal(Y_1 + Y_2) &\subset balY_1 + balY_2 \\ &\subset B\left(Y_1, r_1 + \frac{1}{2}\varepsilon\right) + B\left(Y_2, r_2 + \frac{1}{2}\varepsilon\right) \\ &\subset B(Y_1 + Y_2, r_1 + r_2 + \varepsilon), \end{aligned}$$

which gives $d_{BAL}(D_H)(Y_1 + Y_2) \leq r_1 + r_2$ and consequently

$$d_{BAL}(D_H)(Y_1 + Y_2) \leq d_{BAL}(D_H)(Y_1) + d_{BAL}(D_H)(Y_2).$$

Now, we will prove that for each bounded nonempty set $Y \subset X$ and each $r_0 > 0$, we have $balB(Y, r_0) \subset B(balY, r_0)$. Indeed, let $x \in balB(Y, r_0)$, it follows $x = \lambda z$, where $|\lambda| \leq 1$ and $z \in B(Y, r_0)$, that is $\exists y \in Y$ with $\|z - y\| < r_0$. We get $\lambda y \in balY$ and

$$\|x - \lambda y\| = \|\lambda z - \lambda y\| = |\lambda| \cdot \|z - y\| < |\lambda| \cdot r_0 < r_0,$$

that is $x \in B(balY, r_0)$.
If we denote

$$\begin{aligned} \rho(x, Y) &= \inf\{\|x - y\|; y \in Y\} \\ \rho(Y_1, Y_2) &= \sup\{\rho(y, Y_2); y \in Y_1\} = \inf\{\varepsilon > 0; Y_1 \subset B(Y_2, \varepsilon)\}, \end{aligned}$$

we obviously have

$$D_H(Y_1, Y_2) = \max\{\rho(Y_1, Y_2), \rho(Y_2, Y_1)\}.$$

Let us denote $\rho(Y_1, Y_2) = r$. Then for each $\varepsilon > 0$, we have $Y_1 \subset B(Y_2, r + \varepsilon)$ and $balY_1 \subset balB(Y_2, r + \varepsilon) \subset B(balY_2, r + \varepsilon)$, which gives $\rho(balY_1, balY_2) \leq r + \varepsilon$. Passing to limit with $\varepsilon \to 0$, it follows $\rho(balY_1, balY_2) \leq r = \rho(Y_1, Y_2)$. Completely analogous we obtain $\rho(balY_2, balY_1) \leq \rho(Y_2, Y_1)$, which implies $D_H(balY_1, balY_2) \leq D_H(Y_1, Y_2)$.
For the final relation, we have

$$d_{BAL}(D_H)(Y_1) = D_H(balY_1, Y_1) \leq D_H(balY_1, balY_2)$$

$$+ D_H(balY_2, Y_2) + D_H(Y_2, Y_1) \leq D_H(balY_1, balY_2)$$

$$+ D_H(Y_2, Y_1) + d_{BAL}(D_H)(Y_2),$$

which implies

$$|d_{BAL}(D_H)(Y_1) - d_{BAL}(D_H)(Y_2)| \leq D_H(balY_1, balY_2) + D_H(Y_1, Y_2),$$

which combined with the last inequality, proves

$$|d_{BAL}(D_H)(Y_1) - d_{BAL}(D_H)(Y_2)| \leq 2D_H(Y_1, Y_2).$$

(iii) We have

$$\begin{aligned} d_{ABS}(D_H)(\overline{Y}) &= D_H(\overline{Y}, \overline{B}(0, R_{\overline{Y}})) = d(\overline{B}(0, R_{\overline{Y}}), \overline{Y}) \\ &= d(\overline{B}(0, R_{\overline{Y}}), Y) = d(\overline{B}(0, R_Y), Y) \\ &= d_{ABS}(D_H)(Y), \end{aligned}$$

because $R_{\overline{Y}} = R_Y$ and $d(A, \overline{C}) = d(A, C), \forall A, C \subset X$ (here d appears in definition of D_H). Indeed, firstly it is obvious that $R_Y \leq R_{\overline{Y}}$. Also, by $R_{\overline{Y}} = \sup\{\|y\|; y \in \overline{Y}\}$, if $y \in \overline{Y}$ then $\exists y_n \in Y$ such that $\|y_n - y\| \stackrel{n\to\infty}{\to} 0$, and consequently, for any $\varepsilon > 0$, there is $n_0 \in \mathbf{N}$ with $\|y_n - y\| < \varepsilon, \forall n \geq n_0$ and $\|y\| \leq \|y - y_n\| + \|y_n\| \leq R_Y + \varepsilon$. Therefore $R_{\overline{Y}} \leq R_Y + \varepsilon, \forall \varepsilon > 0$, which implies $R_{\overline{Y}} \leq R_Y$ and consequently $R_{\overline{Y}} = R_Y$. The equality $d(A, \overline{C}) = d(A, C)$ immediately follows by the relation $d(a, \overline{C}) = d(a, C), \forall a \in A$, and by the definition of $d(A, \overline{C})$. Then

$$\begin{aligned} d_{ABS}(D_H)(\lambda Y) &= D_H(\lambda Y, \overline{B}(0, R_{\lambda Y})) = D_H(\lambda Y, \lambda \overline{B}(0, R_Y)) \\ &= \lambda D_H(Y, \overline{B}(0, R_Y)) = \lambda d_{ABS}(D_H)(Y), \forall \lambda > 0, \end{aligned}$$

taking into account that $R_{\lambda Y} = \sup\{\|\lambda y\|; y \in Y\} = \lambda R_Y$.

Finally,

$$d_{ABS}(D_H)(Y_1) = D_H(\overline{B}(0, R_{Y_1}), Y_1)$$

$$\leq D_H(\overline{B}(0, R_{Y_1}), \overline{B}(0, R_{Y_2})) + D_H(\overline{B}(0, R_{Y_2}), Y_2) + D_H(Y_2, Y_1),$$

which implies

$$d_{ABS}(D_H)(Y_1) - d_{ABS}(D_H)(Y_2) \leq D_H(Y_1, Y_2)$$

$$+ D_H(\overline{B}(0, R_{Y_1}), \overline{B}(0, R_{Y_2})),$$

which proves the theorem. □

Remark. The Theorems 6.13 and 6.14 show that $d_{CONV}(D_H), d_{LIN}(D_H)$ and $d_{BAL}(D_H)$ characterize the corresponding properties only on the class of closed bounded subsets of X.

To overcome this shortcoming (and those pointed out by Remark after Theorem 6.13), in what follows we will replace D_H with another distance (see Section 3.5).

Let us define $D^* : \mathcal{P}(X) \times \mathcal{P}(X) \to \mathbf{R}_+ \cup \{+\infty\}$ by

$$D^*(Y_1, Y_2) = \begin{cases} 0, & \text{if } Y_1 = Y_2 \\ \sup\{\|y_1 - y_2\|; y_1 \in Y_1, y_2 \in Y_2\}, & \text{if } Y_1 \neq Y_2. \end{cases}$$

Denoting $\mathcal{P}_b(X) = \{Y \subset X; Y \neq \emptyset, Y \text{ bounded}\}$, we have

Theorem 6.15 *$(\mathcal{P}_b(X), D^*)$ is a metric space.*

Proof. Firstly, $D^*(Y_1, Y_2) = 0$ if and only if $Y_1 = Y_2$, is obvious. Also, $D^*(Y_1, Y_2) = D^*(Y_2, Y_1)$ is immediate.
Let $Y_1, Y_2, Y_3 \in \mathcal{P}_b(X)$. If $Y_1 = Y_3$ or $Y_2 = Y_3$ then

$$D^*(Y_1, Y_2) = D^*(Y_1, Y_3) + D^*(Y_3, Y_2).$$

Let us suppose $Y_1 \neq Y_3$ and $Y_2 \neq Y_3$. We have

$$\|y_1 - y_2\| \leq \|y_1 - y_3\| + \|y_3 - y_1\|, \forall y_i \in Y_i, i \in \{1, 2, 3\},$$

and passing to supremum we easily obtain

$$D^*(Y_1, Y_2) \leq D^*(Y_1, Y_3) + D^*(Y_3, Y_2)$$

which proves the theorem. □

Other properties of D^* are given by the following.

Theorem 6.16 $D^* : \mathcal{P}_b(X) \times \mathcal{P}_b(X) \to \mathbf{R}_+$ *also satisfies:*

(i) $D^*(Y_1 + Y_1', Y_2 + Y_2') \leq D^*(Y_1, Y_2) + D^*(Y_1', Y_2'), \forall Y_1 \neq Y_2, \forall Y_1' \neq Y_2'$.

(ii) $Y_1 \subseteq Z_1, Y_2 \subseteq Z_2, Z_1 \neq Z_2$ *implies* $D^*(Y_1, Y_2) \leq D^*(Z_1, Z_2)$.

(iii) $D^*(Y_1, Y_2) = D^*(\overline{Y}_1, \overline{Y}_2), \forall \overline{Y}_1 \neq \overline{Y}_2$.

(iv) $D^*(\lambda Y_1, \lambda Y_2) = \lambda D^*(Y_1, Y_2), \forall \lambda > 0$.

Proof. *(i)* If $Y_1 + Y_1' = Y_2 + Y_2'$ then the inequality is obvious. Therefore let us suppose $Y_1 + Y_1' \neq Y_2 + Y_2'$. The inequality for D^* follows immediately from

$$\begin{aligned} & \|(y_1 + y_1') - (y_2 + y_2')\| \\ \leq\ & \|y_1 - y_2\| + \|y_1' - y_2'\| \\ \leq\ & D^*(Y_1, Y_2) + D^*(Y_1', Y_2'), \forall y_i \in Y_i, y_i' \in Y_i', i \in \{1,2\}. \end{aligned}$$

(*ii*) If $Y_1 = Y_2$ then the inequality is obvious. If $Y_1 \neq Y_2$ and by $Z_1 \neq Z_2$, we get

$$\begin{aligned} & \sup\{\|y_1 - y_2\|; y_1 \in Y_1, y_2 \in Y_2\} \\ \leq\ & \sup\{\|y_1 - y_2\|; y_1 \in Z_1, y_2 \in Y_2\} \\ \leq\ & \sup\{\|y_1 - y_2\|; y_1 \in Z_1, y_2 \in Z_2\}. \end{aligned}$$

(*iii*) Firstly, by (*ii*) we get

$$D^*(Y_1, Y_2) \leq D^*(\overline{Y}_1, \overline{Y}_2) \text{ if } \overline{Y}_1 \neq \overline{Y}_2.$$

Conversely,

$$D^*(\overline{Y}_1, \overline{Y}_2) = \sup\{\|y_1 - y_2\|; y_1 \in \overline{Y}_1, y_2 \in \overline{Y}_2\}.$$

By $y_i \in \overline{Y}_i$, there exist $y_i^{(n)} \in Y_i, i \in \{1,2\}, n \in \mathbf{N}$ such that $\left\|y_i^{(n)} - y_i\right\| \overset{n\to\infty}{\to} 0, i \in \{1,2\}$. We get

$$\begin{aligned} \|y_1 - y_2\| &\leq \left\|y_1 - y_1^{(n)}\right\| + \left\|y_1^{(n)} - y_2^{(n)}\right\| + \left\|y_2^{(n)} - y_2\right\| \\ &\leq 2\varepsilon + D^*(Y_1, Y_2), \end{aligned}$$

for any $\varepsilon > 0$, which implies finally

$$D^*(\overline{Y}_1, \overline{Y}_2) \leq D^*(Y_1, Y_2).$$

Combined with the converse inequality we obtain the equality

$$D^*(\overline{Y}_1, \overline{Y}_2) = D^*(Y_1, Y_2),$$

if $\overline{Y}_1 \neq \overline{Y}_2$.

(*iv*) It is immediate. □

Replacing D in Definition 6.8 by D^*, we can state the following

Theorem 6.17 *Let $(X, \|\cdot\|)$ be a real normed space and let $Y \in \mathcal{P}_b(X)$. We have:*

(*i*) $d_{CONV}(D^*)(Y) = 0$ *if and only if Y is convex;*

(ii) $d_{LIN}(D^*)(Y) = 0$ *if and only if* Y *is linear subspace and* $d_{LIN}(D^*)(Y) = +\infty$ *if* Y *is not linear;*

(iii) $d_{BAL}(D^*)(Y) = 0$ *if and only if* Y *is balanced;*

(iv) $d_{ABS}(D^*)(Y) = 0$ *if and only if* Y *is absorbent.*

Proof. Are immediate by Theorem 6.15 and by the Remark after Definition 6.7. □

Concerning these new defects we present other properties too.

Theorem 6.18 *Let* $(X, \|\cdot\|)$ *be a real normed space and* $Y, Y_1, Y_2 \in \mathcal{P}_b(X)$. *We have:*

(i) $d_\sharp(D^*)(\overline{Y}) = d_\sharp(D^*)(Y)$, *if* $\overline{Y} \neq \sharp\overline{Y}$ *and* $\overline{Y} \neq \overline{\sharp Y}$.

(ii) $d_\sharp(D^*)(\lambda Y) = \lambda d_\sharp(D^*)(Y), \forall\lambda > 0$.

Here $\sharp$ *represents any element of the set* $\{CONV, BAL, ABS\}$.

In addition, for $\sharp$ *any element of* $\{CONV, BAL\}$ *we have*

(iii) $d_\sharp(D^*)(Y_1 + Y_2) \leq d_\sharp(D^*)(Y_1) + d_\sharp(D^*)(Y_2)$, *if* $Y_1 \neq \sharp Y_1, Y_2 \neq \sharp Y_2$ *and* $Y_1 + Y_2 \neq \sharp Y_1 + \sharp Y_2$.

Proof. *(i)* First let us consider the case $CONV$. We have $conv\overline{Y} \subset \overline{convY}$ and

$$\begin{aligned} d_{CONV}(D^*)(\overline{Y}) &= D^*\left(conv\overline{Y}, \overline{Y}\right) \leq D^*\left(\overline{convY}, \overline{Y}\right) \\ &= \text{(see Theorem 6.16, } (iii)\text{)} \\ &= D^*(convY, Y) = d_{CONV}(D^*)(Y). \end{aligned}$$

Conversely, by $convY \subset conv\overline{Y}, Y \subset \overline{Y}$ and taking into account Theorem 6.16, *(ii)*, we immediately get

$$d_{CONV}(D^*)(Y) = D^*(Y, convY) \leq D^*\left(\overline{Y}, conv\overline{Y}\right) = d_{CONV}(D^*)(\overline{Y}),$$

which implies $d_{CONV}(D^*)(Y) = d_{CONV}(D^*)(\overline{Y})$.

Now take the case BAL. We get $bal\overline{Y} \subset \overline{balY}$ and

$$\begin{aligned} d_{BAL}(D^*)(\overline{Y}) &= D^*\left(bal\overline{Y}, \overline{Y}\right) \leq D^*\left(\overline{balY}, \overline{Y}\right) \\ &= D^*(balY, Y) = d_{BAL}(D^*)(Y). \end{aligned}$$

Conversely, reasoning as for the case $CONV$, we get finally

$$d_{BAL}(D^*)(\overline{Y}) = d_{BAL}(D^*)(Y).$$

For the case ABS, we have $R_{\overline{Y}} = R_Y$ and

$$d_{ABS}(D^*)(\overline{Y}) = D^*\left(\overline{B}(0, R_{\overline{Y}}), \overline{Y}\right)$$

$$\begin{aligned}&= \sup\left\{\|y_1 - y_2\| ; y_1 \in \overline{B}(0, R_Y), y_2 \in \overline{Y}\right\}\\ &\geq \sup\left\{\|y_1 - y_2\| ; y_1 \in \overline{B}(0, R_Y), y_2 \in Y\right\}\\ &= d_{ABS}(D^*)(Y).\end{aligned}$$

Conversely, for any $y_2 \in \overline{Y}$, let $y_2^{(n)} \in Y, n \in \mathbf{N}$, be with $\left\|y_2^{(n)} - y_2\right\| \overset{n\to\infty}{\to} 0.$ We get for $y_1 \in \overline{B}(0, R_Y)$,

$$\|y_1 - y_2\| \leq \left\|y_1 - y_2^{(n)}\right\| + \left\|y_2^{(n)} - y_2\right\| \leq d_{ABS}(D^*)(Y) + \varepsilon, \forall n \geq n_0(\varepsilon).$$

¿From here we immediately get $d_{ABS}(D^*)(\overline{Y}) = d_{ABS}(D^*)(Y)$.

(*ii*) It is immediate for all the cases $CONV, BAL$ and ABS.

(*iii*) By $conv(Y_1 + Y_2) \subset convY_1 + convY_2$, we get

$$d_{CONV}(D^*)(Y_1 + Y_2) = D^*(conv(Y_1 + Y_2), Y_1 + Y_2)$$

$$\begin{aligned}&\leq D^*(convY_1 + convY_2, Y_1 + Y_2) \leq (\text{see Theorem 6.16, } (i))\\ &\leq D^*(convY_1, Y_1) + D^*(convY_2, Y_2)\\ &= d_{CONV}(D^*)(Y_1) + d_{CONV}(D^*)(Y_2).\end{aligned}$$

Also, by $bal(Y_1 + Y_2) \subset balY_1 + balY_2$ we get

$$d_{BAL}(D^*)(Y_1 + Y_2) = D^*(bal(Y_1 + Y_2), Y_1 + Y_2)$$

$$\begin{aligned}&\leq D^*(balY_1 + balY_2, Y_1 + Y_2) \leq (\text{see Theorem 6.16, } (i))\\ &\leq D^*(balY_1, Y_1) + D^*(balY_2, Y_2)\\ &= d_{BAL}(D^*)(Y_1) + d_{BAL}(D^*)(Y_2).\end{aligned}$$

□

Remarks. 1) The defects in Definition 6.8 can be combined in various ways. For example, taking $d_1(D)(Y) = d_{CONV}(D)(Y) + d_{BAL}(D)(Y)$, then we can call d_1 as defect of convexity-balancing of Y, because, for example, if we take $D \equiv D^*$, then by Theorem 6.17 we get that $Y \in \mathcal{P}_b(X)$ is convex and balanced if and only if $d_1(D)(Y) = 0$. Similarly, we can combine any defect in Definition 6.8 with the measure of noncompactness in Section 3.1. Note that the defect of compactness-convexity (called measure of noncompactness-nonconvexity) appears in *e.g.* Rus [184].

2) It is well-known (see *e.g.* Banas-Goebel [32], Rus [184]) that the measures of noncompactness and of nonconvexity can be introduced by

axioms in abstract spaces. Similarly, by using the properties in Theorems 6.13-6.18 we can introduce in abstract spaces by axioms other defects too, like the defect of balancing, the defect of linearity and so on.

At the end of this section we present applications.

For $Y \in \mathcal{P}_b(X)$, let us denote

$$\begin{aligned} E_{CONV}(Y) &= \inf\{D_H(Y,A); A \in \mathcal{P}_b(X), A \text{ convex}\}, \\ E_{BAL}(Y) &= \inf\{D_H(Y,A); A \in \mathcal{P}_b(X), A \text{ balanced}\}, \end{aligned}$$

the best approximation of a bounded set by convex bounded and by balanced bounded sets, respectively.

Theorem 6.19 *Let $(X, \|\cdot\|)$ be a real normed space. Then for any $Y \in \mathcal{P}_b(X)$ we get*

$$\begin{aligned} E_{CONV}(Y) &\geq \frac{1}{2} d_{CONV}(D_H)(Y), \\ E_{BAL}(Y) &\geq \frac{1}{2} d_{BAL}(D_H)(Y). \end{aligned}$$

Proof. We have, for any $A \in \mathcal{P}_b(X)$, A convex

$$\begin{aligned} d_{CONV}(D_H)(Y) &= D_H(Y, convY) \leq D_H(Y,A) + D_H(A, convA) \\ +D_H(convA, convY) &\leq 2D_H(Y,A). \end{aligned}$$

(We have used the well-known property $D_H(convA, convY) \leq D_H(A,Y)$). Passing to infimum we obtain the desired inequality.

Similarly, for any $A \in \mathcal{P}_b(X)$, A balanced, we have

$$d_{BAL}(D_H)(Y) \leq D_H(Y,A) + D_H(A, balA)$$

$$+D_H(balA, balY) \leq 2D_H(Y,A).$$

□

6.3 Defect of Property for Functionals

Let $(X, +, \cdot)$ be a real linear space. In this section we deal with the defects of various properties for functionals $f : Y \to \mathbf{R}, Y \subset X$. Firstly, let us recall some well-known properties of functionals.

Definition 6.9 Let $(X, +, \cdot)$ be a real linear space.

(i) Let $Y \subset X$ be convex. Then $f : Y \to \mathbf{R}$ is called convex if

$$f(\alpha_1 y_1 + \alpha_2 y_2) \leq \alpha_1 f(y_1) + \alpha_2 f(y_2), \forall y_i \in Y, \alpha_i \geq 0, i = 1, 2, \alpha_1 + \alpha_2 = 1$$

(It is well-known that this is equivalent with $f(\sum_{i=1}^n \alpha_i y_i) \leq \sum_{i=1}^n \alpha_i f(y_i)$, $\forall n \in \mathbf{N}$, $\forall y_i \in Y, \alpha_i \geq 0, i \in \{1, ..., n\}, \sum_{i=1}^n \alpha_i = 1$);

(ii) $f : X \to \mathbf{R}$ is called subadditive if

$$f(x+y) \leq f(x) + f(y), \forall x, y \in X;$$

(iii) $f : X \to \mathbf{R}$ is called positive homogeneous if

$$f(\lambda x) = \lambda f(x), \forall \lambda \geq 0, \forall x \in X;$$

(iv) $f : X \to \mathbf{R}$ is called absolute homogeneous if

$$f(\lambda x) = |\lambda| f(x), \forall \lambda \in \mathbf{R}, \forall x \in X;$$

(v) $f : X \to \mathbf{R}$ is called sublinear if f is subadditive and positive homogeneous;

(vi) $f : X \to \mathbf{R}$ is called quasi-seminorm if f is sublinear and $f(x) \geq 0, \forall x \in X$;

(vii) $f : X \to \mathbf{R}$ is seminorm if f is subadditive and absolute homogeneous;

$(viii)$ $f : X \to \mathbf{R}$ is norm if f is seminorm and $f(x) = 0$ implies $x = 0$.

Concerning these properties we can introduce the following defects.

Definition 6.10 Let $(X, +, \cdot)$ be a real linear space.

(i) If $Y \subset X$ is convex and $f : Y \to \mathbf{R}$, then the n-defect of convexity of f on Y, $n \geq 2$, is defined by

$$\begin{aligned} d_{CONV}^{(n)}(f)(Y) &= \sup \left\{ f\left(\sum_{i=1}^n \alpha_i y_i\right) - \sum_{i=1}^n \alpha_i f(y_i); y_i \in Y, \alpha_i \geq 0, \right. \\ &\qquad \left. i \in \{1, ..., n\}, \sum_{i=1}^n \alpha_i = 1 \right\}. \end{aligned}$$

(ii) Let $Y \subset X$ be a linear subspace of X and $f : Y \to \mathbf{R}$. The defect of subadditivity of f on Y is defined by

$$d_{SADD}(f)(Y) = \sup \{ f(y_1 + y_2) - (f(y_1) + f(y_2)); y_1, y_2 \in Y \}.$$

The defect of absolute homogeneity of f on Y is defined by

$$d_{AH}(f)(Y) = \sup\{|f(\lambda y) - |\lambda| f(y)|; \lambda \in \mathbf{R}, y \in Y\}.$$

We present

Theorem 6.20 *Let $(X, +, \cdot)$ be a real linear space.*

(i) Let $Y \subset X$ be convex and $f : Y \to \mathbf{R}$. Then for all $n \geq 2$ we have

$$d_{CONV}^{(n)}(f)(Y) \leq d_{CONV}^{(n+1)}(f)(Y) \leq d_{CONV}^{(n)}(f)(Y) + d_{CONV}^{(2)}(f)(Y).$$

Also, f is convex on Y if and only if for any fixed $n \geq 2$, we have $d_{CONV}^{(n)}(f)(Y) = 0$.
For $f, g : Y \to \mathbf{R}, \alpha \geq 0$ and $n \geq 2$ we have

$$d_{CONV}^{(n)}(f+g)(Y) \leq d_{CONV}^{(n)}(f)(Y) + d_{CONV}^{(n)}(g)(Y)$$

and $d_{CONV}^{(n)}(\alpha f)(Y) = \alpha d_{CONV}^{(n)}(f)(Y)$.

(ii) Let $Y \subset X$ be linear subspace of X and $f : Y \to \mathbf{R}$. If $f(0) = 0$ then f is subadditive (on Y) if and only if $d_{SADD}(f)(Y) = 0$. f is absolute homogeneous (on Y) if and only if $d_{AH}(f)(Y) = 0$. Moreover, if $f, g : Y \to \mathbf{R}$ then

$$\begin{aligned} d_{SADD}(f+g)(Y) &\leq d_{SADD}(f)(Y) + d_{SADD}(g)(Y), \\ d_{AH}(f+g)(Y) &\leq d_{AH}(f)(Y) + d_{AH}(g)(Y), \\ d_{SADD}(\alpha f)(Y) &= \alpha d_{SADD}(f)(Y), \forall \alpha \geq 0, \\ d_{AH}(\alpha f)(Y) &= |\alpha| d_{AH}(f)(Y), \forall \alpha \in \mathbf{R}. \end{aligned}$$

Proof. (*i*) For any $n \geq 2, \alpha_i \geq 0, y_i \in Y, i \in \{1, ..., n+1\}, \sum_{i=1}^{n+1} \alpha_i = 1$, we have the equality

$$\sum_{i=1}^{n+1} \alpha_i y_i = (1 - \alpha_{n+1}) \left(\sum_{i=1}^{n} \frac{\alpha_i}{1 - \alpha_{n+1}} y_i \right) + \alpha_{n+1} y_{n+1}.$$

Then

$$\begin{aligned} & f\left(\sum_{i=1}^{n+1} \alpha_i y_i\right) - \sum_{i=1}^{n+1} \alpha_i f(y_i) \\ = & \left[f\left((1 - \alpha_{n+1}) \left(\sum_{i=1}^{n} \frac{\alpha_i}{1 - \alpha_{n+1}} y_i \right) + \alpha_{n+1} y_{n+1} \right) \right. \end{aligned}$$

$$- \left((1-\alpha_{n+1}) f \left(\sum_{i=1}^{n} \frac{\alpha_i}{1-\alpha_{n+1}} y_i \right) + \alpha_{n+1} f(y_{n+1}) \right) \Bigg]$$

$$+ \left[(1-\alpha_{n+1}) \left(f \left(\sum_{i=1}^{n} \frac{\alpha_i}{1-\alpha_{n+1}} y_i \right) - \sum_{i=1}^{n} \frac{\alpha_i}{1-\alpha_{n+1}} f(y_i) \right) \right]$$

$$\leq d_{CONV}^{(2)}(f)(Y) + (1-\alpha_{n+1}) d_{CONV}^{(n)}(f)(Y)$$

$$\leq d_{CONV}^{(2)}(f)(Y) + d_{CONV}^{(n)}(f)(Y),$$

and passing to supremum, we get

$$d_{CONV}^{(n+1)}(f)(Y) \leq d_{CONV}^{(n)}(f)(Y) + d_{CONV}^{(2)}(f)(Y).$$

Also, for $\alpha_{n+1} = 0$ and arbitrary $\alpha_i \geq 0, i \in \{1, ..., n\}, y_i \in Y, i \in \{1, ..., n+1\}, \sum_{i=1}^{n} \alpha_i = 1$, we have

$$f\left(\sum_{i=1}^{n} \alpha_i y_i\right) - \sum_{i=1}^{n} \alpha_i f(y_i) = f\left(\sum_{i=1}^{n+1} \alpha_i y_i\right) - \sum_{i=1}^{n+1} \alpha_i f(y_i) \leq d_{CONV}^{(n+1)}(f)(Y),$$

and passing to supremum, we obtain

$$d_{CONV}^{(n)}(f)(Y) \leq d_{CONV}^{(n+1)}(f)(Y).$$

As a conclusion

$$d_{CONV}^{(n)}(f)(Y) \leq d_{CONV}^{(n+1)}(f)(Y) \leq d_{CONV}^{(n)}(f)(Y) + d_{CONV}^{(2)}(f)(Y), \forall n \geq 2.$$

Let $n_0 \geq 2$ be fixed. If f is convex on Y then obviously $d_{CONV}^{(n_0)}(f)(Y) = 0$. Conversely, let us suppose $d_{CONV}^{(n_0)}(f)(Y) = 0$. Then by the above inequalities it easily follows that $d_{CONV}^{(2)}(f)(Y) = 0$, which immediately implies that f is convex on Y.

Then, for any $n \geq 2$ we get

$$d_{CONV}^{(n)}(f+g)(Y) = \sup \left\{ f\left(\sum_{i=1}^{n} \alpha_i y_i\right) - \sum_{i=1}^{n} \alpha_i f(y_i) + g\left(\sum_{i=1}^{n} \alpha_i y_i\right) \right.$$

$$\left. - \sum_{i=1}^{n} \alpha_i g(y_i); y_i \in Y, \alpha_i \geq 0, i \in \{1, ..., n\}, \sum_{i=1}^{n} \alpha_i = 1 \right\}$$

$$\leq \sup \left\{ f\left(\sum_{i=1}^{n} \alpha_i y_i\right) - \sum_{i=1}^{n} \alpha_i f(y_i); y_i \in Y, \alpha_i \geq 0, \right.$$

$$i \in \{1, ..., n\}, \sum_{i=1}^{n} \alpha_i = 1\Bigg\}$$

$$+\sup\left\{g\left(\sum_{i=1}^{n}\alpha_i y_i\right) - \sum_{i=1}^{n}\alpha_i g(y_i); y_i \in Y, \alpha_i \geq 0,\right.$$

$$\left. i \in \{1, ..., n\}, \sum_{i=1}^{n} \alpha_i = 1\right\}$$

$$= d_{CONV}^{(n)}(f)(Y) + d_{CONV}^{(n)}(g)(Y).$$

Similarly, for all $\alpha > 0$ we get

$$\begin{aligned} d_{CONV}^{(n)}(\alpha f)(Y) &= \sup\left\{\alpha\left(f\left(\sum_{i=1}^{n}\alpha_i y_i\right) - \sum_{i=1}^{n}\alpha_i f(y_i)\right);\right. \\ &\qquad \left. y_i \in Y, \alpha_i \geq 0, i \in \{1, ..., n\}, \sum_{i=1}^{n}\alpha_i = 1\right\} \\ &= \alpha \sup\left\{f\left(\sum_{i=1}^{n}\alpha_i y_i\right) - \sum_{i=1}^{n}\alpha_i f(y_i);\right. \\ &\qquad \left. y_i \in Y, \alpha_i \geq 0, i \in \{1, ..., n\}, \sum_{i=1}^{n}\alpha_i = 1\right\} \\ &= \alpha d_{CONV}^{(n)}(f)(Y). \end{aligned}$$

(*ii*) Let $Y \subset X$ be linear subspace of X and $f : Y \to \mathbf{R}$. If $f(0) = 0$ then by choosing $y_1 = y_2 = 0$ we get $d_{SADD}(f)(Y) \geq 0$. Then, it is immediate that f is subadditive if and only if $d_{SADD}(f)(Y) = 0$.
Also, it is easily follows that $d_{AH}(f)(Y) = 0$ if and only if f is absolute homogeneous on Y.

The other four inequalities can easily be proved reasoning exactly as for $d_{CONV}^{(n)}$, which proves the theorem. □

Immediate applications of the defects in Definition 6.10 are to lower estimates of best approximation for bounded functionals, by bounded subadditive or convex functionals.

Indeed, for $Y \subset X$ linear subspace or convex set, let us denote

$$\begin{aligned} B_0(Y) &= \{f : Y \to \mathbf{R}; f \text{ is bounded on } Y \text{ and } f(0) = 0\}, \\ E_{SADD}(f)(Y) &= \inf\{\|f - g\|; g \in B_0(Y), g \text{ is subadditive on } Y\}, \end{aligned}$$

$f \in B_0(Y)$, and

$$B(Y) = \{f : Y \to \mathbf{R}; f \text{ is bounded on } Y\},$$
$$E_{CONV}(f)(Y) = \inf\{\|f-g\|; g \in B(Y), g \text{ is convex on } Y\},$$

$f \in B(Y)$, respectively, where $\|f\| = \sup\{|f(x)|; x \in Y\}$.

We present

Theorem 6.21 *(i) Let $Y \subset X$ be linear subspace of X and $f \in B_0(Y)$. Then*

$$E_{SADD}(f)(Y) \geq \frac{d_{SADD}(f)(Y)}{3}.$$

(ii) Let $Y \subset X$ be convex and $f \in B(Y)$. Then

$$E_{CONV}(f)(Y) \geq \frac{d^{(n)}_{CONV}(f)(Y)}{2}, \forall n \geq 2.$$

Proof. (i) We have, for all $g \in B_0(Y)$ and g subadditive,

$$f(x+y) - f(x) - f(y) \leq f(x+y) - f(x) - f(y)$$

$$\begin{aligned} -(g(x+y) - g(x) - g(y)) &= f(x+y) - g(x+y) \\ +g(x) - f(x) + g(y) - f(y) &\leq 3\|f-g\|, \end{aligned}$$

for all $x, y \in Y$. Passing to supremum with $x, y \in Y$, we get

$$0 \leq d_{SADD}(f)(Y) \leq 3\|f-g\|, \forall g \in B_0(Y), g \text{ subadditive}.$$

Now, passing to infimum with g, we obtain the desired inequality.

(ii) For all $g \in B(Y), g$ convex, $\forall n \geq 2, y_i \in Y, \alpha_i \geq 0, i \in \{1, ..., n\}$, $\sum_{i=1}^{n}\alpha_i = 1$, we have

$$\begin{aligned} f\left(\sum_{i=1}^{n}\alpha_i y_i\right) - \sum_{i=1}^{n}\alpha_i f(y_i) &\leq f\left(\sum_{i=1}^{n}\alpha_i y_i\right) - \sum_{i=1}^{n}\alpha_i f(y_i) \\ -\left(g\left(\sum_{i=1}^{n}\alpha_i y_i\right) - \sum_{i=1}^{n}\alpha_i g(y_i)\right) &= f\left(\sum_{i=1}^{n}\alpha_i y_i\right) - g\left(\sum_{i=1}^{n}\alpha_i y_i\right) \\ +\sum_{i=1}^{n}\alpha_i(g(y_i) - f(y_i)) &\leq \|f-g\| + \sum_{i=1}^{n}\alpha_i\|f-g\| \\ &= 2\|f-g\|. \end{aligned}$$

Passing to supremum with α_i and y_i, we get

$$d_{CONV}^{(n)}(f)(Y) \leq 2\|f-g\|, \forall n \geq 2, \forall g \in B(Y), g \text{ convex}.$$

Now passing to infimum with g, we get the desired inequality, which proves the theorem. □

Remark. By Theorems 6.20, (i) and 6.21, (ii), there exists the limit $\lim_{n\to\infty} d_{CONV}^{(n)}(f)(Y)$ and moreover,

$$\frac{1}{2}\lim_{n\to\infty} d_{CONV}^{(n)}(f)(Y) \leq E_{CONV}(f)(Y).$$

6.4 Defect of Property for Linear Operators on Normed Spaces

Everywhere in this section $(X, \langle\cdot,\cdot\rangle)$ will be a Hilbert space over $\mathbf{R}$ or $\mathbf{C}$ and

$$LC(X) = \{A : X \to X; A \text{ is linear and continuous on } X\}.$$

For any $A \in LC(X)$, we define the norm $|||\cdot||| : LC(X) \to \mathbf{R}_+$ by $|||A||| = \sup\{\|A(x)\|; \|x\| \leq 1\}$, where $\|x\| = \sqrt{\langle x, x\rangle}$.

The following concepts are well-known in functional analysis (see *e.g.* Muntean [155] and Ionescu-Tulcea [103]).

Definition 6.11 The operator $A \in LC(X)$ is called:

(i) symmetric (or Hermitean) if $\langle A(x), y\rangle = \langle x, A(y)\rangle, \forall x, y \in X$;

(ii) normal, if $AA^* = A^*A$, where A^* is the adjoint operator of A defined by $\langle A(x), y\rangle = \langle x, A^*(y)\rangle, \forall x, y \in X$ and $AA^*(x) = A(A^*(x)), \forall x \in X$;

(iii) idempotent, if $A^2 = A$, where $A^2(x) = A(A(x)), \forall x \in X$;

(iv) isometry, if $\langle A(x), A(y)\rangle = \langle x, y\rangle, \forall x, y \in X$ (or equivalently, if $\|A(x)\| = \|x\|, \forall x \in X$);

Also, two operators $A, B \in LC(X)$ are called permutable if $AB = BA$.

Suggested by these properties, we can introduce the following

Definition 6.12 Let $A, B \in LC(X)$.

(i) The defect of symmetry of A is given by

$$d_{SIM}(A) = \sup\{|\langle A(x), y\rangle - \langle x, A(y)\rangle|; \|x\|, \|y\| \leq 1\};$$

(*ii*) The defect of normality of A is given by

$$d_{NOR}(A) = \sup\{\|A(A^*(x)) - A^*(A(x))\|; \|x\| \leq 1\} = |||AA^* - A^*A|||;$$

(*iii*) The defect of idempotency of A is defined by

$$d_{IDEM}(A) = \sup\left\{\left\|A^2(x) - A(x)\right\|; \|x\| \leq 1\right\} = \left|\left|\left|A^2 - A\right|\right|\right|;$$

(*iv*) The defect of isometry of A is given by

$$d_{ISO}(A) = \sup\{\|A(x) - x\|; \|x\| \leq 1\};$$

(*v*) The defect of permutability of A and B is given by

$$d_{PERM}(A, B) = \sup\{\|A(B(x)) - B(A(x))\|; \|x\| \leq 1\} = |||AB - BA|||.$$

Also, if $Y \subset LC(X)$, then we can introduce the corresponding defects for Y, by

$$\begin{aligned} D_{SIM}(Y) &= \sup\{d_{SIM}(A); A \in Y\}; \\ D_{NOR}(Y) &= \sup\{|||AA^* - A^*A|||; A \in Y\}; \\ D_{IDEM}(Y) &= \sup\left\{\left|\left|\left|A^2 - A\right|\right|\right|; A \in Y\right\}; \\ D_{ISO}(Y) &= \sup\{d_{ISO}(A); A \in Y\}; \end{aligned}$$

and

$$d_{PERM}(Y) = \sup\{|||AB - BA|||; A, B \in Y\}.$$

Remark. Obviously d_{IDEM}, d_{ISO} and d_{PERM} have sense only if $(X, \|\cdot\|)$ is a normed space.

We have

Theorem 6.22 *Let $A, B \in LC(X)$.*
(*i*) *A is symmetric if and only if $d_{SIM}(A) = 0$;*
(*ii*) *A is normal if and only if $d_{NOR}(A) = 0$;*
(*iii*) *A is idempotent if and only if $d_{IDEM}(A) = 0$;*
(*iv*) *A is isometric if and only if $d_{ISO}(A) = 0$;*
(*v*) *A and B are permutable if and only if $d_{PERM}(A, B) = 0$.*

Proof. Because $(ii)-(v)$ are immediate from the Definitions 6.11 and 6.12, let us prove (i).

If A is symmetric then obviously $d_{SIM}(A)=0$. Conversely, let us suppose $d_{SIM}(A)=0$. It follows

$$\langle A(x_0), y_0\rangle = \langle x_0, A(y_0)\rangle, \forall \|x_0\|, \|y_0\| \le 1.$$

Let $x, y \in X$ be with $\|x\|, \|y\| > 1$. Obviously there exist x_0, y_0 and $\alpha, \beta \neq 0$, such that $\|x_0\|, \|y_0\| \le 1$ and $x=\alpha x_0, y=\beta y_0$ (actually $\alpha \ge \|x\|, \beta \ge \|y\|$). We get

$$\begin{aligned}\langle A(x), y\rangle &= \langle A(\alpha x_0), \beta y_0\rangle = \alpha\overline{\beta}\langle A(x_0), y_0\rangle \\ &= \alpha\overline{\beta}\langle x_0, A(y_0)\rangle = \langle \alpha x_0, A(\beta y_0)\rangle = \langle x, A(y)\rangle,\end{aligned}$$

which proves (i) and the theorem. □

In what follows, we present some properties and applications of the above concepts of defects.

Theorem 6.23 *Let $A, B \in LC(X)$. We have*

(i) $d_{SIM}(\lambda A) = |\lambda| d_{SIM}(A), \forall \lambda \in \mathbf{R}$;
$d_{NOR}(\lambda A) = |\lambda|^2 d_{NOR}(A), \forall \lambda \in \mathbf{R}$ *or* $\mathbf{C}$.

(ii) $d_{SIM}(A+B) \le d_{SIM}(A) + d_{SIM}(B)$;
$d_{SIM}(AB) \le 2|||AB - I|||$, *where* $I(x) = x, \forall x \in X$;
$d_{NOR}(A+B) \le d_{NOR}(A) + d_{NOR}(B) + d_{PERM}(A, B^*)$
$+ d_{PERM}(A^*, B)$.

(iii) $d_{SIM}(A^{-1}) = d_{SIM}(A)$, *if there exists* A^{-1} *and* A *is isometry.*

(iv) $|d_{SIM}(A) - d_{SIM}(A^*)| \le 2|||A - A^*|||$;
$d_{NOR}(A^*) = d_{NOR}(A)$;
$d_{NOR}(A + A^*) \le d_{NOR}(A) + d_{NOR}(A^*)$;
$|d_{IDEM}(A) - d_{IDEM}(A^*)| \le |||A - A^*||| + \left|\left|\left|A^2 - (A^*)^2\right|\right|\right|$;
$d_{IDEM}(I - A) = d_{IDEM}(A)$;
$d_{PERM}(A, A^*) = d_{NOR}(A)$.

Proof. (i) We have

$$\begin{aligned}d_{SIM}(\lambda A) &= \sup\{|\langle \lambda A(x), y\rangle - \langle x, \lambda A(y)\rangle|; \|x\|, \|y\| \le 1\} \\ &= \sup\{|\lambda(\langle A(x), y\rangle - \langle x, A(y)\rangle)|; \|x\|, \|y\| \le 1\} \\ &= |\lambda| d_{SIM}(A), \forall \lambda \in \mathbf{R},\end{aligned}$$

and

$$d_{NOR}(\lambda A) = |||(\lambda A)(\lambda A^*) - (\lambda A^*)(\lambda A)||| = |\lambda|^2 d_{NOR}(A).$$

(*ii*) It follows

$$d_{SIM}(A+B) = \sup\{|\langle A(x) + B(x), y\rangle - \langle x, A(y) + B(y)\rangle|; \|x\|, \|y\| \le 1\}$$

$$\begin{aligned} &\le \sup\{|\langle A(x), y\rangle - \langle x, A(y)\rangle| + |\langle B(x), y\rangle - \langle x, B(y)\rangle|; \|x\|, \|y\| \le 1\} \\ &\le d_{SIM}(A) + d_{SIM}(B). \end{aligned}$$

Then, by

$$|\langle AB(x), y\rangle - \langle x, AB(y)\rangle| = |\langle (AB - I)(x), y\rangle - \langle x, (I - AB)(y)\rangle|$$

$$\begin{aligned} &\le |\langle (AB - I)(x), y\rangle| + |\langle x, (I - AB)(y)\rangle| \\ &\le |||AB - I||| \cdot \|x\| \cdot \|y\| + |||AB - I||| \cdot \|x\| \cdot \|y\|, \end{aligned}$$

and passing to supremum with $\|x\|, \|y\| \le 1$, we get $d_{SIM}(AB)$ $\le 2\,|||AB - I|||$.
Also,

$$d_{NOR}(A+B) = |||(A+B)(A^* + B^*) - (A^* + B^*)(A+B)|||$$

$$\begin{aligned} &= |||(AA^* - A^*A) + (BB^* - B^*B) + (BA^* - A^*B) + (AB^* - B^*A)||| \\ &\le d_{NOR}(A) + d_{NOR}(B) + d_{PERM}(A^*, B) + d_{PERM}(A, B^*). \end{aligned}$$

(*iii*) If A is isometry then we get $\left\|A^{-1}(x)\right\| = \|x\|$ and

$$\begin{aligned} d_{SIM}\left(A^{-1}\right) &= \sup\left\{\left|\left\langle A^{-1}(x), y\right\rangle - \left\langle x, A^{-1}(y)\right\rangle\right|; \|x\|, \|y\| \le 1\right\} \\ &= \sup\left\{\left|\left\langle A^{-1}(x), A\left(A^{-1}(y)\right)\right\rangle - \left\langle A\left(A^{-1}(x)\right), A^{-1}(y)\right\rangle\right|;\right. \\ &\qquad \left.\left\|A^{-1}x\right\|, \left\|A^{-1}y\right\| \le 1\right\} \\ &= d_{SIM}(A). \end{aligned}$$

(*iv*) We have

$$\begin{aligned} d_{SIM}(A^*) &= \sup\{|\langle A^*(x), y\rangle - \langle A(x), y\rangle + \langle A(x), y\rangle - \langle x, A(y)\rangle \\ &\qquad + \langle x, A(y)\rangle - \langle x, A^*(y)\rangle|; \|x\|, \|y\| \le 1\} \\ &\le \sup\{|\langle (A^* - A)(x), y\rangle|; \|x\|, \|y\| \le 1\} \\ &\qquad + \sup\{|\langle A(x), y\rangle - \langle x, A(y)\rangle|; \|x\|, \|y\| \le 1\} \end{aligned}$$

$$
\begin{aligned}
&+\sup\{|\langle x,(A^{*}-A)(y)\rangle|\,;\|x\|,\|y\|\le 1\}\\
\le\;& |||A^{*}-A|||+d_{SIM}(A)+|||A-A^{*}|||,
\end{aligned}
$$

which implies

$$d_{SIM}(A^{*})-d_{SIM}(A)\le 2\,|||A^{*}-A|||.$$

By symmetry, we get

$$d_{SIM}(A)-d_{SIM}(A^{*})\le 2\,|||A^{*}-A|||,$$

which implies

$$|d_{SIM}(A)-d_{SIM}(A^{*})|\le 2\,|||A^{*}-A|||.$$

Then,

$$d_{NOR}(A^{*})=\left|\left|\left|A^{*}(A^{*})^{*}-(A^{*})^{*}A^{*}\right|\right|\right|=|||A^{*}A-AA^{*}|||=d_{NOR}(A),$$

and taking $B=A^{*}$ in the second inequality of above (ii), we immediately obtain

$$d_{NOR}(A+A^{*})\le d_{NOR}(A)+d_{NOR}(A^{*})\,(=2d_{NOR}(A)).$$

Also,

$$
\begin{aligned}
d_{IDEM}(A^{*}) &= \left|\left|\left|(A^{*})^{2}-A^{*}\right|\right|\right|=\left|\left|\left|(A^{*})^{2}-A^{2}+A^{2}-A+A-A^{*}\right|\right|\right|\\
&\le \left|\left|\left|(A^{*})^{2}-A^{2}\right|\right|\right|+\left|\left|\left|A^{2}-A\right|\right|\right|+|||A-A^{*}|||,
\end{aligned}
$$

which implies

$$d_{IDEM}(A^{*})-d_{IDEM}(A)\le |||A-A^{*}|||+\left|\left|\left|A^{2}-(A^{*})^{2}\right|\right|\right|.$$

By symmetry,

$$d_{IDEM}(A)-d_{IDEM}(A^{*})\le |||A-A^{*}|||+\left|\left|\left|A^{2}-(A^{*})^{2}\right|\right|\right|,$$

which concludes

$$|d_{IDEM}(A)-d_{IDEM}(A^{*})|\le |||A-A^{*}|||+\left|\left|\left|A^{2}-(A^{*})^{2}\right|\right|\right|.$$

Then

$$
\begin{aligned}
\left\|(I-A)^{2}(x)-(I-A)(x)\right\| &= \left\|(I-2A+A^{2})(x)-(I-A)(x)\right\|\\
&= \left\|A^{2}(x)-A(x)\right\|,
\end{aligned}
$$

which immediately proves that

$$d_{IDEM}(I-A)=d_{IDEM}(A).$$

Finally, it is obvious that

$$d_{PERM}(A,A^*)=|||AA^*-A^*A|||=d_{NOR}(A),$$

which proves the theorem. □

For given $A\in LC(X)$, let us introduce the following quantities of best approximation:

$$E_{SIM}(A)=\inf\{|||A-B|||\,;B\in LC(X)\,,B\text{ is symmetric}\},$$

$$\begin{aligned}E_{NOR}(A) &= \inf\{|||A-B|||+|||A^*-B^*|||\,;B\in LC(X)\,,\\ &\quad |||B|||\le 1, B\text{ is normal}\},\end{aligned}$$

$$\begin{aligned}E_{IDEM}(A) &= \inf\left\{|||A-B|||+\left|\left|\left|A^2-B^2\right|\right|\right|\,;B\in LC(X)\,,\right.\\ &\quad \left. B\text{ is idempotent}\right\},\end{aligned}$$

$$E_{ISO}(A)=\inf\{|||A-B|||\,;B\in LC(X)\,,B\text{ is isometry}\}.$$

The following lower estimates for these quantities hold.

Theorem 6.24 *Let $A\in LC(X)$. We have:*

$$E_{SIM}(A)\ge\frac{d_{SIM}(A)}{2},$$
$$E_{NOR}(A)\ge\frac{d_{NOR}(A)}{2},\ \textit{if}\ |||A|||\le 1,$$
$$E_{IDEM}(A)\ge d_{IDEM}(A),$$
$$E_{ISO}(A)\ge d_{ISO}(A).$$

Proof. Let $A,B\in LC(X)$. If B is symmetric then we obtain

$$|\langle A(x),y\rangle-\langle x,A(y)\rangle|=$$

$$|\langle A(x),y\rangle-\langle B(x),y\rangle+\langle B(x),y\rangle-\langle x,B(y)\rangle+\langle x,B(y)\rangle-\langle x,A(y)\rangle|$$

$$\le|\langle(A-B)(x),y\rangle|+|\langle B(x),y\rangle-\langle x,B(y)\rangle|+|\langle x,(B-A)(y)\rangle|$$

$$\leq |||A - B||| \cdot \|x\| \cdot \|y\| + |||B - A||| \cdot \|x\| \cdot \|y\| .$$

Passing to supremum with $\|x\|, \|y\| \leq 1$ we get $d_{SIM}(A) \leq 2\,|||A - B|||, \forall B \in LC(X), B$ symmetric. Passing to infimum with B, we get the first lower estimate.

If B is normal and $|||B||| \leq 1$, then

$$|||AA^* - A^*A||| = |||AA^* - BB^* + BB^* - A^*A|||$$

$$\leq |||AA^* - BA^* + BA^* - BB^*||| + |||BB^* - B^*A + B^*A - A^*A|||$$

$$\leq |||AA^* - BA^*||| + |||BA^* - BB^*||| + |||B^*B - B^*A||| + |||B^*A - A^*A|||$$

$$\leq |||A - B||| \cdot |||A^*||| + |||B||| \cdot |||A^* - B^*||| + |||B^*||| \cdot |||B - A|||$$

$$+ |||B^* - A^*||| \cdot |||A||| \leq 2\left(|||A - B||| + |||A^* - B^*|||\right).$$

Passing to infimum with B normal, $|||B||| \leq 1$, we immediately get the desired conclusion.

If $B \in LC(X)$ is idempotent then

$$\left|\left|\left|A^2 - A\right|\right|\right| = \left|\left|\left|A^2 - B^2 + B^2 - B + B - A\right|\right|\right| \leq \left|\left|\left|A^2 - B^2\right|\right|\right| + |||A - B|||,$$

and passing to infimum with B, we get

$$d_{IDEM}(A) \leq E_{IDEM}(A).$$

If $B \in LC(X)$ is isometry, then

$$\begin{aligned} |\,\|A(x)\| - \|x\|\,| &= |\,\|A(x)\| - \|B(x)\| + \|B(x)\| - \|x\|\,| \\ &\leq |\,\|A(x)\| - \|B(x)\|\,| + |\,\|B(x)\| - \|x\|\,| \\ &\leq \|A(x) - B(x)\| + |\,\|B(x)\| - \|x\|\,|. \end{aligned}$$

Passing to supremum with $\|x\| \leq 1$, it follows $d_{ISO}(A) \leq |||A - B|||, \forall B$ isometry, and passing to infimum with B, we obtain

$$d_{ISO}(A) \leq E_{ISO}(A),$$

which proves the theorem. □

Remarks. 1) Taking into account the classical development of spectral theory for symmetric linear operators (see *e.g.* Muntean [155], p. 155-166) we think that the quantity $d_{SIM}(A)$ can play an important role to the study of the spectrum of a linear operator A, not necessarily symmetric.

2) Let $A, B \in LC(X)$ be two symmetric operators. Then according to *e.g.* Bohm [42], p. 206, relations (13) and (24)), the defect $d_{PERM}(A, B)$ can have applications to quantum theory, at the minimization of uncertainties for A and B.

At the end of this section, starting from the remark that for a subset $Y \subset LC(X), d_{PERM}(Y)$ coincides in fact with the defect of commutativity of Y with respect to the binary operation of composition on $LC(X)$, denoted by $e^d_{COM}(\circ)(Y)$ (where $d(x,y) = \|x - y\|$), we will transfer here some corresponding results in the next Chapter 7.

Let us consider two fixed $f, g \in LC(X)$ such that $f \circ g = I$ and define the (f, g)-dual of $\circ$ by (see Definition 7.2)

$$(A \odot B)(x) = g[(f \circ A) \circ (f \circ B)](x), \forall x \in X$$

(or shortly $A \odot B = gfAfB$). According to Theorem 7.2, (i), if moreover $\|g(x) - g(y)\| \le \varepsilon \|x - y\|, \forall x, y \in X$, where $0 < \varepsilon < 1$, then

$$e^d_{COM}(\odot)(Y) \le \varepsilon \cdot e^d_{COM}(\circ)(f(Y)).$$

Remark. Of course, the above relations are not trivial for $Y \subset LC(X), Y$ bounded (that is $\exists M > 0$ such that $|||A||| \le M, \forall A \in Y$). Let $Y \subset LC(X)$ be bounded. Denoting by

$$COM(LC(X)) = \{F; F \text{ is a commutative binary operation on } LC(X)\},$$

defining a distance (between F and the composition law $\circ$)

$$D(\circ, F)(Y) = \sup\{|||AB - F(A, B)|||; A, B \in Y\}$$

and the quantity of best approximation

$$E_{COM}(\circ)(Y) = \inf\{D(\circ, F)(Y); F \in COM(LC(X))\},$$

by Theorem 7.6 we get

$$\frac{1}{2} d_{PERM}(Y) \le E_{COM}(\circ)(Y).$$

6.5 Defect of Fixed Point

Given a set X in a Banach space and $T : X \to X$ that has no fixed points, it is natural to see how near T has fixed points, that is estimation of the quantity $\inf\{\|x - T(x)\| : x \in X\}$ is required. The study of this problem called minimal displacement of points under mappings, was started in the papers of Franchetti [77], Furi-Martelli [81], Goebel [91], Reich [171] and [172] and in the book Goebel-Kirk [92], p. 210-218. In this section, new contributions to this topic are obtained. Many examples illustrating the concepts are presented. Firstly, we give definitions and examples concerning the concepts of defect of fixed point and of best almost-fixed point for a mapping. Then, for various classes of mappings, we study the introduced concepts. Finally, we deal with some applications to various kinds of equations that have no solutions. Also, we use the concept of fixed point to introduce and study the concept of defect of property of fixed point for topological spaces.

Let us introduce the following.

Definition 6.13 (see Goebel-Kirk [92], p.210). Let (X, d) be a metric space and $M \subseteq X$. The defect of fixed point of $f : M \to X$ is defined by $e_d(f; M) = \inf\{d(x, f(x)); x \in M\}$.
If there exists $x_0 \in M$ with $e_d(f; M) = d(x_0, f(x_0))$, then x_0 will be called best almost-fixed point for f on M.

Remarks. 1) If $f : M \to X$ has a fixed point, *i.e.* $\exists x_0 \in M$ with $f(x_0) = x_0$, then $e(f; M) = 0$. On the other hand, the defect can be 0 without f to have fixed point. For example, if $M = X = [1, +\infty)$ and $f(x) = x + \frac{1}{x}$ then $e_d(f; [1, +\infty)) = 0$ (where the metric d is generated by absolute value$|\cdot|$) but f has no fixed point on M.

2) Let us suppose that $e_d(f; X) = 0$. Then obviously there exists a sequence $(x_n)_{n\in\mathbf{N}}$ in X, such that $d(x_n, f(x_n)) \overset{n\to\infty}{\to} 0$, sequence which is called asymptotically f-regular. Although this condition is not sufficient for the existence of fixed points for f, imposing some additional assumptions on f can be derived fixed point theorems (see *e.g.* Engl [73], Guay-Singh [97], Rhoades-Sessa-Khan-Swaleh [173]).

3) If $M \subseteq X$ is compact and $f : M \to X$ is continuous, then introducing $F : M \to \mathbf{R}_+$ defined by $F(x) = d(x, f(x))$, it easily follows that F is continuous on M and as a consequence, there exists $x_0 \in M$ with $e_d(f; M) = d(x_0, f(x_0))$. Also, in this case $e_d(f; M) = 0$ iff f has a fixed point.

4) If M is not compact or f is not continuous on M, then in general does not exist $x_0 \in M$ with $e_d(f; M) = d(x_0, f(x_0))$. The following two simple counterexamples show us the above statement: $M = X = (0,1), d(x,y) = |x - y|, f : X \to X, f(x) = x^2$ and $M = X = [0,1], d(x,y) = |x - y|, f : X \to X, f(x) = 1$ if $x = 0, f(x) = 0$ if $x \in (0,1]$, respectively.

5) Let $(X, \|\cdot\|)$ be a normed space and $M \subseteq X$ be nonempty compact set. In Ky Fan [75] it is proved that for any continuous map $f : M \to X$, there exists a point $x_0 \in M$ such that $\|x_0 - f(x_0)\| = \inf\{\|f(x_0) - y\|; y \in M\}$. It follows that if moreover $f : M \to (X \setminus M)$, then $0 < e_d(f; M) \leq \|x_0 - f(x_0)\|$ (where the metric d is generated by norm $\|\cdot\|$). Indeed, let us suppose that $e_d(f; M) = 0$. It follows that there exists $x_n \in M, n \in \mathbf{N}$, such that $\|x_n - f(x_n)\| \overset{n\to\infty}{\to} 0$. Because M is compact too there is $y_0 \in M$ such that $\|y_0 - f(y_0)\| = 0$, *i.e.* $y_0 = f(y_0)$, which is a contradiction because $y_0 \in M$ and $f(y_0) \in X \setminus M$.

Example 6.1 Let B denote the unit ball in the space

$$C[-1,1] = \{x : [-1,1] \to \mathbf{R} \,|x \text{ is continuous on } [-1,1]\}$$

with the uniform metric $d(x,y) = \sup\{|x(t) - y(t)| : t \in [-1,1]\}$, and for fixed $k > 1$, set

$$\alpha(t) = \begin{cases} -1, & \text{if } -1 \leq t \leq -\frac{1}{k} \\ kt, & \text{if } -\frac{1}{k} \leq t \leq \frac{1}{k} \\ 1, & \text{if } \frac{1}{k} \leq t \leq 1. \end{cases}$$

If the mapping $T : B \to B$ is defined by

$$(Tx)(t) = \alpha(\max\{-1, \min\{1, x(t) + 2t\}\}),$$

then $d(x, Tx) \geq 1 - \frac{1}{k}$ for each $x \in C[-1,1]$ (see Goebel-Kirk [92], p.212). This implies $e_d(T; B) \geq 1 - \frac{1}{k}$.

Example 6.2 Let $X = c_0$ be the Banach space of the sequences that converge to 0, endowed with the norm $\|x\| = \sup\{|x_n|; n \in \mathbf{N}\}, x = (x_n)_{n\in\mathbf{N}}$.

If $f : X \to X$ is defined by

$$f(x_1, x_2, ..., x_n, ...) = \left(1, |x_2|^{\frac{1}{2}}, ..., |x_n|^{\frac{1}{2}}, ...\right) + \left(1, \frac{1}{2}, ..., \frac{1}{n}, ...\right),$$

then f is continuous and for every $k \in \mathbf{R}$, the equation $f(x) = kx$ has no solutions (see Rădulescu-Rădulescu [170], p.67).

Let $M_0 \subset X$,

$$M_0 = \left\{x = (x_1, x_2, ..., x_n, ...) \in c_0 : (|x_n|)_{n\geq 1} \text{ is decreasing}\right\}$$

and $f_0 : M_0 \to X, f_0(x) = f(x)+x$. Then $d(f_0(x), x) = \max\left\{2, |x_2|^{\frac{1}{2}} + \frac{1}{2}\right\}$, where $d(f_0(x), x) = ||f_0(x) - x||$ and

$$e_d(f_0; M_0) = \inf\{d(f_0(x), x) : x \in M_0\} = 2.$$

The set of the best almost-fixed points for f_0 on M_0 is given by

$$\left\{x = (x_1, x_2, ..., x_n, ...) \in M_0 : |x_2| \leq \frac{9}{4}\right\}.$$

Let $M_1 \subset X$,

$$M_1 = \left\{x = (x_1, x_2, ..., x_n, ...) \in c_0 : (x_n)_{n\geq 1} \text{ is increasing}\right\},$$

and $f_1 : M_1 \to X, f_1(x) = f(x)$. Because $x = (x_1, x_2, ..., x_n, ...) \in M_1$ implies $x_n \leq 0, \forall n \geq 1$, we obtain that $\left(|x_n|^{\frac{1}{2}} - x_n\right)_{n\geq 2}$ is a decreasing and positive sequence, therefore

$$d(f_1(x), x) = \max\left\{2 - x_1, |x_2|^{\frac{1}{2}} + \frac{1}{2} - x_2\right\}.$$

The defect of fixed point of f_1 on M_1 is

$$e_d(f_1; M_1) = \inf\{d(f_1(x), x) : x \in M_1\} = 2$$

and the unique best almost-fixed points for f_1 on M_1 is $x_0 = (0, 0, ..., 0, ...)$.

Example 6.3 Let $F : C[0,1] \to C[0,1]$ be defined by

$$(Fx)(t) = (\max\{t, |x(t) - x(0)|\})^{\frac{1}{2}}.$$

The function F is continuous and for every $k \in \mathbf{R}$, the equation $Fx = kx$ has no solutions (see Rădulescu-Rădulescu [170], p.67).
Let $A : C[0,1] \to C[0,1]$ be defined by

$$(Ax)(t) = (Fx)(t) + (1-k)x(t)$$

and let $x \in C[0,1]$. We denote $m = \inf_{t\in[0,1]} x(t)$, $M = \sup_{t\in[0,1]} x(t)$. We get

$$\begin{aligned} |(Ax)(t) - x(t)| &= |(Fx)(t) - kx(t)| \leq |(Fx)(t)| + |kx(t)| \\ &= (\max(t, |x(t) - x(0)|))^{\frac{1}{2}} + |k|\,|x(t)| \end{aligned}$$

for every $t \in [0,1]$. Because the above inequalities become equalities if $x(t) \geq 0, \forall t \in [0,1], k \leq 0$ and 1 is the maximum point of x or if $x(t) \leq 0, \forall t \in [0,1], k \geq 0$ and 1 is the minimum point of x, denoting

$$\begin{aligned} C_+[0,1] &= \{x \in C[0,1] : x(t) \geq 0, \forall t \in [0,1]\}, \\ C_-[0,1] &= \{x \in C[0,1] : x(t) \leq 0, \forall t \in [0,1]\}, \\ C_+^*[0,1] &= \{x \in C_+[0,1] : x(1) \geq x(t), \forall t \in [0,1]\}, \\ C_-^*[0,1] &= \{x \in C_-[0,1] : x(1) \leq x(t), \forall t \in [0,1]\}, \end{aligned}$$

and considering the uniform metric d, we obtain

$$e_d\left(A; C_+^*[0,1]\right) = \inf\left\{(\max(1, M - x(0)))^{\frac{1}{2}} - kM : x \in C_+^*[0,1]\right\} = 1,$$

for all $k \leq 0$, and

$$e_d\left(A; C_-^*[0,1]\right) = \inf\left\{(\max(1, x(0) - m))^{\frac{1}{2}} - km : x \in C_-^*[0,1]\right\} = 1,$$

for all $k \geq 0$. A best almost-fixed point for F on $C_+^*[0,1]$ and on $C_-^*[0,1]$ is the same, namely the constant zero function (if $k \neq 0$ then it is unique).
Now, let us consider the metric $d : C[0,1] \to \mathbf{R}$ defined by

$$d(x,y) = \left(\int_0^1 (x(t) - y(t))^2 \,\mathrm{d}t\right)^{\frac{1}{2}}.$$

We observe that $x(t) \geq 0, \forall t \in [0,1]$ and $k \leq 0$ or $x(t) \leq 0, \forall t \in [0,1]$ and $k \geq 0$ implies

$$(Fx)(t) - kx(t) = (\max(t, |x(t) - x(0)|))^{\frac{1}{2}} - kx(t) \geq t^{\frac{1}{2}} - kx(t) \geq t^{\frac{1}{2}},$$

for all $t \in [0,1]$ with equalities when $x(t) = 0, \forall t \in [0,1]$. By using the above notations, we have

$$e_d(A; C_+[0,1]) = \inf\left\{\left(\int_0^1 ((Fx)(t) - kx(t))^2 \mathrm{d}t\right)^{\frac{1}{2}} : x \in C_+[0,1]\right\}$$
$$= \left(\int_0^1 t\mathrm{d}t\right)^{\frac{1}{2}} = 1, \forall k \le 0$$

and

$$e_d(A; C_-[0,1]) = \inf\left\{\left(\int_0^1 ((Fx)(t) - kx(t))^2 \mathrm{d}t\right)^{\frac{1}{2}} : x \in C_-[0,1]\right\}$$
$$= \left(\int_0^1 t\mathrm{d}t\right)^{\frac{1}{2}} = 1, \forall k \ge 0.$$

A best almost-fixed point for A is the constant function zero.

Taking into account the Remarks 1)-3) after Definition 6.13, we consider some results for mappings f that satisfy $e_d(f; M) > 0$, *i.e.* for mappings that have not neither fixed points nor asymptotically regular sequences. The first result is the following.

Theorem 6.25 *Let (X,d) be a metric space and $f : X \to X$ be a non-expansive mapping, i.e. $d(f(x), f(y)) \le d(x,y)$, for all $x, y \in X$. Then $e_d(f^n; X) \le n e_d(f; X)$ and $\inf\{d(f^n(x), f^{n+1}(x)); x \in X\} = e_d(f; X)$, $\forall n \in \{1, 2, ...\}$, where f^n denotes the n-th iterate of f.*

Proof. Firstly, by

$$d(x, f^n(x)) \le d(x, f(x)) + d(f(x), f^2(x)) + \ldots$$
$$+ d(f^{n-1}(x), f^n(x)) \le nd(x, f(x)),$$

we obtain the inequality. Secondly, by

$$e_d(f; X) \le d(f^n(x), f^{n+1}(x)) \le d(f^{n-1}(x), f^n(x)) \le \ldots \le d(x, f(x)),$$

passing to infimum, we obtain the desired equality. □

Remarks. 1) If (X,d) is compact and $f : X \to X$ is contractive, *i.e.* $d(f(x), f(y)) < d(x,y)$, for all $x, y \in X, x \ne y$, then it is known that f has a fixed point $x_0 = f(x_0)$ and in this case obviously we have $e_d(f; X) =$

$e_d(f^n; X) = 0, \forall n \in \{1, ..., n\}$. But a nonexpansive mapping even on a compact metric space (X, d), has no in general a fixed point, so the inequality in Theorem 6.25 is not a trivial one.

2) If $f : X \to X$ is α-Lipschitz with $\alpha > 1$, *i.e.* $d(f(x), f(y)) \leq \alpha d(x, y)$, $\forall x, y \in X$, then reasoning as in the proof of Theorem 6.25 we obtain the inequality

$$e_d(f; X) \leq \inf\left\{d\left(f^n(x), f^{n+1}(x)\right); x \in X\right\} \leq \alpha^n e_d(f; X), \forall n \geq 1.$$

Let us denote

$$\mathcal{F} = \{g : X \to X; g \text{ has fixed point in } X\},$$

where (X, d) is supposed to be compact. A natural question is to find the best approximation of a function $f : X \to X, f \notin \mathcal{F}$, by elements in $\mathcal{F}$. In this sense, we can define

$$E_{\mathcal{F}}(f) = \inf\{D(f, g); g \in \mathcal{F}\},$$

where $D(f, g) = \sup\{d(f(x), g(x)); x \in X\}$.

The following lower estimate for $E_{\mathcal{F}}(f)$ holds.

Theorem 6.26 *We have $e_d(f; X) \leq E_{\mathcal{F}}(f)$, for any $f : X \to X$.*

Proof. Let $g \in \mathcal{F}$ be with the fixed point y. We get

$$d(y, f(y)) \leq d(y, g(y)) + d(g(y), f(y)) \leq D(f, g),$$

i.e.

$$e_d(f; X) \leq D(f, g)$$

for all $g \in \mathcal{F}$. As a consequence, $e_d(f; X) \leq E_{\mathcal{F}}(f)$, which proves the theorem. □

Given a metric space (X, d) no necessarily compact and $f : X \to X$, an important problem is to establish the existence of points $x_0 \in X$ with $e_d(f; X) = d(x_0, f(x_0))$. Notice that Franchetti [77], Furi-Martelli [81], Goebel [91], Goebel-Kirk [92], Reich [171] and [172], do not treat this problem.

In this sense might be useful the following results.

Theorem 6.27 *Let (X,d) be a complete metric space and $f : X \to X$ be continuous on X with $e_d(f;X) > 0$. Let $\varepsilon > 0$ be arbitrary and $x \in X$ with*

$$e_d(f;X) \leq d(x, f(x)) \leq e_d(f;X) + \varepsilon.$$

Then there exists $x_\varepsilon \in X$ such that $d(x_\varepsilon, x) \leq 1, 0 < d(x_\varepsilon, f(x_\varepsilon)) \leq d(x, f(x))$ and $0 < d(x_\varepsilon, f(x_\varepsilon)) < d(y, f(y)) + \varepsilon d(x_\varepsilon, y)$, for all $y \in X, y \neq x_\varepsilon$.

Proof. Let us define $F : X \to \mathbf{R}_+, F(x) = d(x, f(x))$. By hypothesis, F is continuous on X and bounded from below. Applying the well-known Ekeland's variational principle to F (see *e.g.* Ekeland [71] and [72] or Barbu [33], p. 33, Th. 3.1) we obtain the statement in theorem. □

By replacing d with $\varepsilon^{-\frac{1}{2}}d$, in Theorem 6.27 we immediately obtain:

Theorem 6.28 *Let (X,d) be a complete metric space and $f : X \to X$ be continuous on X, with $e_d(f;X) > 0$. Let $\varepsilon > 0$ be arbitrary and $x \in X$ be with*

$$0 < e_d(f;X) \leq d(x, f(x)) \leq e_d(f;X) + \varepsilon.$$

Then there exists $x_\varepsilon \in X$ such that $d(x_\varepsilon, x) \leq \sqrt{\varepsilon}, 0 < d(x_\varepsilon, f(x_\varepsilon)) \leq d(x, f(x))$ and $0 < d(x_\varepsilon, f(x_\varepsilon)) < d(y, f(y)) + \sqrt{\varepsilon} d(x_\varepsilon, y)$, for all $y \in X, y \neq x_\varepsilon$.

Deeper results can be obtained if we consider $(X, \|\cdot\|)$ as a real Banach space, because in this case we can use the differential calculus too in normed spaces.

Firstly we need some well-known concepts.

Definition 6.14 Let $(X, \|\cdot\|_1), (Y, \|\cdot\|_2)$ be normed spaces and $f : X \to Y$. We say that f is Gâteaux differentiable at a point $x \in X$, if there exists the limit $\lim_{t\to 0, t\neq 0} \frac{f(x+th)-f(x)}{t}$, for all $h \in X$.

We say that f is Gâteaux derivable on $M \subseteq X$ if for each $x \in X$ there exists a mappings denoted $\nabla f(x) \in L(X,Y) = \{G : X \to Y; G\text{-linear}\}$, such that

$$\lim_{t\to 0, t\neq 0} \frac{f(x+th) - f(x)}{t} = (\nabla f(x))(h),$$

for all $h \in X$. The mapping $\nabla f(x)$ is called the gradient of f at the point x.

We present

Theorem 6.29 *Let $(X, \langle , \rangle)$ be a real Hilbert space and $f : X \to X$ be Gâteaux derivable on X with $e_d(f; X) > 0$. Then for each $\varepsilon > 0$ there exists $x_\varepsilon \in X$ such that*

$$0 < e_d(f; X) \le \|x_\varepsilon - f(x_\varepsilon)\| \le e_d(f; X) + \varepsilon$$

and

$$\left\langle \frac{x_\varepsilon - f(x_\varepsilon)}{\|x_\varepsilon - f(x_\varepsilon)\|}, (1_X - \nabla f(x_\varepsilon))(h) \right\rangle \le \sqrt{\varepsilon},$$

where $1_X(h) = h$, for all $h \in X$, $\|x\| = \sqrt{\langle x, x \rangle}, x \in X$ and the metric d is generated by the norm $\|\cdot\|$.

Proof. Let us denote $F(x) = \|x - f(x)\| > 0, x \in X$. Obviously $F : X \to \mathbf{R}_+$ is bounded from below. Then

$$\begin{aligned} & \lim_{t \to 0, t \neq 0} \frac{F(x+th) - F(x)}{t} \\ = & \lim_{t \to 0, t \neq 0} \left(\frac{F^2(x+th) - F^2(x)}{t} \cdot \frac{1}{F(x+th) + F(x)} \right) \\ = & \left(\nabla F^2(x)\right)(h) \cdot \frac{1}{2F(x)}. \end{aligned}$$

But because $F^2(x) = \langle x - f(x), x - f(x) \rangle$, simple calculations show us that
$\left(\nabla F^2(x)\right)(h) = 2\langle x - f(x), (1_X - \nabla f(x))(h) \rangle, \forall h \in X.$

As a consequence, $\exists\ (\nabla F(x))(h) = \left\langle \frac{x - f(x)}{\|x - f(x)\|}, (1_X - \nabla f(x))(h) \right\rangle$, for all $x \in X$ and all $h \in X$. Then taking in the last inequality of Theorem 6.28, $y = x_\varepsilon \pm th, h \in X, t \neq 0$, and passing with $t \to 0$ (see also *e.g.* Barbu [33], Theorem 3.4, p.35), we obtain

$$0 < e_d(f; X) \le \|x_\varepsilon - f(x_\varepsilon)\| \le e_d(f; X) + \varepsilon$$

and

$$|(\nabla F(x))(h)| \le \sqrt{\varepsilon},$$

which proves the theorem. □

Because the main problem is to prove the existence of the minimum points for the nonlinear functional $F(x) = \|x - f(x)\|$, obviously we can use well-known variational methods.

In this sense, it is immediate the following.

Theorem 6.30 *Let $(X, \langle , \rangle)$ be a real Hilbert space, $\|x\| = \sqrt{\langle x, x \rangle}, x \in X$, the metric generated by norm $\|\cdot\|$, denoted by d and $M \subseteq X$.*

(i) Let $f : M \to X$ be Gâteaux derivable on $x_0 \in M, x_0$ interior point of M, with $e_d(f; M) > 0$. If $\|x_0 - f(x_0)\| = e_d(f; M), x_0 \in M$, then

$$\langle x_0 - f(x_0), (1_X - \nabla f(x_0))(h) \rangle = 0,$$

for all $h \in X$.

(ii) If $F(x) = \|x - f(x)\|, x \in X$ is moreover Gâteaux derivable on $M \subset X$ open convex, then $\|x_0 - f(x_0)\| = e_d(f; M)$ if and only if

$$\langle x_0 - f(x_0), (1_X - \nabla f(x_0))(h) \rangle = 0, \forall h \in X.$$

Remark. As a consequence, the best almost-fixed points of f on M must be among the solutions of the equation $\langle x - f(x), (1_X - \nabla f(x))(h) \rangle = 0, \forall h \in X$.

The concepts and results obtained in this section allow us to approach the study of equations that have no solutions.

Indeed, because each equation $E(x) = 0_X$ in a normed space $(X, \|\cdot\|)$ can be written as $f(x) = x$, where $f(x) = E(x) + x$, for the case when $E(x) = 0_X$ has no solutions in X, a best almost-fixed point y of f will also be called best almost-solution of the equation $E(x) = 0_X$, satisfying $|E(y)| = \inf\{|E(x)|; x \in X\}$.

Firstly, we consider the case of algebraic equations.

Theorem 6.31 *Let $P_m(x)$ be an algebraic polynomial of degree $m \in \mathbf{N}$, with real coefficients such that the equation $P_m(x) = 0$ has no real solutions. Then m is even, there exists at least one and at most $\frac{m}{2}$ best almost-real solutions of the equation $P_m(x) = 0$.*

Proof. The fact that m must be even is obvious. Denote $f(x) = P_m(x) + x$. Let $x_0 \in \mathbf{R}$ be such that $e_d(f) = |x_0 - f(x_0)| = |P_m(x_0)| > 0$ (the metric d is generated by absolute value $|\cdot|$). Because the scalar product on $\mathbf{R}$ is the usual one, $(1_X - \nabla f(x_0))(h) = (1 - f'(x_0)) \cdot h$, by Theorem

6.30, (i), it easily follows that $f'(x_0) = 1$, *i.e.* $P'_m(x_0) = 0$. But the degree of $P'_m(x)$ is odd, so the equation $P'_m(x) = 0$ has at least one solution. We are interested in the maximum number of points $x_k \in \mathbf{R}$ that satisfy

$$e_d(f;\mathbf{R}) = \inf\{|P_m(x)|; x \in \mathbf{R}\} = |P_m(x_k)|, k \in \{1, ..., p\}.$$

Obviously that $P'_m(x_k) = 0, k \in \{1, ..., p\}$, where $x_k, k \in \{1, ...p\}$ are considered in increasing order.

We show that if $e_d(f;\mathbf{R}) = |P_m(x_k)|$, then $e_d(f;\mathbf{R}) \neq |P_m(x_{k+1})|$, which will prove the theorem. Indeed, because $P_m(x)$ has no real solutions, it follows that $P_m(x) > 0$, for all $x \in \mathbf{R}$ or $P_m(x) < 0$, for all $x \in \mathbf{R}$. Let us suppose, for example, that $P_m(x) > 0, \forall x \in \mathbf{R}$ (the case $P_m(x) < 0, \forall x \in \mathbf{R}$ is similar). If $e_d(f;\mathbf{R}) = P_m(x_k) = P_m(x_{k+1})$, then we get that there is $\xi \in (x_k, x_{k+1})$ with $P'_m(\xi) = 0$, *i.e.* x_k and x_{k+1} are not consecutive, a contradiction. The theorem is proved. □

Now, we will consider the following integral equation which appears in statistical mechanics

$$u(x) = 1 + \lambda \int_x^1 u(s-x)\, u(s) \mathrm{d}s, x \in [0,1],$$

where λ is a real parameter, $\lambda > -1$. In Rus [182], p. $236 - 237$ it is proved that for $\lambda > \frac{1}{2}$, the above equation has no solutions $u \in C[0,1]$.

Let us consider $C[0,1]$ endowed with the uniform metric $d(f,g) = \sup\{|f(x) - g(x)| : x \in [0,1]\}$. Denoting $A : C[0,1] \to C[0,1]$ by

$$(Au)(x) = 1 + \lambda \int_x^1 u(s-x)\, u(s) \mathrm{d}s,$$

where $\lambda > \frac{1}{2}$, it follows that A has no fixed points in $C[0,1]$.

Theorem 6.32 *Let $A : M \to C[0,1]$, where $\lambda > \frac{1}{2}, A$ defined as above and*

$$M = \{u \in C[0,1] : u \textit{ is derivable, } u'(x) \geq 0, \forall x \in [0,1], \\ 0 \leq u(0) \leq u(1) \leq 1\}.$$

Then $e_d(A;M) = \frac{4\lambda - 1}{4\lambda}$ and a best almost-fixed point for A on M is $u : [0,1] \to \mathbf{R}, u(x) = \frac{1}{2\lambda}, \forall x \in [0,1]$.

Proof. We have

$$e_d(A;M) = \inf\{\sup\{|(Au)(x) - u(x)| : x \in [0,1]\} : u \in M\}$$

$$= \inf\left\{\sup\left\{\left|\lambda \int_x^1 u(s-x)\,u(s)\mathrm{d}s - u(x) + 1\right| : x \in [0,1]\right\} : u \in M\right\}.$$

The function $g : [0,1] \to \mathbf{R}$ defined by

$$g(x) = \lambda \int_x^1 u(s-x)\,u(s)\mathrm{d}s - u(x) + 1$$

is monotone decreasing. Indeed,

$$g'(x) = -\lambda \int_x^1 u'(s-x)\,u(s)\mathrm{d}s - u(0)\,u(x) - u'(x) \le 0$$

for every $u \in M$. Then

$$\begin{aligned} & \sup\{|g(x)| : x \in [0,1]\} \\ = \; & \max\left\{\left|\sup_{x\in[0,1]} g(x)\right|, \left|\inf_{x\in[0,1]} g(x)\right|\right\} \\ = \; & \max\left\{\left|\lambda \int_0^1 u^2(s)\mathrm{d}s - u(0) + 1\right|, |1 - u(1)|\right\} \end{aligned}$$

and the conditions $0 \le u(0) \le u(1) \le 1$ imply

$$\sup\{|g(x)| : x \in [0,1]\} = \lambda \int_0^1 u^2(s)\mathrm{d}s - u(0) + 1.$$

This means that the defect of fixed point of A is

$$\begin{aligned} e_d(A;M) &= \inf\left\{\lambda \int_0^1 u^2(s)\mathrm{d}s - u(0) + 1 : u \in X\right\} \\ &= \inf\left\{\lambda u^2(0) - u(0) + 1 : u \in X, u(x) = u(0), \forall x \in [0,1]\right\} \\ &= \frac{4\lambda - 1}{4\lambda} \end{aligned}$$

and a best almost-fixed point for A on M is the function $u : [0,1] \to \mathbf{R}$, defined by $u(x) = \frac{1}{2\lambda}$. □

If we replace the uniform metric on $C[0,1]$ with the metric $d : C[0,1] \times C[0,1] \to \mathbf{R}$ defined by $d(u,v) = \left(\int_0^1 (u(x) - v(x))^2 \, dx\right)^{\frac{1}{2}}$, we obtain

Corollary 6.3 *For the operator A defined above, $\lambda > \frac{1}{2}$, we have $e_d(A; M) = 1$ and a best almost-fixed point for A on M is $u(x) = 0, \forall x \in [0,1]$.*

Proof. We get

$$d(u, Au) = \left(\int_0^1 \left(\lambda \int_x^1 u(s-x)\, u(s) \mathrm{d}s - u(x) + 1\right)^2 \mathrm{d}x\right)^{\frac{1}{2}} \geq -u(1) + 1$$

by using the monotony of g (as above). The inequality becomes equality if $\lambda \int_x^1 u(s-x)\, u(s)\mathrm{d}s - u(x) = -u(1)$, almost everywhere $x \in [0,1]$. It follows $\lambda \int_x^1 u(s-x)\, u(s)\mathrm{d}s = u(x) - u(1)$, almost everywhere $x \in [0,1], u \in M$. But $u(x) - u(1) \leq 0$ and $\lambda \int_x^1 u(s-x)\, u(s)\mathrm{d}s \geq 0$, which implies $u(x) = u(1)$, almost everywhere $x \in [0,1]$. Replacing in the above equation, it easily follows $u(x) = 0$, almost everywhere $x \in [0,1]$, which by $u \in C[0,1]$ implies $u(x) = 0, \forall x \in [0,1]$. Therefore

$$e_d(A; M) = \inf\{d(u, Au) : u \in X\} = 1,$$

and a best almost-fixed points for A on M is $u(x) = 0, \forall x \in [0,1]$, which proves the corollary. □

Finally, we will use the defect of fixed point to introduce a similar concept for topological spaces. Firstly, we recall the following definition, well-known in the theory of fixed point (see for example Goebel-Kirk [92], Rus [182]).

Definition 6.15 The topological space $(X, \mathcal{T})$ has the property of fixed point if any continuous function $f : X \to X$ has fixed point.

The concept of defect of fixed point suggests the following

Definition 6.16 Let (X, d) be a metric space and $\mathcal{T}_d$ be the topology on X generated by d. The defect of fixed point of topological space $(X, \mathcal{T}_d)$ is defined by

$$t(X, \mathcal{T}_d) = \sup\{e_d(f; X) \,|\, f : X \to X \text{ is continuous}\}.$$

Remark. We can reformulate the above definition in a more general frame, considering $(X, \mathcal{T})$ a metrizable space and d the corresponding metric.

Remark. If $(X, \mathcal{T}_d)$ has the property of fixed point then $e_d(f; X) = 0$ for every continuous function $f : X \to X$, therefore $t(X, \mathcal{T}_d) = 0$.

Example 6.4 We consider the Euclidean metric δ on $\mathbf{R}^2$ and we denote by $\mathcal{T}_\delta$ the Euclidean topology on $\mathbf{R}^2$. Because the defect of fixed point of the continuous function $f : \mathbf{R}^2 \to \mathbf{R}^2$ defined by $f(x, y) = (x + a, y + b)$ is $e_\delta(f; \mathbf{R}^2) = \sqrt{a^2 + b^2}$, we obtain $t(\mathbf{R}^2, \mathcal{T}_\delta) = +\infty$.
Nevertheless, there is a subset X dense in the topological space $(\mathbf{R}^2, \mathcal{T}_\delta)$ such that $t(X, \mathcal{T}_\delta|_X) = 0$. Indeed, denoting $C = \{(x, y) \in \mathbf{R}^2 : x^2 + y^2 = 1\}$ and $\{z_1, ..., z_n, ...\}$ a dense subset of C then the set $X = \bigcup_{n \in \mathbf{N}^*} X_n$, where $X_n = \{tnz_n : 0 \leq t \leq 1\}$, is dense in $(\mathbf{R}^2, \mathcal{T}_\delta)$ and any continuous function $f : X \to X$ has at least one fixed point (see Rădulescu-Rădulescu [170], p.145).

Example 6.5 There exist topological spaces that have as defect of fixed point any real positive numbers. Indeed, let us consider $a, b, c, d \in \mathbf{R}, a < b < c < d$, $a + d = b + c, ad - bc \neq 0$ and the continuous function $f_0 : [a, b] \cup [c, d] \to [a, b] \cup [c, d]$ defined by $f_0(x) = x + \frac{ad-bc}{a-b}$ if $x \in [a, b]$ and $f_0(x) = x + \frac{bc-ad}{c-d}$ if $x \in [c, d]$. We have $e_d(f_0; [a, b] \cup [c, d]) = \frac{|ad-bc|}{b-a}$ (with respect to the metric d generated by absolute value $|\cdot|$), therefore $t([a, b] \cup [c, d], \mathcal{T}_d) \geq \frac{|ad-bc|}{b-a}$. On the other hand, $e_d(f; [a, b] \cup [c, d]) \leq d - a$ for every function f defined on $[a, b] \cup [c, d]$ with values in $[a, b] \cup [c, d]$, which implies $t([a, b] \cup [c, d], \mathcal{T}_d) \leq d - a$.

By using the Remark after Definition 6.16 and Remark 3) after Definition 6.13, we obtain the following characterization of topological spaces with the property of fixed point.

Theorem 6.33 *Let (X, d) be a compact metric space and $\mathcal{T}_d$ the topology on X generated by d. Then $(X, \mathcal{T}_d)$ has the property of fixed point if and only if $t(X, \mathcal{T}_d) = 0$.*

The property of fixed point is invariant by homeomorphisms, in other words, it is a topological property. An analogous result can be proved for the defect of this property.

Theorem 6.34 *(i) If (X, d) and (X', d') are isometric then $t(X, \mathcal{T}_d) = t\left(X', \mathcal{T}_{d'}\right)$;*

(*ii*) $t(X, \mathcal{T}_d) \leq diam\ X$;

Proof. Let $f : X' \to X'$ be a continuous function and $i : X \to X'$ the isometry between (X, d) and (X', d'), that is the function i is bijective and $d(x, y) = d'(i(x), i(y))$, for every $x, y \in X$. Because i and i^{-1} are continuous (see Kelley [116], p. 123), the function $i^{-1} \circ f \circ i : X \to X$ is continuous and

$$d\left(\left(i^{-1} \circ f \circ i\right)(x), x\right) = d'(f(i(x)), i(x)), \forall x \in X.$$

We have

$$\begin{aligned} e_{d'}(f; X') &= \inf\{d'(f(x'), x') : x' \in X'\} \\ &= \inf\{d'(f(i(x)), i(x)) : x \in X\} \\ &= \inf\left\{d\left(\left(i^{-1} \circ f \circ i\right)(x), x\right) : x \in X\right\} \\ &= e_d\left(i^{-1} \circ f \circ i; X\right), \end{aligned}$$

therefore $t(X', \mathcal{T}_{d'}) \leq t(X, \mathcal{T}_d)$. We analogously obtain the converse inequality and the property is proved.

(*ii*) By

$$e_d(f; X) = \inf\{d(f(x), x) : x \in X\} \leq \text{diam } X$$

for any continuous function $f : X \to X$, we obtain the inequality. □

6.6 Bibliographical Remarks and Open Problems

Definition 6.2, Theorems 6.1-6.12, Corollary 6.1, Lemma 6.1 are in Ban-Gal [29]. Definitions 6.8, 6.12 and 6.10, Theorems 6.13-6.19, 6.20, 6.21, 6.22, 6.24 appear for the first time in this book. Examples 6.1-6.5, Theorems 6.25-6.34, Definition 6.16, Corollary 6.3 are in Ban-Gal [27]. Completely new are Open problems 6.1 and 6.2.

Open problem 6.1 If $Y \subset X$ is an absorbent subset of the linear space $(X, +, \cdot)$ then the well-known Minkowski's functional attached to Y is defined by

$$p_Y(x) = \inf\{\alpha > 0; x \in \alpha Y\}, x \in X.$$

This functional characterizes the quasi-seminorms and the seminorms as follows (see *e.g.* Muntean [154], p. 43-45):

(i) $p : X \to \mathbf{R}$ is quasi-seminorm if and only if there exists $Y \subset X$, absorbent and convex such that $p = p_Y$;

(ii) $p : X \to \mathbf{R}$ is seminorm if and only if there exists $Y \subset X$, absorbent convex and balanced such that $p = p_Y$.

In the proof of (i), given an absorbent subset $Y \subset X$, the subadditivity of p_Y is essentially a consequence of the convexity of Y (because p_Y is always positive homogeneous if Y is absorbent). Let us suppose, in addition, that $(X, \|\cdot\|)$ is a real normed space. Then, would be natural to search for a relationship between the defect of subadditivity $d_{SADD}(p_Y)(X)$ and the defect of convexity $d_{CONV}(D)(Y)$ (where $D \equiv D_H$ or $D \equiv D^*$) in such a way that for absorbent $Y \subset X, d_{SADD}(p_Y)(X) = 0$ if and only if $d_{CONV}(D)(Y) = 0$. Similarly, in the proof of (ii), given an absorbent and convex subset $Y \subset X$, the absolute homogeneity of p_Y is essentially a consequence of the fact that Y is balanced. Therefore, in this case would be natural to search for a relationship between the defect of absolute homogeneity $d_{AH}(p_Y)(X)$ and the defect of balancing $d_{BAL}(D)(Y)$, (where $D \equiv D_H$ or $D \equiv D^*$) in such a way that for absorbent and convex $Y \subset X, d_{AH}(p_Y)(X) = 0$ if and only if $d_{BAL}(D)(Y) = 0$.

Open problem 6.2 Firstly let us recall some known facts about algebras of operators. Let E denote a real or complex Jordan-Banach algebra, that is a non associative algebra whose product satisfies $a \cdot b = b \cdot a, (a \cdot b) \cdot a^2 = a \cdot (b \cdot a^2), \forall a, b \in E$ and whose underlying vector space is endowed with a complete norm $\|\cdot\|$ with the property $\|a \cdot b\| \leq \|a\| \cdot \|b\|, \forall a, b \in E$. By a derivation of E we mean a linear operator $D : E \to E$, satisfying

$$D(a \cdot b) = D(a) \cdot b + a \cdot D(b), \forall a, b \in E.$$

It is known that every derivation on a semisimple Jordan-Banach algebra is continuous. We define the operator of right multiplication by an element $a \in E$, as the operator $R_a : E \to E$, given by

$$R_a(x) = a \cdot x, \forall x \in E.$$

We also define the operator $U_a = 2R_a^2 - R_{a^2}$. The multiplication algebra of E, denoted by $M(E)$ is defined as the subalgebra of $L(E)$ (the algebra of all linear operators on E) generated by all multiplication operators on E. Note that $M(E) \subset LC(E)$-the algebra of all bounded linear operators on E. If D is a derivation on E, then for every $a \in E$ we have $DR_a - R_aD = R_{D(a)}$, and so the subalgebra of those elements T in $M(E)$ for which $DT - TD$

lies in $M(E)$, equals to $M(E)$ and consequently we can define a derivation on $M(E)$ by

$$D^*(T) = DT - TD, \forall T \in M(E).$$

(For all the above concepts and results see *e.g.* Villena [215]). Now, given $A \in LC(X)$, we can introduce the defect of derivation of A on unit ball $B(0;1)$, by

$$d_{DER}(A)(B(0;1)) = \sup\{\|A(a \cdot b) - (A(a) \cdot b + a \cdot A(b))\|; a, b \in B(0;1)\}.$$

Also, it is obvious that

$$|||D^*(T)||| = d_{PERM}(D, T).$$

It would be interesting to study the properties of $d_{DER}(A)$ and $d_{PERM}(D,T)$ and their possible implications in operator theory.

Open problem 6.3 Let $(E, \|\cdot\|)$ be real normed space and

$$\mathcal{P}_b(E) = \{X \subset E; X \text{ is bounded}\}.$$

For any $\varepsilon \geq 0$ and $X \in \mathcal{P}_b(E)$, characterize/study the following problem of best approximation with restrictions of X:

$$E_\varepsilon^{\perp_*}(X) = \inf\{D_H(X,Y); Y \in \mathcal{P}_b(E), d_*^\perp(X,Y) \leq \varepsilon\},$$

where $D_H(X,Y)$ denotes the Hausdorff-Pompeiu distance and $d_*^\perp(X,Y)$ is any from the defects of orthogonalities introduced by Definition 6.2.

Chapter 7

Defect of Property in Algebra

Let (X, d) be a metric space and $F : X \times X \to X$ be a binary operation on X. If F is not commutative, or is not associative, or is not distributive (with respect to another binary operation $G : X \times X \to X$), or has no identity element, or not every element has an inverse, so on, it is natural to look for a concept of defect of F with respect to these properties.

It is the main aim of this chapter to introduce and study the concept of defect of F with respect to the above properties.

In Section 7.1 we study this problem in general context and a method that decrease these defects is presented. Section 7.2 deals with the calculation of these defects for various concrete examples and Section 7.3 is devoted to a particular class of binary operations called triangular norms. Finally, in Section 7.4 we give some applications of introduced defects.

7.1 Defects of Property for Binary Operations

We begin with the following basic definitions.

Definition 7.1 Let (X, d) be a metric space, $Y \subseteq X$ and $F : X \times X \to X$ be a binary operation on X.

(*i*) The quantity

$$e_{COM}^{d}(F)(Y) = \sup \{d(F(x, y), F(y, x)); x, y \in Y\}$$

is called defect of commutativity of F on Y with respect to the metric d.

(*ii*) The quantity

$$e_{AS}^{d}(F)(Y) = \sup \{d(F(F(x, y), z), F(x, F(y, z))); x, y, z \in Y\}$$

is called defect of associativity of F on Y with respect to the metric d.

(iii) The quantity $\sup\{d(F(x,y),x); x \in Y\}$ is called defect of identity element at right of y (with respect to d) and

$$e_{IDR}^{d}(F)(Y) = \inf\{\sup\{d(F(x,y),x); x \in Y\}; y \in Y\}$$

is called defect of identity element at right of F on Y (with respect to d). Similarly, we can define the defect of identity element at left,

$$e_{IDL}^{d}(F)(Y) = \inf\{\sup\{d(F(y,x),x); x \in Y\}; y \in Y\}.$$

If there exists $e_R \in Y$ such that

$$e_{IDR}^{d}(F)(Y) = \sup\{d(F(x,e_R),x); x \in Y\} > 0,$$

then e_R will be called best almost-identity element at right of F on Y. Analogously, if there exists $e_L \in Y$ such that

$$e_{IDL}^{d}(F)(Y) = \sup\{d(F(e_L,x),x); x \in Y\} > 0,$$

then e_L will be called best almost-identity element at left of F on Y.

(iv) Let us suppose that there exists $a \in X$ such that $F(x,a) = F(a,x) = x, \forall x \in X$. Then

$$\inf\{d(F(x,y),a); y \in Y\}$$

and

$$\inf\{d(F(y,x),a); y \in Y\}$$

are called defects of invertibility of x at right and at left, respectively. The quantities

$$e_{INR}^{d}(F)(Y) = \sup\{\inf\{d(F(x,y),a); y \in Y\}; x \in Y\}$$

and

$$e_{INL}^{d}(F)(Y) = \sup\{\inf\{d(F(y,x),a); y \in Y\}; x \in Y\}$$

are called defect of invertibility of F on Y, at right and at left, respectively. The quantity $e_{IN}^{d}(F)(Y) = \max\left\{e_{INL}^{d}(F)(Y), e_{INR}^{d}(F)(Y)\right\}$ is called defect of invertibility of F on Y (with respect to d).
If there exists $x_R^* \in Y$ such that

$$0 < \inf\{d(F(x,y),a); y \in Y\} = d(F(x,x_R^*),a),$$

then x_R^* will be called best almost-inverse at right of x on Y with respect to d. Similarly, if there is $x_L^* \in Y$ with

$$0 < \inf \{d(F(y,x),a); y \in Y\} = d(F(x_L^*,x),a),$$

then x_L^* will be called best almost-inverse at left of x on Y with respect to d.

(v) The quantity

$$e_{REL}^d(F)(Y) = \sup \{d(x,y); x,y \in Y, F(z,x) = F(z,y)\}$$

is called defect of regularity at left of z on Y. Analogously,

$$e_{RER}^d(F)(Y) = \sup \{d(x,y); x,y \in Y, F(x,z) = F(y,z)\}$$

is called defect of regularity at right of z on Y.

(vi) The quantity

$$e_{IDEM}^d(F)(Y) = \sup \{d(F(x,x),x); x \in Y\}$$

is called defect of idempotency of F on Y (with respect to d).

(vii) If $G : X \times X \to X$ is another binary operation on X then

$$e_{DISL}^d(F;G)(Y) = \sup \quad \{d(F(x,G(y,z)),G(F(x,y),F(x,z))); \\ x,y,z \in Y\}$$

is called defect of left-distributivity of F with respect to G on Y. Similarly,

$$e_{DISR}^d(F;G)(Y) = \sup \quad \{d(F(G(y,z),x,),G(F(y,x),F(z,x))); \\ x,y,z \in Y\}$$

is called defect of right-distributivity of F with respect to G on Y. If $F \equiv G$, then $e_{DISL}^d(F;F)(Y)$ and $e_{DISR}^d(F;F)(Y)$ are called defects of autodistributivity (left and right) of F on Y. Also,

$$e_{AB}^d(F;G)(Y) = \max \{\sup \{d(F(a,G(a,b)),a); a,b \in Y\},$$

$$\sup \{d(G(a,F(a,b)),a); a,b \in Y\}\}$$

is called defect of absorption of (F,G) on Y.

Remark. If (X,d) is of finite diameter (*i.e.* d is bounded on X) obviously all the quantities (defects) in Definition 7.1 are real nonnegative numbers. If d is not bounded on X, then it is known, for example, that $d_1 = \frac{d}{1+d}$ is bounded on X and equivalent to d.

It is immediate the following

Lemma 7.1 *With the notations in Definition 7.1 we have:*

(i) F is commutative on Y if and only if $e^d_{COM}(F)(Y) = 0$.

(ii) F is associative on Y if and only if $e^d_{AS}(F)(Y) = 0$.

(iii) If F has identity element at right in Y, i.e. there is $a \in Y$ such that $F(x,a) = x, \forall x \in Y$, then $e^d_{IDR}(F)(Y) = 0$. Conversely, if (Y,d) is compact and F is continuous on $Y \times Y$ (with respect to the box metric on $Y \times Y$) then $e^d_{IDR}(F)(Y) = 0$ implies that F has identity element at right in Y. Similar results hold in the case of identity element at left.

(iv) If F has identity element in Y and each $x \in Y$ has inverse at right, then $e^d_{INR}(F)(Y) = 0$. Conversely, if (Y,d) is compact and F is continuous on $Y \times Y$ (with respect to the box metric on $Y \times Y$), then $e^d_{INR}(F)(Y) = 0$ implies that each $x \in Y$ has inverse at right. Similar results hold for invertibility at left.

(v) A set $Y \subseteq X$ is called regular at left if each $z \in Y$ is regular at left (i.e. $F(z,x) = F(z,y)$ implies $x = y$). Then Y is regular at left with respect to F if and only if $e^d_{REL}(F)(Y) = 0$. Similar results hold for regularity at right.

(vi) A set $Y \subseteq X$ is called idempotent if each $x \in Y$ is idempotent (with respect to F). Then Y is idempotent with respect to F if and only if $e^d_{IDEM}(F)(Y) = 0$.

(vii) F is left-distributive (right-distributive) with respect to G on Y if and only if $e^d_{DISL}(F;G)(Y) = 0$ ($e^d_{DISR}(F;G)(Y) = 0$). Also, the pair (F,G) has the property of absorption on Y if and only if $e^d_{AB}(F,G)(Y) = 0$.

Proof. The proofs of $(i), (ii), (v), (vi), (vii)$ are immediate. It remains to prove (iii) and (iv).

(iii) The first part is obvious. Conversely, by the obvious inequalities,

$$\begin{aligned} d(F(x,y_n),x) &\leq d(F(x,y_n),F(x,y_0)) + d(F(x,y_0);x), \\ d(F(x,y_0),x) &\leq d(F(x,y_0),F(x,y_n)) + d(F(x,y_n),x), \end{aligned}$$

passing to supremum after $x \in Y$ (here $y_n, y_0 \in Y, n \in \mathbf{N}$), we immediately get

$$|f(y_n) - f(y_0)| = |\sup\{d(F(x, y_n), x); x \in Y\}$$

$$-\sup\{d(F(x, y_0), x); x \in Y\}| \leq \sup\{d(F(x, y_0), F(x, y_n)); x \in Y\},$$

where we have denoted $f : Y \to \mathbf{R}$, $f(y) = \sup\{d(F(x, y), x); x \in Y\}$.

Let $y_n \stackrel{n\to\infty}{\to} y_0$ in the metric d. Because $F : Y \times Y \to X$ is continuous on the compact $Y \times Y$ (with respect to the metric $D((x_1, y_1), (x_2, y_2)) = \max\{d(x_1, x_2), d(y_1, y_2)\}$) it follows that is uniformly continuous on $Y \times Y$, which implies

$$\sup\{d(F(x, y_0), F(x, y_n)); x \in Y\} \stackrel{n\to\infty}{\to} 0$$

that is f is continuous on Y. Therefore, by the equality $e^d_{IDR}(F)(Y) = \inf\{f(y); y \in Y\}$, there exists $a \in Y$ such that $f(a) = e^d_{IDR}(F)(Y)$. Now, by $e^d_{IDR}(F)(Y) = 0$, it easily follows that $d(F(x, a), x) = 0, \forall x \in Y$, *i.e.* $F(x, a) = x, \forall x \in Y$.

(iv) The first part is obvious. Let us suppose now $e^d_{INR}(F)(Y) = 0$. It follows that $\inf\{d(F(x, y), a); y \in Y\} = 0, \forall x \in Y$. Let $x \in Y$ be fixed. By hypothesis, there exists $y \in Y$ such that $d(F(x, y), a) = 0$, that is $F(x, y) = a$ and y is the inverse of x at right. The lemma is proved. □

A very natural question is that starting from a binary operation G, to construct another binary operation F that improves G, that is decreases the defects of properties in Definition 7.1. To give an answer to the question, we introduce the following.

Definition 7.2 Let $f, g : X \to X$ be with $f \circ g = i$, where i is the identity function on X. If G is a binary operation on X, then $F : X \times X \to X$ given by $F = g \circ G(f, f)$ is called the (f, g)-dual of G (*i.e.* $F(x, y) = g(G(f(x), f(y))), \forall (x, y) \in X \times X$).

Remark. Actually, the above definition was introduced in Mayor-Calvo [143] only for $X = [0, 1] \subset \mathbf{R}$.

Concerning this concept, firstly we present

Theorem 7.1 (i) *F is associative on X if and only if G is associative on X.*

(ii) *F is commutative on X if and only if G is commutative on X.*

(iii) If F has identity element at right (left) then G has identity element at right (left). Conversely, if in addition f is injective, then the fact that G has identity element at right (left) implies that F has identity element at right (left). If a is identity for F, then $f(a)$ is identity for G in all the cases.

(iv) If F has identity element a and $Y \subset X$ is invertible at left (right) with respect to F (that is each element of Y is invertible at left (right)), then $f(a)$ is identity element for G and $f(Y)$ is invertible at left (right) with respect to G. Conversely, if in addition f is injective, then $Y \subset X$ is invertible with respect to G implies that $f^{-1}(Y)$ is invertible with respect to F, at left and right, respectively.

(v) Let us suppose, in addition, that f is injective. If $Y \subset X$ is regular at left with respect to F (that is each element of Y is regular at left), then $f(Y)$ is regular at left with respect to G. Conversely, if $Y \subset X$ is regular at left with respect to G implies that $f^{-1}(Y)$ is regular at left with respect to F. Similar results for regularity at right hold.

(vi) If $Y \subset X$ is idempotent with respect to F, then $f(Y)$ is idempotent with respect to G. Conversely, if in addition f is injective, then $Y \subset X$ is idempotent with respect to G implies that $f^{-1}(Y)$ is idempotent with respect to F.

(vii) Let F_i be the (f,g)-duals of $G_i, i \in \{1,2\}$. F_1 is left-distributive (right-distributive) with respect to F_2 if and only if G_1 is left-distributive (right-distributive) with respect to G_2. Also, if the pair (F_1, F_2) satisfies the axioms of absorption, then (G_1, G_2) satisfies the axioms of absorption. If, in addition, f is injective, then the converse of the last property holds.

Proof. (*i*) See Mayor-Calvo [143], Proposition 1.1.

(*ii*) If G is commutative then we get

$$F(x,y) = g(G(f(x), f(y))) = g(G(f(y), f(x))) = F(y,x).$$

Conversely, if F is commutative, since f is surjective, denoting $x = f(x')$, $y = f(y')$, we get

$$\begin{aligned} G(x,y) &= G(f(x'), f(y')) = f(F(x',y')) \\ &= f(F(y',x')) = G(f(y'), f(x')) = G(y,x). \end{aligned}$$

(*iii*) Let $a \in X$ be identity element at right of F, that is $F(x,a) = x$, for all $x \in X$. We get

$$f(x) = f(F(x,a)) = G(f(x), f(a)),$$

for all $x \in X$. Since f is surjective we get $f(X) = X$, which implies that $f(a)$ is identity element at right of G.

Conversely, let us suppose that f is injective, *i.e.* f is bijection on X and let $a' \in X$ be the identity element at right of G, that is $G(x', a') = x'$, for all $x' \in X$. Since f is bijection, let $x' = f(x), a' = f(a)$. We get

$$G(f(x), f(a)) = f(F(x,a)) = f(x),$$

which implies $F(x,a) = x$, for all $x \in X$.

The case of identity element at left is similar.

(*iv*) Let $a \in X$ be identity element of F and let us suppose that $Y \subset X$ is invertible at left with respect to F, that is, for all $y \in Y$, there is $y^* \in Y$ such that $F(y^*, y) = a$. By the above point (*iii*), $f(a)$ is identity element of G. Then

$$f(a) = f(F(y^*, y)) = G(f(y^*), f(y)),$$

that is $f(y)$ has left inverse, $f(y^*)$.

Conversely, let us suppose that f is injective and that $Y \subset X$ is invertible at left with respect to G. Let $a' \in X$ be the identity element with respect to G. That is, for all $y' \in Y$, there is y'_* such that $G(y'_*, y') = a'$. Since f is bijection, let $y' = f(x), y'_* = f(x^*), a' = f(a)$. We get

$$G(f(x^*), f(x)) = f(F(x^*, x)) = f(a),$$

which implies $F(x^*, x) = a$, with $x \in f^{-1}(Y)$ and a identity element with respect to F.

The case of invertibility at right is similar.

(*v*) Let $Y \subset X$ be regular at left with respect to F, that is for all $z \in Y$, from $F(z,x) = F(z,y)$ it follows $x = y$. Let $z' \in f(Y)$ be certain, $z' = f(z), z \in Y$, and let us consider the equality $G(z', b') = G(z', c')$. Since f is surjective, let us denote $b' = f(b), c' = f(c)$. The above equality becomes

$$f(F(z,b)) = f(F(z,c)),$$

which by injectivity implies $F(z,b) = F(z,c)$, which by hypothesis implies $b = c$, that is $b' = f(b) = f(c) = c'$.

Conversely, let $z \in f^{-1}(Y)$ and let us consider the equality $F(z,b) = F(z,c)$. It follows

$$G(f(z), f(b)) = G(f(z), f(c)),$$

with $f(z) \in Y$, which by hypothesis implies $f(b) = f(c)$, *i.e.* $b = c$.

(vi) Let $x \in Y$ be with $F(x,x) = x$. Applying f we get $G(f(x), f(x)) = f(x)$, *i.e.* $f(Y)$ is idempotent with respect to G. Conversely, let $z \in f^{-1}(Y)$, *i.e.* $f(z) \in Y$, which implies $G(f(z), f(z)) = f(z)$. It follows $f(F(z,z)) = f(z)$, which by injectivity of f implies $F(z,z) = z$, for all $z \in f^{-1}(Y)$.

(vii) Firstly, we consider the case of distributivity. Let us suppose that F_1 is left-distributive with respect to F_2. Since f is surjective we get $x = f(x'), y = f(y'), z = f(z')$ and

$$G_1(x, G_2(y,z)) = G_1(f(x'), G_2(f(y'), f(z')))$$

$$\begin{aligned}
&= G_1(f(x'), f(F_2(y',z'))) = f(F_1(x', F_2(y',z'))) \\
&= f(F_2(F_1(x',y'), F_1(x',z'))) = f(F_2(g(G_1(x,y)), g(G_1(x,z)))) \\
&= G_2(f(g(G_1(x,y))), f(g(G_1(x,z)))) = G_2(G_1(x,y), G_1(x,z)).
\end{aligned}$$

Conversely, let us suppose that G_1 is left-distributive with respect to G_2. We get

$$\begin{aligned}
F_1(x, F_2(y,z)) &= g(G_1(f(x), G_2(f(y), f(z)))) \\
&= g(G_2(G_1(f(x), f(y)), G_1(f(x), f(z)))) \\
&= F_2(F_1(x,y), F_1(x,z)).
\end{aligned}$$

Similar results hold for right-distributivity.

Secondly, let us suppose that the pair (F_1, F_2) satisfies the axioms of absorption, *i.e.* $F_1(a, F_2(a,b)) = a$ and $F_2(a, F_1(a,b)) = a$. Because f is surjective, we have

$$G_1(a, G_2(a,b)) = G_1(f(a'), G_2(f(a'), f(b')))$$

$$= G_1(f(a'), f(F_2(a',b'))) = f(F_1(a', F_2(a',b'))) = f(a') = a,$$

where $a = f(a')$ and $b = f(b')$. Similarly, $G_2(a, G_1(a,b)) = a$.

Conversely, let us suppose that the pair (G_1, G_2) satisfies the axioms of absorption. We get

$$f(F_1(a, F_2(a,b))) = G_1(f(a), f(F_2(a,b)))$$

$$= G_1(f(a), G_2(f(a), f(b))) = f(a),$$

which by the injectivity of f implies $F_1(a, F_2(a,b)) = a$. Similarly, we get $F_2(a, F_1(a,b)) = a$. The theorem is proved. □

The next theorem needs the following

Definition 7.3 Let (X, d) be a metric space and $F, G : X \times X \to X$. We say that F is less than G (we write $F < G$) if F is the (f,g)-dual of G, where $d(g(x), g(y)) \leq kd(x,y), \forall x, y \in X$ and $0 < k < 1$ is a constant.

Remark. Definition 7.3 in the case $X = [0,1]$ and $d(x,y) = |x - y|$ was considered in Mayor-Calvo [143].

Theorem 7.2 *With the notations in Definitions 7.1 and 7.3, the condition $F < G$ implies:*

(i) $e^d_{COM}(F)(Y) \leq ke^d_{COM}(G)(f(Y)), \forall Y \subseteq X$. *In particular,* $e^d_{COM}(F)(X) \leq ke^d_{COM}(G)(X)$.

(ii) $e^d_{AS}(F)(Y) \leq ke^d_{AS}(G)(f(Y)), \forall Y \subseteq X$. *In particular,* $e^d_{AS}(F)(X) \leq ke^d_{AS}(G)(X)$.

(iii) Let F_i be the (f,g)-duals of $G_i, i \in \{1,2\}$, such that $F_i < G_i, i \in \{1,2\}$. Then $e^d_{DISL}(F_1;F_2)(Y) \leq ke^d_{DISL}(G_1;G_2)(f(Y)), \forall Y \subseteq X$, and in particular $e^d_{DISL}(F_1;F_2)(X) \leq ke^d_{DISL}(G_1;G_2)(X)$. Similar results hold for e^d_{DISR}.

Proof. *(i)* It is immediate by the relations

$$\begin{aligned} d(F(x,y), F(y,x)) &= d(g(G(f(x), f(y))), g(G(f(y), f(x)))) \\ &\leq kd(G(f(x), f(y)), G(f(y), f(x))) \end{aligned}$$

and by $f(X) = X$.

(ii) It is immediate by the relations

$$\begin{aligned} & d(F(F(x,y), z), F(x, F(y,z))) \\ = \;& d(g(G(f(F(x,y)), f(z))), g(G(f(x), f(F(y,z))))) \\ \leq \;& kd(G(G(f(x), f(y)), f(z)), G(f(x), G(f(y), f(z)))) \end{aligned}$$

and by $f(X) = X$.

(*iii*) It follows by $f(X) = X$ and by the relations

$$d(F_1(x, F_2(y,z)), F_2(F_1(x,y), F_1(x,z)))$$

$$= d(g(G_1(f(x), f(F_2(y,z)))),$$

$$g(G_2(G_1(f(x), f(y)), G_1(f(x), f(y)))))$$

$$\leq kd(G_1(f(x), G_2(f(y), f(z))),$$

$$G_2(G_1(f(x), f(y)), G_1(f(x), f(y)))),$$

which proves the theorem. □

Remark. Because $0 < k < 1$, Theorem 7.2 shows that the (f,g)-duals improve the properties of commutativity, associativity and distributivity.

Reasoning similarly with the proof of Lemma 7.1, (*iii*) and (*iv*), we get the following

Theorem 7.3 *Let us suppose that $Y \subseteq X$ is compact in (X,d).*

(i) If $F : X \times X \to X$ is continuous on $Y \times Y$ with respect to the box metric on $X \times X$ (i.e. $D((x_1,y_1),(x_2,y_2)) = \max\{d(x_1,x_2), d(y_1,y_2)\}$), $e^d_{IDR}(F)(Y) > 0, e^d_{IDL}(F)(Y) > 0$, then there exist $e_R, e_L \in Y$, best almost-identity on Y, at right and at left, respectively.

(ii) If F is continuous on Y with respect to the second variable for each fixed $x \in Y$, then each $x \in Y$ has best almost-inverse at right in Y. Similarly, if F is continuous on Y with respect the first variable for each fixed $x \in Y$, then each $x \in Y$ has best almost-inverse at left in Y.

In what follows we present some connections between the defects of properties in Definition 7.1 and some classical concepts in the theory of algebraic structures. The first result is the following

Theorem 7.4 *Let $(X_i, d_i), i \in \{1,2\}$, be two metric spaces and $F_i : X_i \times X_i \to X_i, i \in \{1,2\}$.*

(i) If there exists isometric homomorphism $h : X_1 \to X_2$ (i.e. $d_1(x,y) = = d_2(h(x), h(y))$ and $h(F_1(x,y)) = F_2(h(x), h(y)), \forall x,y \in X_1$), then $e^{d_1}_{COM}(F_1)(X_1) = e^{d_2}_{COM}(F_2)(X_2)$ and $e^{d_1}_{AS}(F_1)(X_1) = e^{d_2}_{AS}(F_2)(X_2)$.

(ii) Let us define $F : (X_1 \times X_2) \times (X_1 \times X_2) \to X_1 \times X_2$ *by* $F((x_1, x_2), (y_1, y_2)) = (F_1(x_1, y_1), F_2(x_2, y_2))$ *and* $d((x_1, x_2), (y_1, y_2)) = \max\{d_1(x_1, y_1), d_2(x_2, y_2)\}$ *the box metric on* $X = X_1 \times X_2$. *Then we have*

$$\begin{aligned} e^{d}_{COM}(F)(X) &= \max\left\{e^{d_1}_{COM}(F_1)(X_1), e^{d_2}_{COM}(F_2)(X_2)\right\}, \\ e^{d}_{AS}(F)(X) &= \max\left\{e^{d_1}_{AS}(F_1)(X_1), e^{d_2}_{AS}(F_2)(X_2)\right\}. \end{aligned}$$

Proof. The proof is immediate. □

Concerning the factor group we present

Theorem 7.5 *Let* $(X, \cdot, d)$ *be a metrizable non-commutative group,* N *a closed normal divisor of* X *and* d *a metric that is left invariant. If* $\widetilde{d}$ *denotes the induced metric on* X/N *and* $\odot$ *the induced operation on* X/N*, then*

$$e^{\widetilde{d}}_{COM}(\odot)(X/N) \leq e^{d}_{COM}(\cdot)(X).$$

Proof. According to *e.g.* Meghea [144], p.621, $\widetilde{d}$ is defined by

$$\widetilde{d}(\alpha, \beta) = \inf\{d(ux, vy); u, v \in N, x \in \alpha, y \in \beta\}, \forall \alpha, \beta \in X/N,$$

and it is left invariant. We have $\alpha = Nx, \beta = Ny$ and we get

$$e^{\widetilde{d}}_{COM}(\odot)(X/N)$$

$$= \sup\left\{\widetilde{d}((Nx_1) \odot (Nx_2), (Nx_2) \odot (Nx_1)); x_1, x_2 \in X\right\}$$

$$= \sup\left\{\widetilde{d}(N(x_1x_2), N(x_2x_1)); x_1, x_2 \in X\right\}$$

$$= \sup\{\inf\{d(u(x_1x_2), v(x_2x_1)); u, v \in N\}; x_1, x_2 \in X\}$$

$$\leq (\text{choosing } u = v = 1_X) \leq e^{d}_{COM}(\cdot)(X),$$

which proves the theorem. □

Remark. Let $(X, \cdot)$ be a non-commutative group and d a metric on X. Let us denote the center of X by $Z(X) = \{x \in X; \forall y \in X, xy = yx\}$. Then it is obvious that $e^{d}_{COM}(\cdot)(X) = e^{d}_{COM}(\cdot)(X \backslash Z(X))$.

Now, for a metric space (X, d), let us denote

$$COM(X) = \{F : X \times X \to X; F \text{ is commutative on } X\}$$

and

$$AS(X) = \{F : X \times X \to X; F \text{ is associative on } X\}.$$

For $F, G : X \times X \to X$, we define the distance between F and G by

$$D(F, G) = \sup\{d(F(x, y), G(x, y)); x, y \in X\}.$$

Also we define

$$E_{COM}(F) = \inf\{D(F, G); G \in COM(X)\}$$

and

$$E_{AS}(F) = \inf\{D(F, G); G \in AS(X)\},$$

the best approximation of a binary operation on X by commutative operations and by associative operations, respectively.

We present

Theorem 7.6 *Let (X, d) be a metric space with d bounded on X. We have $\frac{1}{2}e^d_{COM}(F) \leq E_{COM}(F)$ and $\frac{1}{2}e^d_{AS}(F) \leq E_{AS}(F)$, for any binary operation $F : X \times X \to X$.*

Proof. We have

$$\begin{aligned} d(F(x, y), F(y, x)) &\leq d(F(x, y), G(x, y)) + d(G(x, y), G(y, x)) \\ +d(G(y, x), F(y, x)) &\leq 2D(F, G), \forall G \in COM(X), x, y \in X. \end{aligned}$$

Passing to supremum after $x, y \in X$ and then to infimum after $G \in COM(X)$, we get the first inequality.

To prove the second inequality, we start from

$$d(F(x, F(y, z)), F(F(x, y), z)) \leq d(F(x, F(y, z)), G(x, G(y, z)))$$

$$+d(G(x, G(y, z)), G(G(x, y), z)) + d(G(G(x, y), z), F(F(x, y), z)),$$

for all $G \in AS(X)$ and we follow the above reasonings, which proves the theorem. □

7.2 Calculations of the Defect of Property

In this section we calculate and estimate the defects of property in Definition 7.1 for various concrete examples. A possible application would be that if we know, for example, the defect of commutativity of a binary operation F and if we know $F(x,y)$, then we can obtain an estimate of $F(y,x)$ in terms of $F(x,y)$ and $e^d_{COM}(F)(X)$.

Example 7.1 Take $X = [0,1] \subset \mathbf{R}, d(x,y) = |x-y|$ and $F(x,y) = (1-\lambda)x + \lambda y$, where $\lambda \in (0,1)$ is fixed. F is non-associative, has no identity element and if $\lambda \neq \frac{1}{2}$ then it is non-commutative. Because

$$|F(x,y) - F(y,x)| = |(1-\lambda)(x-y) + \lambda(y-x)|$$

$$\leq |1-2\lambda| \cdot |x-y| \leq |1-2\lambda|$$

for all $x, y \in [0,1]$, it follows

$$e^d_{COM}(F)([0,1]) = |1-2\lambda|.$$

Then,

$$\begin{aligned}
& |F(x, F(y,z)) - F(F(x,y), z)| \\
= \; & |(1-\lambda)x + \lambda((1-\lambda)y + \lambda z) - ((1-\lambda)((1-\lambda)x + \lambda y) + \lambda z)| \\
= \; & \left| x\left((1-\lambda) - (1-\lambda)^2\right) - z\left(\lambda - \lambda^2\right)\right| = (1-\lambda)|\lambda x - \lambda z| \\
= \; & (1-\lambda)\lambda|x-z|, \forall x,y,z \in [0,1],
\end{aligned}$$

which immediately implies $e^d_{AS}(F)([0,1]) = \lambda(1-\lambda)$.
Let $y \in [0,1]$ be fixed and let us consider

$$|F(x,y) - x| = |(1-\lambda)x + \lambda y - x| = \lambda|x-y|.$$

Passing to supremum, we get

$$\sup\{|F(x,y) - x|; x \in [0,1]\} = \lambda \max\{y, 1-y\},$$

which represents the defect of identity element at right of y. Passing now to infimum, we get

$$e^d_{IDR}(F)([0,1]) = \inf\{\lambda \max\{y, 1-y\}; y \in [0,1]\} = \frac{\lambda}{2},$$

and $\frac{1}{2}$ is the best almost-identity element at right of F on $[0,1]$. Similarly,

$$|F(y,x) - x| = |(1-\lambda)y + \lambda x - x| = (1-\lambda)|x-y|$$

and reasoning as above, we get $e^{d}_{IDL}(F)([0,1]) = \frac{1-\lambda}{2}$ and $y = \frac{1}{2}$ is the best almost-identity element at left of F on $[0,1]$.

Remark. Given $\varepsilon \in (0,1)$, we can define f and g in Definition 7.2 and Definition 7.3, by $g(x) = (1-\varepsilon)x, \forall x \in [0,1]$ and $f(x) = \frac{x}{1-\varepsilon}, x \in [0, 1-\varepsilon], f(x) = 1, x \in [1-\varepsilon, 1]$. Then the (f,g)-dual of F defined by

$$F^{*}(x,y) = (1-\varepsilon)((1-\lambda)f(x) + \lambda f(y))$$

satisfies Theorem 7.2, (i) and (ii), with $k = 1-\varepsilon$.

Example 7.2 If $X = Y = [0,1], d(x,y) = |x-y|, \forall x,y \in [0,1]$, then the binary operation $F(x,y) = x^2 \wedge y$, where $a \wedge b = \min\{a,b\}$, has the defect of associativity equal to $\frac{1}{4}$. Indeed, denoting

$$\begin{aligned} E(x,y,z) &= |F(F(x,y),z) - F(x,F(y,z))| \\ &= \left|\left((x^2 \wedge y)^2 \wedge z\right) - \left(x^2 \wedge (y^2 \wedge z)\right)\right|, \end{aligned}$$

the following situations are possible:
Case 1: $y \le x^2$. Then $y^2 \le y \le x^2$ and $E(x,y,z) = 0$;
Case 2: $y \ge x^2$.
If $y \ge x^2 \ge y^2$ then $E(x,y,z) = \left|(x^4 \wedge z) - (y^2 \wedge z)\right| \le (x^2 \wedge z) - (x^4 \wedge z)$. In the case $z \le y^4 \le y^2$ we get $E(x,y,z) = 0$. If $y^4 \le y^2 \le z$ then $E(x,y,z) = y^2 - y^4 \le \frac{1}{4}$, with equality if $y = \frac{\sqrt{2}}{2}$. If $y^4 \le z \le y^2$ then $E(x,y,z) = z - y^4 \le \frac{1}{4}$, with equality if $z = y^2 = \frac{1}{2}$.
If $y \ge y^2 \ge x^2$ then $E(x,y,z) = \left|(x^4 \wedge z) - (x^2 \wedge z)\right|$. In the case $z \le x^4 \le x^2$ we get $E(x,y,z) = 0$. If $x^4 \le z \le x^2$ then $E(x,y,z) = z - x^4 \le x^2 - x^4 \le \frac{1}{4}$, with equality if $x = \frac{\sqrt{2}}{2}$. If $x^4 \le x^2 \le z$ then $E(x,y,z) = x^2 - x^4 \le \frac{1}{4}$, with equality if $x = \frac{\sqrt{2}}{2}$.
As a conclusion,

$$e^{|\cdot|}_{AS}(F)[0,1] = \sup\left\{\left|\left((x^2 \wedge y)^2 \wedge z\right) - \left(x^2 \wedge y^2 \wedge z\right)\right|; x,y,z \in [0,1]\right\} = \frac{1}{4}.$$

Example 7.3 Let $H = \{q = a_0 + ia_1 + ja_2 + ka_3; a_p \in \mathbf{R}, p \in \{0,1,2,3\}\}$ be the quaternionic division ring, where $i^2 = j^2 = k^2 = -1, ij = -ji = k, jk = -kj = i, ki = -ik = j$. It is known that the multiplication denoted by $\cdot$ is non-commutative. If $q_1 = a_0+ia_1+ja_2+ka_3, q_2 = b_0+ib_1+jb_2+kb_3$, then simple calculations show that

$$q_1 \cdot q_2 = (a_0b_0 - a_1b_1 - a_2b_2 - a_3b_3) + i(a_0b_1 + a_1b_0 + a_2b_3 - a_3b_2)$$

$$+j(a_0b_2 + a_2b_0 + a_3b_1 - a_1b_3) + k(a_0b_3 + a_3b_0 + a_1b_2 - a_2b_1),$$

$$q_2 \cdot q_1 = (a_0b_0 - a_1b_1 - a_2b_2 - a_3b_3) + i(a_0b_1 + a_1b_0 - (a_2b_3 - a_3b_2))$$

$$+j(a_0b_2 + a_2b_0 - (a_3b_1 - a_1b_3)) + k(a_0b_3 + a_3b_0 - (a_1b_2 - a_2b_1)).$$

Let us choose on $\mathbf{R}^4$ the norm $\|x\| = \max\{|x_p|; p \in \{0,1,2,3\}\}$, for every $x = (x_1, x_2, x_3, x_4) \in \mathbf{R}^4$ and the metric on H generated by this norm, that is

$$d(q_1, q_2) = \|(a_0 - b_0, a_1 - b_1, a_2 - b_2, a_3 - b_3)\|.$$

We have

$$d(q_1 \cdot q_2, q_2 \cdot q_1) = 2\max\{|a_2b_3 - a_3b_2|, |a_3b_1 - a_1b_3|, |a_1b_2 - a_2b_1|\}.$$

Let us choose $Y = \{x \in \mathbf{R}^4; \|x\| \leq r\}$ and let us suppose $q_1, q_2 \in Y$. It follows $|a_p|, |b_p| \leq r, \forall p \in \{0,1,2,3\}$ and

$$\max\{|a_2b_3 - a_3b_2|, |a_3b_1 - a_1b_3|, |a_1b_2 - a_2b_1|\} = 2r^2.$$

As a consequence, in this case we get

$$e^d_{COM}(\cdot)(Y) = 4r^2.$$

Example 7.4 Let us consider a particular case of Lie group called the Heisenberg group (see *e.g.* Howe [102]), defined by

$$H = \{g = (t, z), t \in \mathbf{R}, z \in \mathbf{C}\},$$

where the multiplication is given by

$$(t_1, z_1) * (t_2, z_2) = \left(t_1 + t_2 + \frac{1}{2}\mathrm{Im}(\overline{z_1}z_2), z_1 + z_2\right).$$

It is known that $*$ is non-commutative. Denoting $z_p = x_p + iy_p \in \mathbf{C}, p \in \{1, 2\}$, we have

$$(t_1, z_1) * (t_2, z_2) = \left(t_1 + t_2 + \frac{x_1y_2 - x_2y_1}{2}, (x_1 + x_2) + i(y_1 + y_2)\right),$$

$$(t_2, z_2) * (t_1, z_1) = \left(t_1 + t_2 - \frac{x_1 y_2 - x_2 y_1}{2}, (x_1 + x_2) + i (y_1 + y_2) \right).$$

Let us consider on H the metric $D : H \times H \to \mathbf{R}_+$ defined by

$$D ((t_1, z_1), (t_2, z_2)) = \max \left\{ |t_1 - t_2|, |z_1 - z_2|^* \right\},$$

for every $(t_p, z_p) \in \mathbf{R} \times \mathbf{C}, p \in \{1, 2\}$, where

$$|z_1 - z_2|^* = \max \{ |x_1 - x_2|, |y_1 - y_2| \},$$

$z_1 = x_1 + i y_1, z_2 = x_2 + i y_2$. We get

$$D ((t_1, z_1) * (t_2, z_2), (t_2, z_2) * (t_1, z_1)) = |x_1 y_2 - x_2 y_1|.$$

Let $Y = \mathbf{R} \times U_r$, where $U_r = \left\{ z \in \mathbf{C}; |z|^* \leq r \right\}$. We easily obtain

$$e^d_{COM} (*) (Y) = 2r^2.$$

Example 7.5 Let $c > 0, U_c = \{ z \in \mathbf{C}; |z| < c \}$ and $\oplus : U_c \times U_c \to U_c$ given by $z_1 \oplus z_2 = \frac{z_1 + z_2}{1 + \frac{\overline{z_1} z_2}{c^2}}, \forall z_1, z_2 \in U_c$ (see *e.g.* Ungar [213], p.1410). The binary operation $\oplus$ is called Einstein's velocity addition. Obviously $\oplus$ is non-commutative and non-associative. In fact $(U_c, \oplus)$ forms a grouplike object called gyrogroup (see *e.g.* Ungar [213], p.1410). For simplicity, let us take $c = 1$ and let us consider on U_c the Euclidean metric $d (z_1, z_2) = |z_1 - z_2|$. We have

$$d (z_1 \oplus z_2, z_2 \oplus z_1) = \left| \frac{(z_1 \overline{z_2} - \overline{z_1} z_2)(z_1 + z_2)}{(1 + \overline{z_1} z_2)(1 + z_1 \overline{z_2})} \right|$$

$$\leq (|z_1| + |z_2|) \left| \frac{u - \overline{u}}{(1 + u)(1 + \overline{u})} \right| = (|z_1| + |z_2|) \frac{2b}{(1 + a)^2 + b^2},$$

where $u = z_1 \overline{z_2} = a + ib$.
Let us choose $Y = \left\{ z \in \mathbf{C}; |z| < \frac{1}{4} \right\}$, for example. Then, for all $z_1, z_2 \in Y$ we get

$$d (z_1 \oplus z_2, z_2 \oplus z_1) \leq \frac{b}{(1 + a)^2 + b^2},$$

where $|u| \leq \frac{1}{16}$ and therefore $|a|, |b| \leq |u| \leq \frac{1}{16}$. By $|1+a| \geq |1-|a|| \geq \frac{15}{16}$, we obtain

$$d(z_1 \oplus z_2, z_2 \oplus z_1) \leq \frac{b}{(1+a)^2} \leq \frac{1}{16} \cdot \frac{256}{225} = \frac{16}{225}$$

and $e^d_{COM}(\oplus)(Y) \leq \frac{16}{225} = 0,07(1)$.

7.3 Defect of Idempotency and Distributivity of Triangular Norms

In 1960 Schweizer and Sklar [191] defined the triangular norm as a binary operation on the unit interval $[0,1]$,(*i.e.*, a function $T : [0,1] \times [0,1] \to [0,1]$) which is commutative, associative, monotone in each variable and $T(x,1) = x, \forall x \in [0,1]$. Similarly, a triangular conorm is a commutative, associative and monotone (in each variable) function $S : [0,1] \times [0,1] \to [0,1]$ which satisfies $S(x,0) = x, \forall x \in [0,1]$ (see also Klement-Mesiar [119]).

The triangular norms and conorms are especially used in probabilistic analysis and in fuzzy mathematics (fuzzy measure theory, fuzzy logic, *etc.*). For example, the most important operations on fuzzy sets are extensions of triangular norms and conorms. That is, if X is a nonempty set, denoting by $FS(X) = \{A \,|A : X \to [0,1]\}$ the family of all fuzzy sets on X, then for every $A, B \in FS(X)$ we can define

$$\begin{aligned} (A \cap_T B)(x) &= T(A(x), B(x)), \forall x \in X, \\ (A \cup_S B)(x) &= S(A(x), B(x)), \forall x \in X. \end{aligned}$$

Also, the fuzzy propositional logic is described as a $[0,1]$ -valued logic in which the disjunction symbol $\vee$ and the conjunction symbol $\wedge$ are interpreted with the help of triangular conorms and triangular norms, respectively (see Butnariu-Klement-Zafrany [51]).

Concerning the defect of idempotency of triangular norms and triangular conorms, we present the following properties (the metric d on $[0,1]$ is generated by the absolute value $|\cdot|$ and, for short, we denote $e_{IDEM}(F) = e^d_{IDEM}(F)([0,1])$ for a triangular norm or conorm F):

Theorem 7.7 *Let T be a triangular norm and S be a triangular conorm. We have:*

(i) $e_{IDEM}(T) = \sup\{x - T(x,x); x \in [0,1]\}$;

(ii) $e_{IDEM}(S) = \sup\{S(x,x) - x; x \in [0,1]\}$;

(iii) If S is the dual of T, that is $S(x,y) = 1 - T(1-x, 1-y), \forall x, y \in [0,1]$, then $e_{IDEM}(T) = e_{IDEM}(S)$;

(iv) If $T_1 \leq T_2$, that is $T_1(x,y) \leq T_2(x,y), \forall x, y \in [0,1]$, then $e_{IDEM}(T_2) \leq e_{IDEM}(T_1)$;

(v) If $S_1 \leq S_2$, that is $S_1(x,y) \leq S_2(x,y), \forall x, y \in [0,1]$, then $e_{IDEM}(S_1) \leq e_{IDEM}(S_2)$;

(vi) If T^* is the reverse of triangular norm T, that is $T^*(x,y) = \max(0, x + y - 1 + T(1-x, 1-y))$ (see Kimberling [117] or Šabo [186]), then $e_{IDEM}(T^*) \leq e_{IDEM}(T)$;

Proof. The proofs of *(i)* and *(ii)* are immediate because $T(x,x) \leq x, \forall x \in [0,1]$ and $x \leq S(x,x), \forall x \in [0,1]$.

(iii) By using *(i)*, *(ii)* and relation $S(x,x) = 1 - T(1-x, 1-x), \forall x \in [0,1]$, we obtain

$$\begin{aligned} e_{IDEM}(S) &= \sup\{S(x,x) - x; x \in [0,1]\} \\ &= \sup\{1 - x - T(1-x, 1-x); x \in [0,1]\} \\ &= \sup\{y - T(y,y); y \in [0,1]\} = e_{IDEM}(T). \end{aligned}$$

(iv), *(v)* Are immediate by using *(i)* and *(ii)*.

(vi) Denoting by S the dual conorm of T, we have

$$\begin{aligned} x - T^*(x,x) &= x - \max(0, x + x - 1 + T(1-x, 1-x)) \\ &= x - \max(0, 2x - S(x,x)) \leq x - 2x + S(x,x) \\ &= S(x,x) - x, \forall x \in [0,1], \end{aligned}$$

which implies $e_{IDEM}(T^*) \leq e_{IDEM}(S) = e_{IDEM}(T)$. □

Remark. Due to Theorem 7.7, *(iii)*, we will study the defect of idempotency only for triangular norms.

The basic triangular norms together with their duals triangular conorms are the following (see Klement-Mesiar [119]):

(i) Minimum T_M and Maximum S_M given by $T_M(x,y) = \min(x,y)$ and $S_M(x,y) = \max(x,y)$.

(ii) Product T_P and Probabilistic Sum S_P given by $T_P(x,y) = xy$ and $S_P(x,y) = x + y - xy$.

(*iii*) Lukasiewicz triangular norm T_L and Lukasiewicz triangular conorm S_L given by $T_L(x,y) = \max(x+y-1,0)$ and $S_L(x,y) = \min(x+y,1)$.

(*iv*) The weakest triangular norm T_W and the strongest triangular conorm S_W given by

$$T_W(x,y) = \begin{cases} \min(x,y), & \text{if } \max(x,y) = 1, \\ 0, & \text{otherwise,} \end{cases}$$

$$S_W(x,y) = \begin{cases} \max(x,y), & \text{if } \min(x,y) = 0, \\ 1, & \text{otherwise.} \end{cases}$$

Also, other important families of triangular norms together with their corresponding families of triangular conorms are the following (see Klement-Mesiar [119]):

(*i*) The family of Frank triangular norms $(T_\lambda)_{\lambda\in[0,\infty]}$ given by

$$T_\lambda(x,y) = \log_\lambda\left(1 + \frac{(\lambda^x - 1)(\lambda^y - 1)}{\lambda - 1}\right)$$

if $\lambda \in (0,1)\cup(1,\infty)$, $T_0 = T_M$, $T_1 = T_P$, $T_\infty = T_L$ and Frank triangular conorms $(S_\lambda)_{\lambda\in[0,\infty]}$ given by

$$S_\lambda(x,y) = 1 - \log_\lambda\left(1 + \frac{(\lambda^{1-x} - 1)(\lambda^{1-y} - 1)}{\lambda - 1}\right)$$

if $\lambda \in (0,1)\cup(1,\infty)$, $S_0 = S_M$, $S_1 = S_P$, $S_\infty = S_L$.

(*ii*) The family of Yager triangular norms $(T^\lambda)_{\lambda\in[0,\infty]}$ given by

$$T^\lambda(x,y) = \max\left(0, 1 - \left((1-x)^\lambda + (1-y)^\lambda\right)^{\frac{1}{\lambda}}\right)$$

if $\lambda \in (0,\infty)$, $T^0 = T_W$, $T^\infty = T_M$ and Yager triangular conorms $(S^\lambda)_{\lambda\in[0,\infty]}$ given by

$$S^\lambda(x,y) = \min\left(1, (x^\lambda + y^\lambda)^{\frac{1}{\lambda}}\right)$$

if $\lambda \in (0,\infty)$, $S^0 = S_W$, $S^\infty = S_M$.

For these families of triangular norms (and implicitly for their dual triangular conorms, see Theorem 7.7, (*iii*)) we present

Theorem 7.8 (i) $e_{IDEM}(T_\lambda) = \begin{cases} 0, & \text{if } \lambda = 0 \\ \frac{1}{4}, & \text{if } \lambda = 1 \\ \frac{1}{2}, & \text{if } \lambda = \infty \\ \log_\lambda \frac{\sqrt{\lambda}+1}{2}, & \text{if } \lambda \in (0,1) \cup (1,\infty) \end{cases}$

(ii) $e_{IDEM}(T^\lambda) = \begin{cases} 1, & \text{if } \lambda = 0 \\ 0, & \text{if } \lambda = \infty \\ 1 - 2^{-\frac{1}{\lambda}}, & \text{if } \lambda \in (0,\infty). \end{cases}$

Proof. (i) If $\lambda = 0$ then $x - T_\lambda(x,x) = 0, \forall x \in [0,1]$. If $\lambda = 1$ then $e_{IDEM}(T_\lambda) = \sup\{x - T_\lambda(x,x) : x \in [0,1]\} = \sup\{x - x^2 : x \in [0,1]\} = \frac{1}{4}$. If $\lambda = \infty$ then $x - T_\lambda(x,x) = x - T_L(x,x)$ is equal to x if $x \le \frac{1}{2}$ and to $1 - x$ if $x \ge \frac{1}{2}$, therefore $e_{IDEM}(T_\lambda) = \frac{1}{2}$. If $\lambda \in (0,1) \cup (1,\infty)$ then

$$x - T_\lambda(x,x) = x - \log_\lambda\left(1 + \frac{(\lambda^x - 1)(\lambda^x - 1)}{\lambda - 1}\right).$$

Because the function $f : [0,1] \to \mathbf{R}$ defined by

$$f(x) = x - \log_\lambda\left(1 + \frac{(\lambda^x - 1)(\lambda^x - 1)}{\lambda - 1}\right)$$

has the derivative $f'(x) = \frac{\lambda - \lambda^{2x}}{\lambda - 1 + (\lambda^x - 1)^2}$ positive on $\left[0, \frac{1}{2}\right]$ and negative on$\left[\frac{1}{2}, 1\right]$, we have

$$\begin{aligned} e_{IDEM}(T_\lambda) &= \sup\left\{x - \log_\lambda\left(1 + \frac{(\lambda^x - 1)(\lambda^x - 1)}{\lambda - 1}\right) : x \in [0,1]\right\} \\ &= f\left(\frac{1}{2}\right) = \log_\lambda \frac{\sqrt{\lambda}+1}{2}. \end{aligned}$$

(ii) If $\lambda = 0$ then

$$x - T^\lambda(x,x) = x - T_W(x,x) = \begin{cases} 0, & \text{if } x = 1 \\ x, & \text{if } x \in [0,1) \end{cases},$$

therefore $e_{IDEM}(T^\lambda) = 1$. Because

$$x - T^\lambda(x,x) = \begin{cases} x, & \text{if } x \le 1 - 2^{-\frac{1}{\lambda}} \\ (1-x)\left(2^{\frac{1}{\lambda}} - 1\right), & \text{if } x \ge 1 - 2^{-\frac{1}{\lambda}} \end{cases}$$

for every $\lambda \in (0,\infty)$, we obtain $e_{IDEM}(T^\lambda) = 1 - 2^{-\frac{1}{\lambda}}$. □

We also obtain a general result for the defect of idempotency of ordinal sums of triangular norms. Firstly, we recall the following

Theorem 7.9 *(see Klement-Mesiar [119]) Let $(T_\alpha)_{\alpha\in A}$ be a family of triangular norms and $(]a_\alpha, b_\alpha[)_{\alpha\in A}$ be a family of pairwise disjoint open subintervals of $[0,1]$. Then the function $T:[0,1]\times[0,1]\to[0,1]$ defined by*

$$T(x,y)=\begin{cases} a_\alpha+(b_\alpha-a_\alpha)T_\alpha\left(\frac{x-a_\alpha}{b_\alpha-a_\alpha},\frac{y-a_\alpha}{b_\alpha-a_\alpha}\right), & \text{if } (x,y)\in[a_\alpha,b_\alpha]^2 \\ \min(x,y), & \text{otherwise,}\end{cases}$$

is a triangular norm. It is called the ordinal sum of the summands $\langle a_\alpha, b_\alpha, T_\alpha\rangle$ $\alpha\in A$, and we shall write $T\approx(\langle a_\alpha,b_\alpha,T_\alpha\rangle)_{\alpha\in A}$.

Theorem 7.10 *If $T\approx(\langle a_\alpha,b_\alpha,T_\alpha\rangle)_{\alpha\in A}$ then*

$$e_{IDEM}(T)\le\sup\{e_{IDEM}(T_\alpha):\alpha\in A\}.$$

Proof. Let $\alpha\in A$. We obtain

$$\begin{aligned} x-T(x,x) &= x-a_\alpha-(b_\alpha-a_\alpha)T_\alpha\left(\frac{x-a_\alpha}{b_\alpha-a_\alpha},\frac{x-a_\alpha}{b_\alpha-a_\alpha}\right) \\ &= \left(\frac{x-a_\alpha}{b_\alpha-a_\alpha}-T_\alpha\left(\frac{x-a_\alpha}{b_\alpha-a_\alpha},\frac{x-a_\alpha}{b_\alpha-a_\alpha}\right)\right)(b_\alpha-a_\alpha) \\ &\le \frac{x-a_\alpha}{b_\alpha-a_\alpha}-T_\alpha\left(\frac{x-a_\alpha}{b_\alpha-a_\alpha},\frac{x-a_\alpha}{b_\alpha-a_\alpha}\right)\le e_{IDEM}(T_\alpha), \end{aligned}$$

for every $x\in[a_\alpha,b_\alpha]$, therefore

$$e_{IDEM}(T)=\sup\{x-T(x,x):x\in X\}\le\sup\{e_{IDEM}(T_\alpha):\alpha\in A\}.$$

□

Example 7.6 If we take into account that each triangular norm T can be written as a trivial ordinal sum with one summand $\langle 0,1,T\rangle$ (see Klement-Mesiar [119]), then in the previous theorem we obtain equality.
The inequality in the same theorem can be strict. Indeed, if we consider the triangular norm T as ordinal sum of the summands $\langle\frac{1}{4},\frac{1}{2},T_P\rangle$ and $\langle\frac{2}{3},\frac{3}{4},T_L\rangle$, that is (see Klement-Mesiar [119])

$$T(x,y)=\begin{cases}\frac{1}{4}(1+(4x-1)(4y-1)), & \text{if } (x,y)\in[1/4,1/2]\times[1/4,1/2] \\ \frac{2}{3}+\max\left(0,x+y-\frac{17}{12}\right), & \text{if } (x,y)\in[2/3,3/4]\times[2/3,3/4] \\ \min(x,y), & \text{otherwise,}\end{cases}$$

then

$$x - T(x,x) = \begin{cases} -4x^2 + 3x - \frac{1}{2}, & \text{if } x \in [1/4, 1/2] \\ x - \frac{2}{3}, & \text{if } x \in [2/3, 17/24] \\ -x + \frac{3}{4}, & \text{if } x \in [17/24, 3/4] \\ 0, & \text{otherwise,} \end{cases}$$

therefore the defect of idempotency of T is equal to $\frac{1}{16}$. The defect of idempotency of summands are $e_{IDEM}(T_P) = \frac{1}{4}$ and $e_{IDEM}(T_L) = \frac{1}{2}$.

In the sequel, we study the defect of distributivity of a triangular norm or conorm with respect to another triangular norm or conorm. Because triangular norms and triangular conorms are commutative, we have $e^d_{DISR}(F;G)([0,1]) = e^d_{DISL}(F;G)([0,1])$ for every F, G triangular norms or conorms. We denote the common value by $e_{DIS}(F;G)$, the metric d on $[0,1]$ being generated by absolute value $|\cdot|$. Concerning this defect, we present

Theorem 7.11 *(i)* $0 \le e_{DIS}(F;G) \le 1$ *for every triangular norms or conorms* F *and* G;

(ii) $e_{DIS}(T;S_M) = 0$, *for every triangular norm* T;

(iii) $e_{DIS}(S;T_M) = 0$, *for every triangular conorm* S;

(iv) $e_{DIS}(F;F) = e_{IDEM}(F)$, *for every triangular norm or conorm* F;

(v) If S *is the dual of* T *then* $e_{DIS}(T;T) = e_{DIS}(S;S)$.

Proof. *(i)* It is obvious.

(ii) By using the monotonicity of T in the second variable we get

$$T(x, \max(y,z)) = T(x,y) = \max(T(x,y), T(x,z)),$$

in the hypothesis $y \ge z$. The proof is similar in the case $y \le z$.

(iii) It is similar to *(ii)* because S is monotone too.

(iv) Let F be a triangular norm. Due to associativity and commutativity of F, we get

$$F(F(x,y), F(x,z)) = F(F(x, F(y,z)), x), \forall x, y, z \in [0,1].$$

Denoting $F(x, F(y,z)) = u$, the properties of F imply

$$u = F(x, F(y,z)) \le F(x, F(1,1)) = F(x,1) = x, \forall x, y, z \in [0,1]$$

and the inequality becomes equality if $y = z = 1$.

The above relations together with the monotonicity of F imply

$$e_{DIS}(F;F)$$

$$= \sup\{|F(x,F(y,z)) - F(F(x,y),F(x,z))|; x,y,z \in [0,1]\}$$

$$= \sup\{F(x,F(y,z)) - F(F(x,F(y,z)),x); x,y,z \in [0,1]\}$$

$$= \sup\{u - F(u,x); u,x \in [0,1], u \leq x\}$$

$$= \sup\{u - F(u,u); u \in [0,1]\}$$

$$= e_{IDEM}(F)$$

Analogously, if F is a triangular conorm then we have

$$e_{DIS}(F;F)$$

$$= \sup\{|F(x,F(y,z)) - F(F(x,y),F(x,z))|; x,y,z \in [0,1]\}$$

$$= \sup\{F(F(x,F(y,z)),x) - F(x,F(y,z)); x,y,z \in [0,1]\}$$

$$= \sup\{F(u,x) - u; u,x \in [0,1], x \leq u\}$$

$$= \sup\{F(u,u) - u; u \in [0,1]\}$$

$$= e_{IDEM}(F)$$

(v) It is immediate by Theorem 7.7, (iii). □

Remark. Due to property (iv), the calculus of defect of autodistributivity becomes simple. Also, property (iv) implies other properties of the defect of autodistributivity. For example, if T_1 and T_2 are triangular norms and $T_1 \leq T_2$ then $e_{DIS}(T_2;T_2) \leq e_{DIS}(T_1;T_1)$ (see also Theorem 7.7, (iv)). If S_1 and S_2 are triangular conorms and $S_1 \leq S_2$, then $e_{DIS}(S_1;S_1) \leq e_{DIS}(S_2;S_2)$ (see also Theorem 7.7, (v)).

For some basic triangular norms and conorms we obtain the following

Theorem 7.12 *We have:*

(i) $e_{DIS}(T_M; S_M) = e_{DIS}(T_W; S_M) = e_{DIS}(T_L; S_M)$
$= e_{DIS}(T_P; S_M) = 0$;

(ii) $e_{DIS}(S_M; T_M) = e_{DIS}(S_W; T_M) = e_{DIS}(S_L; T_M)$
$= e_{DIS}(S_P; T_M) = 0$;

(iii) $e_{DIS}(T_L; T_M) = e_{DIS}(S_L; S_M) = 0$;

(iv) $e_{DIS}(S_P; T_P) = e_{DIS}(T_P; S_P) = \frac{1}{4}$.

Proof. (i) and (ii) are consequences of Theorem 7.11, (ii), (iii).

(iii) The distributivity of T_L with respect to T_M and the distributivity of S_L with respect to S_M are given by Butnariu [49].

(iv) Starting from definition we have

$$\begin{aligned} & e_{DIS}(S_P; T_P) \\ = \ & \sup\{|x + yz - xyz - (x + y - xy)(x + z - xz)| ; x, y, z \in [0,1]\} \\ = \ & \sup\{x(1-x)(1-y)(1-z) ; x, y, z \in [0,1]\} = \frac{1}{4} \end{aligned}$$

and

$$\begin{aligned} & e_{DIS}(T_P; S_P) \\ = \ & \sup\left\{\left|x(y + z - yz) - (xy + xz - x^2yz)\right| ; x, y, z \in [0,1]\right\} \\ = \ & \sup\{x(1-x)yz; x, y, z \in [0,1]\} = \frac{1}{4} \end{aligned}$$

□

7.4 Applications

The calculation of defects of idempotency and distributivity for triangular norms and triangular conorms can be useful for some estimations in fuzzy mathematics. For example, if S is a triangular conorm and the fuzzy set $A \in FS(X)$ is given (in fact the function $A : X \to [0,1]$), then the membership function of $A \cup_S A \in FS(X)$ can be estimated by:

$$(A \cup_S A)(x) \leq \min(1, A(x) + e_{IDEM}(S)), \forall x \in X.$$

If, in addition, S' is a triangular conorm and $B, C \in FS(X)$ then we have

$$((A \cup_S B) \cup_{S'} (A \cup_S C))(x)$$

$$\leq \min(1, (A \cup_S (B \cup_{S'} C))(x) + e_{DIS}(S; S')), \forall x \in X.$$

Also, if $\mathcal{A}$ is a σ-ring of subsets of the set X, S is a triangular conorm and $m : \mathcal{A} \to [0,1]$ is a S-decomposable measure (that is $m(\emptyset) = 0$ and $m(A \cup B) = m(A)\, Sm(B), \forall A, B \in \mathcal{A},\ A \cap B = \emptyset$, see *e.g.* Pap [162]), then we have

$$m(A)\, Sm(A) \leq m(A) + e_{IDEM}(S).$$

In what follows, we give an application of the defect of associativity.

Firstly, let us recall the definition of the general fuzzy integral $\int^{\oplus}$ (which includes Choquet and Sugeno integrals, for example) based on a pseudo-addition $\oplus$ and a pseudo-multiplication $\odot$, introduced in Benvenuti-Mesiar [37].

Definition 7.4 A binary operation $\oplus : [0,M]^2 \to [0,M]$, where $M \in]0,\infty]$, is called a pseudo-addition on $[0,M]$ if the following properties are satisfied:

(*i*) $a \oplus b = b \oplus a$;
(*ii*) $a \leq a'$ and $b \leq b'$ implies $a \oplus b \leq a' \oplus b'$;
(*iii*) $(a \oplus b) \oplus c = a \oplus (b \oplus c)$;
(*iv*) $a \oplus 0 = 0 \oplus a = a$;
(*v*) $a_n \to a, b_n \to b$ implies $a_n \oplus b_n \to a \oplus b$.

Definition 7.5 Let $\oplus$ be a given pseudo-addition on $[0,M]$. A binary operation $\odot : [0,M]^2 \to [0,M]$ is called a pseudo-multiplication if the following properties are satisfied:

(*i*) $(a \oplus b) \odot c = (a \odot c) \oplus (b \odot c)$;
(*ii*) $a \leq a'$ and $b \leq b'$ implies $a \odot b \leq a' \odot b'$;
(*iii*) $a \odot 0 = 0 \odot b = 0$;
(*iv*) $\exists u \in]0,M] : u \odot a = a$;
(*v*) $(\sup_{n \in \mathbf{N}} a_n) \odot (\sup_{m \in \mathbf{N}} b_m) = \sup_{n,m \in \mathbf{N}} (a_n \odot b_m)$.

The integral of a basic simple function $U_A, A \in \mathcal{A}$, where $U_A(x) = u$ if $x \in A$ and $U_A(x) = 0$ if $x \notin A$, with respect to measure μ, is introduced by

$$\int^{\oplus} U_A \odot d\mu = u \odot \mu(A) = \mu(A).$$

Then, the integral of a simple function $s = \oplus_{i=1}^{m} b(c_i, C_i)$, is given by

$$\int^{\oplus} s \odot d\mu = \oplus_{i=1}^{m} (c_i \odot \mu(C_i)),$$

where $b(c, C)(x) = c$ if $x \in C$ and $b(c, C)(x) = 0$ if $x \notin C$, and finally the integral of a measurable function f, is given by

$$\int^{\oplus} f \odot d\mu = \sup \left\{ \int^{\oplus} s \odot d\mu, s \leq f, s \text{ simple function} \right\}.$$

If the pseudo-multiplication $\odot$ is associative ($(a \odot b) \odot c = a \odot (b \odot c)$, $\forall a, b, c \in [0, M]$), then the integral above introduced has the property of $\odot$-homogeneity, that is $a \odot \int^{\oplus} f \odot d\mu = \int^{\oplus} (a \odot f) \odot d\mu, \forall a \in [0, M], \forall f$ a $\mathcal{A}$-measurable function (see Benvenuti-Mesiar [37], Theorem 4.5). In the case of non-associativity of $\odot$, the difference between $a \odot \int^{\oplus} f \odot d\mu$ and $\int^{\oplus} (a \odot f) \odot d\mu$ (which can be looked as the defect of homogeneity of the integral) can be estimated with the help of defect of associativity for the operation $\odot$.

For example, if $\oplus \equiv \vee$ on $[0, 1]$ and the pseudo-multiplication is given by $a \odot b = g(a) \wedge b, \forall a, b \in [0, 1]$, where $g : [0, 1] \to [0, 1]$ is an increasing bijection, then

$$a \odot \int^{\oplus} f \odot d\mu - \int^{\oplus} (a \odot f) \odot d\mu \leq e_{AS}^{|\cdot|}(\odot)([0, 1]),$$

for all $a \in [0, 1]$ and f $\mathcal{A}$-measurable. (Here $x \vee y = \max(x, y), x \wedge y = \min(x, y)$ and $|\cdot|$ denotes the absolute value). Indeed,

$$a \odot \int^{\oplus} s \odot d\mu = a \odot \vee_{i=1}^{m} (c_i \odot \mu(C_i)) = \vee_{i=1}^{m} (a \odot (c_i \odot \mu(C_i)))$$

$$\begin{aligned}
&\leq \vee_{i=1}^{m} \left(((a \odot c_i) \odot \mu(C_i)) + e_{AS}^{|\cdot|}(\odot)([0, 1]) \right) \\
&= \vee_{i=1}^{m} ((a \odot c_i) \odot \mu(C_i)) + e_{AS}^{|\cdot|}(\odot)([0, 1]) \\
&= \int^{\oplus} \vee_{i=1}^{m} b(a \odot c_i, C_i) \odot d\mu + e_{AS}^{|\cdot|}(\odot)([0, 1]) \\
&\quad \text{(because } b(a \odot c_i, C_i) \text{ and } a \odot b(c_i, C_i) \text{ are equal)} \\
&= \int^{\oplus} \vee_{i=1}^{m} (a \odot b(c_i, C_i)) \odot d\mu + e_{AS}^{|\cdot|}(\odot)([0, 1]) \\
&= \int^{\oplus} a \odot (\vee_{i=1}^{m} b(c_i, C_i)) \odot d\mu + e_{AS}^{|\cdot|}(\odot)([0, 1]) \\
&= \int^{\oplus} (a \odot s) \odot d\mu + e_{AS}^{|\cdot|}(\odot)([0, 1]),
\end{aligned}$$

for every simple function s such that $s \leq f$. Passing to supremum to the right side with f $\mathcal{A}$-measurable, $f \geq s$, we get (because $s \leq f$ implies $a \odot s \leq a \odot f, \forall a \in [0, M]$)

$$a \odot \int^{\oplus} s \odot d\mu \leq \int^{\oplus} (a \odot f) \odot d\mu + e_{AS}^{|\cdot|}(\odot)([0,1]), \forall s \leq f, s \text{ simple.}$$

Now, passing to supremum with $s \leq f$ to the left side, we have

$$\sup_{s \leq f} \left\{ a \odot \int^{\oplus} s \odot d\mu \right\} = \sup_{s \leq f} \left\{ g(a) \wedge \int^{\oplus} s \odot d\mu \right\}$$

$$\begin{aligned} &= g(a) \wedge \sup_{s \leq f} \left\{ \int^{\oplus} s \odot d\mu \right\} = a \odot \sup_{s \leq f} \left\{ \int^{\oplus} s \odot d\mu \right\} \\ &= a \odot \int^{\oplus} f \odot d\mu \leq \int^{\oplus} (a \odot f) \odot d\mu + e_{AS}^{|\cdot|}(\odot)([0,1]). \end{aligned}$$

For example, if $g(a) = a^2$, that is $a \odot b = a^2 \wedge b$, then (see Example 7.2)

$$a \odot \int^{\oplus} f \odot d\mu - \int^{\oplus} (a \odot f) \odot d\mu \leq \frac{1}{4}, \forall a \in [0,1], \forall f \ \mathcal{A}\text{-measurable.}$$

The right distributivity of the pseudo-multiplication is not, in general, required. Nevertheless, it must be required if we want the integral to be $\oplus$-additive (*i.e.* $\int^{\oplus} (f \oplus g) \odot d\mu = \int^{\oplus} f \odot d\mu \oplus \int^{\oplus} g \odot d\mu, \forall f, g$ $\mathcal{A}$-measurable functions), when the fuzzy measure μ is $\oplus$-additive (*i.e.* $\mu(A \cup B) = \mu(A) \oplus \mu(B), \forall A, B \in \mathcal{A}, A \cap B = \emptyset$). As an example, for basic simple functions,

$$\int^{\oplus} b(a, A) \odot d\mu \oplus \int^{\oplus} b(a, B) \odot d\mu = (a \odot \mu(A)) \oplus (a \odot \mu(B))$$

and

$$\int^{\oplus} (b(a, A) \oplus b(a, B)) \odot d\mu = \int^{\oplus} b(a, A \cup B) \odot d\mu = a \odot (\mu(A) \oplus \mu(B)),$$

such that if we request the $\oplus$-additivity of the integral, one reduces to

$$(a \odot \mu(A)) \oplus (a \odot \mu(B)) = a \odot (\mu(A) \oplus \mu(B)),$$

$\forall a \in [0, M], \forall A, B \in \mathcal{A}, A \cap B = \emptyset$. In the case of right distributivity of the pseudo-multiplication with respect to pseudo-addition, the above equality is true.

By using the definition of defect of distributivity at right (Definition 7.1), we can estimate the difference between $\int^{\oplus} b(a,A) \odot d\mu \oplus \int^{\oplus} b(a,B) \odot d\mu$ and $\int^{\oplus} (b(a,A) \oplus b(a,B)) \odot d\mu$ (which can be considered as defect of additivity of the integral for simple basic functions) as follows:

$$\left| \int^{\oplus} b(a,A) \odot d\mu \oplus \int^{\oplus} b(a,B) \odot d\mu \right.$$

$$\left. - \int^{\oplus} (b(a,A) \oplus b(a,B)) \odot d\mu \right| \leq e_{DISR}^{|\cdot|}(\odot, \oplus)([0,M]).$$

Remark. Similar reasoning with those in Subsection 4.1.1 easily show us that the above inequality can be framed into the general scheme in Section 1.1.

7.5 Bibliographical Remarks

Definition 7.1, $(i), (ii), (vi), (vii)$, Lemma 7.1, $(i), (ii), (vi), (vii)$, Definition 7.2, Theorem 7.1, $(i), (ii), (vi), (vii)$, Definition 7.3, Theorem 7.2, Theorems 7.4, 7.5, 7.6, Examples 7.1, 7.3, 7.4, 7.5, Theorems 7.7, 7.8, 7.9, 7.10, 7.11, 7.12, Example 7.6, and Section 7.4 are from Ban-Gal [28]. All the other results (excepting those where are mentioned the authors) appear for the first time in this book.

Chapter 8

Miscellaneous

In this chapter we study defects of property in Complex Analysis, Geometry, Number Theory and Fuzzy Logic.

8.1 Defect of Property in Complex Analysis

Let $f : D \to \mathbf{C}$, where D is a domain of the field of complex numbers, $f(z) = u(x, y) + iv(x, y), z = x + iy, i = \sqrt{-1}$.

A well-known result of Morera [153] states that the holomorphy of f on D is equivalent to the following two conditions:

(i) f is continuous on D;

(ii) The integral $\int_C f(z)\, dz = 0$, for any closed rectifiable curve in D.

Starting from this definition, Pompeiu [163] notes that if $\int_C f(z)\, dz \neq 0$, then $\left|\int_C f(z)\, dz\right|$ can be considered as a measure of non-holomorphy of f inside of the domain bounded by the closed curve C.

Suggested by this remark we can introduce the following.

Definition 8.1 Let $D \subset \mathbf{C}$ be a bounded domain and $f : D \to \mathbf{C}$ integrable on D. The number

$$d_{HOL}(f)(D) = \sup\left\{\left|\int_C f(z)\, dz\right|; C \subset D, \text{ closed rectifiable curve}\right\}$$

will be called defect of holomorphy of f on D.

Remark. If f is continuous on D (we write $f \in C(D)$), then from Morera's theorem it follows that f is holomorphic on D if and only if $d_{HOL}(f)(D) =$

0.

The following properties are immediate:

$$\begin{aligned} d_{HOL}(f+g)(D) &\leq d_{HOL}(f)(D)+d_{HOL}(g)(D), \forall f,g \in C(D) \\ d_{HOL}(\lambda f)(D) &= |\lambda| d_{HOL}(f)(D), \forall \lambda \in \mathbf{C}, f \in C(D). \end{aligned}$$

Continuing the ideas in Pompeiu [163], let us suppose that $f \in C^1(D)$, that is if $f = u+iv$ then u, v are of C^1 class. In this case, can be introduced the following

Definition 8.2 (Pompeiu [163]) Let $f \in C^1(D), f(z) = u(x,y)+iv(x,y)$ $, z = x+iy \in D$ and $z_0 = x_0 + iy_0 \in D$. The areolar derivative of f at z_0 is given by

$$d_A(f)(z_0) = \lim_{C \to z_0} \frac{\frac{1}{2i}\int_C f(z)\,dz}{m(\Delta)},$$

where the limit is considered for all closed curves C (in D) surrounding z_0, that converge to z_0 by a continuous deformation ($m(\Delta)$ represents the area of the domain Δ closed by C).

In the same paper Pompeiu [163], it is proved the formula

$$d_A(f)(z_0) = \frac{1}{2}\left[\left(\frac{\partial u}{\partial x} - \frac{\partial v}{\partial y}\right) + i\left(\frac{\partial v}{\partial x} + \frac{\partial u}{\partial y}\right)\right](z_0).$$

Remarks. 1). Because for $f \in C^1(D)$ and $z_0 \in D$, it is obvious that (see Pompeiu [164])

$$d_A(f)(z_0) = 0 \text{ if and only if } f \text{ is differentiable at } z_0,$$

we can call $|d_A(f)(z_0)|$ as defect of differentiability of $f \in C^1(D)$ on z_0. In Szu-Hoa Min [209], $|d_A(f)(z_0)|$ is called deviation from analiticity.

2). It is well-known the application of $d_A(f)(z)$ to the classical so-called Cauchy-Pompeiu formula, valid for all f of C^1-class

$$f(z) = \frac{1}{2\pi i}\int_C \frac{f(\xi)}{\xi - z}\mathrm{d}\xi - \frac{1}{\pi}\int\int_\Delta \frac{d_A(f)(v)}{v-z}\mathrm{d}\varpi,$$

where $v = \alpha + i\beta, \mathrm{d}\varpi = \mathrm{d}\alpha \mathrm{d}\beta, z = x+iy$ (see Pompeiu [164]). Obviously, if f is holomorphic it follows $d_A(f) \equiv 0$ and then the above formula one reduces to the Cauchy's formula. The Cauchy-Pompeiu formula can easily be framed into the general scheme in Section 1.1. Indeed, for a fixed closed curve $C \subset D$, we can define

$$\begin{aligned} A_{z,C}(f) &= f(z) - \frac{1}{2\pi i}\int_C \frac{f(\xi)}{\xi - z}\mathrm{d}\xi, \\ B_{z,C}(v) &= \frac{1}{\pi}\int\int_\Delta \frac{v}{v-z}\mathrm{d}\alpha\mathrm{d}\beta \text{ (here } v = \alpha + i\beta), \ z \in \Delta = \operatorname{int}(C), \\ U &= C^1(D) \end{aligned}$$

and

$$P = \text{"the property of differentiability (holomorphy) on } D\text{"}.$$

If we introduce the defect of differentiability of f on D by

$$d_A(f)(D) = \sup\{|d_A(f)(z)|; z \in D\},$$

obviously it is different from $d_{HOL}(f)(D)$. With this notation, from the above formula we get

$$\left| f(z) - \frac{1}{2\pi i}\int_C \frac{f(\xi)}{\xi - z}\mathrm{d}\xi \right| \le \frac{1}{\pi}\left| \int\int_\Delta \frac{d_A(f)(v)}{v - z}\mathrm{d}\varpi \right|,$$

and if we choose $C = \{u \in \mathbf{C}; |u - a| = r\} \subset D, \Delta = int(C) \subset D$ and $z \in D$ such that $|z - a| \ge 2r$, then it follows

$$\begin{aligned} \left| f(z) - \frac{1}{2\pi i}\int_C \frac{f(\xi)}{\xi - z}\mathrm{d}\xi \right| &\le \frac{1}{\pi}\int\int_\Delta \frac{|d_A(f)(v)|}{|v - z|}\mathrm{d}\varpi \\ &\le \frac{1}{\pi}d_A(f)(\Delta)\cdot\int\int_\Delta \frac{1}{|v-z|}\mathrm{d}\alpha\mathrm{d}\beta \le \frac{1}{\pi}d_A(f)(\Delta)\,\pi r^2\frac{1}{r} \\ &= r d_A(f)(\Delta), \end{aligned}$$

because $|v - z| \ge r, \forall v \in \Delta, \forall z \in D$ with $|z - a| \ge 2r$.

As a conclusion, the defect of differentiability $d_A(f)(D)$ appears in the estimate of deviation of $f(z)$ from the Cauchy's integral $\frac{1}{2\pi i}\int_C \frac{f(\xi)}{\xi - z}\mathrm{d}\xi$.

8.2 Defect of Property in Geometry

In geometric language, the curvature can be considered as the deviation (in a point) of a curve, from the right line and the torsion can be considered as the deviation (in a point) of a curve from the plane curve. In what follows, we introduce two global indicators which measure these deviations.

In the Euclidean and non-Euclidean geometries, the concept of curvature and torsion of a curve $\mathcal{C}$ in a point M are introduced by (see *e.g.* Mihăileanu [148], p. 99-100)

$$\gamma(M) = \lim_{\Delta s \to 0} \frac{\Delta \alpha}{\Delta s}$$

and

$$\tau(M) = \lim_{\Delta s \to 0} \frac{\Delta \theta}{\Delta s},$$

respectively, where $\Delta\alpha$ is the angle of tangents in M and $M', \Delta\theta$ is the angle of binormals in M and M' and Δs is the length of the arc MM' $(M, M' \in \mathcal{C})$ when M' tends to M.

Because the curvature of a right line and the torsion of a plane are equal to zero in each point, we can consider the following definition.

Definition 8.3 Let $\mathcal{C}$ be a curve. The quantities

$$\gamma(C) = \sup\{|\gamma(M)|; M \in \mathcal{C}\}$$

and

$$\tau(C) = \sup\{|\tau(M)|; M \in \mathcal{C}\}$$

are called the defect of right line and the defect of plane curve of $\mathcal{C}$, respectively.

An interpretation of the above introduced defects can be given, considering an Euclidean curve $\mathcal{C}$. Are well-known the Frenet's formulas (in canonical form)

$$\begin{aligned} \frac{d\vec{t}(s)}{ds} &= \gamma(s) \cdot \vec{n}(s), \\ \frac{d\vec{b}(s)}{ds} &= -\tau(s) \cdot \vec{n}(s), \\ \frac{d\vec{n}(s)}{ds} &= -\gamma(s) \cdot \vec{t}(s) + \tau(s) \cdot \vec{b}(s), \end{aligned}$$

where $\gamma(s)$ represents the curvature, $\tau(s)$ is the torsion, $\vec{t}(s)$ is the tangent versor, $\vec{n}(s)$ is the normal versor and $\vec{b}(s)$ is the binormal versor, all taken in a variable point on the curve $\mathcal{C}$.

Passing to Euclidean norm $\|\cdot\|_{\mathbf{R}^3}$ and taking into account that $\left\|\overrightarrow{t}(s)\right\|_{\mathbf{R}^3} = \|\overrightarrow{n}(s)\|_{\mathbf{R}^3} = \left\|\overrightarrow{b}(s)\right\|_{\mathbf{R}^3} = 1$, we get

$$\begin{aligned}
\left\|\frac{d\overrightarrow{t}(s)}{ds}\right\|_{\mathbf{R}^3} &= |\gamma(s)|, \\
\left\|\frac{d\overrightarrow{b}(s)}{ds}\right\|_{\mathbf{R}^3} &= |\tau(s)|, \\
\left\|\frac{d\overrightarrow{n}(s)}{ds}\right\|_{\mathbf{R}^3} &\leq |\gamma(s)| + |\tau(s)|.
\end{aligned}$$

Passing to supremum with s, it follows

$$\begin{aligned}
\sup_s \left\{\left\|\frac{d\overrightarrow{t}(s)}{ds}\right\|_{\mathbf{R}^3}\right\} &= \gamma(\mathcal{C}), \\
\sup_s \left\{\left\|\frac{d\overrightarrow{b}(s)}{ds}\right\|_{\mathbf{R}^3}\right\} &= \tau(\mathcal{C})
\end{aligned}$$

and

$$\sup_s \left\{\left\|\frac{d\overrightarrow{n}(s)}{ds}\right\|_{\mathbf{R}^3}\right\} \leq \gamma(\mathcal{C}) + \tau(\mathcal{C}).$$

Because $\frac{d\overrightarrow{t}(s)}{ds}$ is in fact the acceleration, it follows that the defect of right line of a curve, $\gamma(\mathcal{C})$, is, from kinematical viewpoint, the maximum value of the acceleration on the curve $\mathcal{C}$. Similarly, the defect of plane curve, $\tau(\mathcal{C})$, is the maximum value of the speed on the binormal $\overrightarrow{b}$.

In other interpretation, when a solid moves around a fixed point, it is well-known that the Frenet's formulas can be written as

$$\begin{aligned}
\frac{d\overrightarrow{t}}{ds} &= \overrightarrow{\omega} \times \overrightarrow{t}, \\
\frac{d\overrightarrow{n}}{ds} &= \overrightarrow{\omega} \times \overrightarrow{n}, \\
\frac{d\overrightarrow{b}}{ds} &= \overrightarrow{\omega} \times \overrightarrow{b},
\end{aligned}$$

where $\overrightarrow{\omega}$ is the Darboux's vector and represents the angular speed (around a rotational axis). In other words, the defect of plane curve $\tau(\mathcal{C})$ represents the maximum value of the norm (in $\mathbf{R}^3$) of vector $\overrightarrow{\omega} \times \overrightarrow{b}$, *i.e.*

$$\tau(\mathcal{C}) = \sup_{M \in \mathcal{C}} \left\| \overrightarrow{\omega} \times \overrightarrow{b} \right\|_{\mathbf{R}^3}.$$

Remark. For an Euclidean curve given by parametric equations $x = x(t), y = y(t), z = z(t), t \in D$, we have

$$\gamma(\mathcal{C}) = \sup \left\{ \frac{\left((y'z'' - z'y'')^2 + (z'x'' - x'z'')^2 + (x'y'' - y'x'')^2 \right)^{\frac{1}{2}}}{(x'^2 + y'^2 + z'^2)^{\frac{3}{2}}}; t \in D \right\}$$

and

$$\tau(\mathcal{C}) = \sup \left\{ \frac{(y'z'' - z'y'')^2 + (z'x'' - x'z'')^2 + (x'y'' - y'x'')^2}{\begin{vmatrix} x' & y' & z' \\ x'' & y'' & z'' \\ x''' & y''' & z''' \end{vmatrix}}; t \in D \right\}.$$

In the case when $\mathcal{C}$ is a plane curve given in Cartesian coordinates by the equation $y = f(x), x \in [a, b]$, the curvature in a point $M(x, y)$ is (see *e.g.* Ionescu [104], p.87)

$$\gamma(M) = \frac{|f''(x)|}{\sqrt{\left(1 + (f'(x))^2\right)^3}},$$

and consequently,

$$\gamma(\mathcal{C}) = \sup \left\{ \frac{|f''(x)|}{\sqrt{\left(1 + (f'(x))^2\right)^3}}; x \in [a, b] \right\}.$$

Example 8.1 If we consider the Euclidean curve $\mathcal{C}$ given by $x(t) = t, y(t) = t^2, z(t) = \frac{2t^3}{3}, t \in [0, 1]$ then

$$\gamma(\mathcal{C}) = \sup \left\{ \frac{2}{(2t^2 + 1)^3}; t \in [0, 1] \right\} = 2$$

and

$$\tau(\mathcal{C}) = \sup\left\{\frac{4\left(2t^2+1\right)^2}{8}; t \in [0,1]\right\} = \frac{9}{2}.$$

Example 8.2 The above introduced defects can be infinite. Indeed, if $\mathcal{C}$ is the plane curve given by $x = x(t), y = y(t), t \in D$, then the formula of curvature in $M \in \mathcal{C}$ one reduces to

$$\gamma(M) = \frac{x'y'' - y'x''}{\left(x'^2 + y'^2\right)^{\frac{3}{2}}}$$

and the remarkable curve $\mathcal{C} : x = a(t - \sin t), y = a(1 - \cos t), t \in \mathbf{R}$ (called cycloid) has the defect of right line equal to

$$\begin{aligned}
\gamma(\mathcal{C}) &= \sup\left\{\left|\frac{a^2 \cos t(1-\cos t) - a^2 \sin^2 t}{\left(a^2(1-\cos t)^2 + a^2 \sin^2 t\right)^{\frac{3}{2}}}\right|; t \in \mathbf{R}\right\} \\
&= \sup\left\{\left|\frac{1}{4a \sin\frac{t}{2}}\right|; t \in \mathbf{R}\right\} = +\infty.
\end{aligned}$$

Next, we give an example of calculus for defects in the non-Euclidean case.

Example 8.3 Let us consider the curve $\mathcal{C}$ by

$$\begin{aligned}
x_0 &= x_0(t) = \frac{1}{2}\left(t^4 + t^2 + 2\right), \\
x_1 &= x_1(t) = \frac{1}{2}\left(t^4 + t^2\right), \\
x_2 &= x_2(t) = t, \\
x_3 &= x_3(t) = t^2,
\end{aligned}$$

in the hyperbolic space with the absolute given by

$$x_0^2 - x_1^2 - x_2^2 - x_3^2 = 1.$$

Denoting $A = \begin{vmatrix} x_0 & x_1 & x_2 \\ x_0' & x_1' & x_2' \\ x_0'' & x_1'' & x_2'' \end{vmatrix}, B = \begin{vmatrix} x_0 & x_2 & x_3 \\ x_0' & x_2' & x_3' \\ x_0'' & x_2'' & x_3'' \end{vmatrix}, C = \begin{vmatrix} x_0 & x_1 & x_3 \\ x_0' & x_1' & x_3' \\ x_0'' & x_1'' & x_3'' \end{vmatrix}$

and $D = \begin{vmatrix} x_1 & x_2 & x_3 \\ x_1' & x_2' & x_3' \\ x_1'' & x_2'' & x_3'' \end{vmatrix}$, the curvature of $\mathcal{C}$ is (see *e.g.* Mihăileanu [148], p.111-112)

$$\gamma(t) = \frac{\left(A^2+B^2+C^2-D^2\right)^{\frac{1}{2}}}{\left(-x_0'^2+x_1'^2+x_2'^2+x_3'^2\right)^{\frac{3}{2}}} = \frac{\left(64t^6+48t^4+12t^2+5\right)^{\frac{1}{2}}}{\left(1+4t^2\right)^{\frac{3}{2}}}$$
$$= \left(1+\frac{4}{\left(1+4t^2\right)^3}\right)^{\frac{1}{2}},$$

therefore

$$\gamma(\mathcal{C}) = \sup\left\{\left(1+\frac{4}{\left(1+4t^2\right)^3}\right)^{\frac{1}{2}}; t \in \mathbf{R}\right\} = \sqrt{5}.$$

Then, the torsion of $\mathcal{C}$ is given (see also Mihăileanu [148], p.111-112) by

$$\tau(t) = \frac{\begin{vmatrix} x_0 & x_1 & x_2 & x_3 \\ x_0' & x_1' & x_2' & x_3' \\ x_0'' & x_1'' & x_2'' & x_3'' \\ x_0''' & x_1''' & x_2''' & x_3''' \end{vmatrix}}{\left(-x_0'^2+x_1'^2+x_2'^2+x_3'^2\right)^3 \gamma^2(t)} = \frac{24t}{\left(1+4t^2\right)^3+4},$$

therefore

$$\tau(\mathcal{C}) = \sup\left\{\frac{24t}{\left(1+4t^2\right)^3+4}; t \in \mathbf{R}_+\right\} = \frac{132}{25}.$$

Remark. The above introduced defects can be applied to various problems or estimations, as follows.

Firstly, for example, given the plane curve $y = f(x), x \in [a,b], f \in C^2[a,b]$, a problem would be to find out (if there exists) a right line, of equation $y = mx+n$, which minimizes the defect of right line of the difference curve, that is given by the equation $y = f(x) - (mx+n)$. The problem one reduces to find out the quantity

$$\min\left\{\max\left\{\frac{|f''(x)|}{\sqrt{\left(1+\left(f'(x)-m\right)^2\right)^3}}; x \in [a,b]\right\}; m \in \mathbf{R}\right\}.$$

In other order of ideas, if for example, $\mathcal{C}$ is a Ţiţeica curve (in space), that is, satisfies the formula $d^2(M) = c_0 \cdot \tau(M)$, where $\tau(M)$ is the torsion (in the point M), $d(M)$ represents the distance from the origin of coordinate axis to the osculating plane at the curve $\mathcal{C}$ in the point M and c_0 is a constant, then we obviously get

$$\sup \{d(M) ; M \in \mathcal{C}\} = \sqrt{|c_0| \cdot |\tau(C)|}.$$

Another defect of property in geometry can be introduced starting from the following well-known:

Theorem 8.1 *(see e.g. Mihăileanu-Neumann [149], p.42, Vasiu [214], p.46-47). The sum of angles of a triangle in the absolute geometry is less or equal to π (that is the sum of two right angles). The sum of angles of a triangle in the elliptic geometry is greater than π. The sum of angles of a triangle in Euclidean geometry is exactly π.*

Definition 8.4 A triangle ABC (in a geometry) is called Euclidean if $A + B + C = \pi$, where A, B, C denotes the measures of angles.

Definition 8.5 Let ABC be a triangle in a geometry. The quantity

$$D_{\triangle}(ABC) = |A + B + C - \pi|$$

is called defect of Euclidean triangle.

Remark. There exist other concepts too which measure the deviation of the sum of angles of a triangle from π. For example, in Félix [76], p. 521 the difference $\pi - (A + B + C)$ is called deficit of the triangle ABC in the Euclidean geometry and in Lobacevski's geometry. Also, in Mihăileanu-Neumann [149], p.43 the quantity $\pi - (A + B + C)$ is called defect of the triangle ABC in the absolute geometry. In Félix [76], p.543 is introduced the concept of excess of a triangle ABC as being the quantity $A+B+C-\pi$, in the spherical model of geometry.

Example 8.4 If we consider the model of elliptic geometry on a sphere of radius r, then in the Gauss' formula (see Vrănceanu-Teleman [218], p.126) we obtain

$$D_{\triangle}(ABC) = \frac{S}{r^2},$$

where S is the area of triangle ABC.

In what follows, we deal with other two main concepts in geometry: orthogonality and parallelness.

Intuitively, we can accept that two right lines or two planes are "more parallel" or "more orthogonal" than another pair of right lines or planes. Therefore, it is natural to introduce indicators which measure the deviation from parallelness and orthogonality.

Let us assume that the absolute of a non-Euclidean space is given by

$$q^2x_0^2 + x_1^2 + x_2^2 + x_3^2 = 0,$$

where $q^2 \in \mathbf{R}$ and let us denote $\varepsilon = \frac{1}{q}$.

Remark. We have $q^2 > 0$ in the case of elliptic geometry and $q^2 < 0$ in the case of hyperbolic geometry.

We introduce the inner product of two points X and Y with coordinates $x_i, i \in \{0, 1, 2, 3\}$ and $y_i, i \in \{0, 1, 2, 3\}$, respectively, by

$$X \cdot Y = \frac{q^2x_0y_0 + x_1y_1 + x_2y_2 + x_3y_3}{\sqrt{q^2x_0^2 + x_1^2 + x_2^2 + x_3^2}\sqrt{q^2y_0^2 + y_1^2 + y_2^2 + y_3^2}}$$

and the inner product of two planes $\alpha : \alpha_0x_0 + \alpha_1x_1 + \alpha_2x_2 + \alpha_3x_3 = 0$, $\beta : \beta_0x_0 + \beta_1x_1 + \beta_2x_2 + \beta_3x_3 = 0$ by

$$\alpha \cdot \beta = \frac{\varepsilon^2\alpha_0\beta_0 + \alpha_1\beta_1 + \alpha_2\beta_2 + \alpha_3\beta_3}{\sqrt{\varepsilon^2\alpha_0^2 + \alpha_1^2 + \alpha_2^2 + \alpha_3^2}\sqrt{\varepsilon^2\beta_0^2 + \beta_1^2 + \beta_2^2 + \beta_3^2}}.$$

Definition 8.6 (*i*) The quantity

$$d_{ORTH}(X, Y) = |X \cdot Y|$$

is called defect of orthogonality of the points X and Y.

(*ii*) The quantity

$$d_{ORTH}(\alpha, \beta) = |\alpha \cdot \beta|$$

is called defect of orthogonality of the planes α and β.

We have

Theorem 8.2 (*i*) *$d_{ORTH}(X, Y) = 0$ if and only if X and Y are orthogonal; $d_{ORTH}(\alpha, \beta) = 0$ if and only if α and β are orthogonal.*

(*ii*) *$0 \leq d_{ORTH}(X, Y) \leq 1$ and $0 \leq d_{ORTH}(\alpha, \beta) \leq 1$.*

(*iii*) *$d_{ORTH}(X, Y) = d_{ORTH}(Y, X)$; $d_{ORTH}(\alpha, \beta) = d_{ORTH}(\beta, \alpha)$.*

(iv) $d_{ORTH}(X,Y) = \left|\cos\frac{d}{q}\right|$, *where* d *is the distance between* X *and* Y*;* $d_{ORTH}(\alpha,\beta) = |\cos\theta|$, *where* θ *is the angle of the planes* α *and* β*.*

Proof. *(i)* By definition (see Mihăileanu [148], p.16) two points (planes) are orthogonal if and only if their inner product is null.

(ii) , *(iii)* Are immediate.

(iv) By definition (see Mihăileanu [148], p.16), the distance between two points X and Y is the number d $(d < q\pi)$ such that $\cos\frac{d}{q} = X \cdot Y$. Also, the angle of two planes α and β is the positive number θ $(\theta < \pi)$ such that $\cos\theta = \alpha \cdot \beta$ (see also Mihăileanu [148], p.18). □

Remark. In the Euclidean case (that is $q \to +\infty, \varepsilon \to 0$) we have

$$d_{ORTH}(X,Y) = 1$$

and

$$d_{ORTH}(\alpha,\beta) = \frac{|\alpha_1\beta_1 + \alpha_2\beta_2 + \alpha_3\beta_3|}{\sqrt{\alpha_1^2+\alpha_2^2+\alpha_3^2}\sqrt{\beta_1^2+\beta_2^2+\beta_3^2}}.$$

As a conclusion, two points in Euclidean space cannot by orthogonal and the condition of orthogonality of two planes in Euclidean space is $\alpha_1\beta_1 + \alpha_2\beta_2 + \alpha_3\beta_3 = 0$.

Let us consider two right lines u and v with the common point X and let U, V be the orthogonals of the point X on u and v, respectively (that is $U \in u$ and $V \in v, U \cdot X = 0, V \cdot X = 0$). The angle Δ of the right lines u and v is given by (see Mihăileanu [148], p.33)

$$\cos\Delta = U \cdot V.$$

Definition 8.7 The quantity

$$d_{ORTH}(u,v) = |\cos\Delta|$$

is called defect of orthogonality of the right lines u and v.

Theorem 8.3 *(i)* $d_{ORTH}(u,v) = 0$ *if and only if* u *and* v *are orthogonal.*

(ii) $0 \le d_{ORTH}(u,v) \le 1.$

(iii) $d_{ORTH}(u,v) = d_{ORTH}(v,u).$

(iv) $d_{ORTH}(u,v) = \left|\cos\frac{\delta}{q}\right|$, *where* δ *is the distance between* u *and* v*.*

(*v*) *If* $u : ax + by + cz = 0$ *and* $v : a'x + b'y + c'z = 0$ *are two right lines in plane, then* $d_{ORTH}(u,v) = \frac{|aa'+bb'+cc'|}{\sqrt{a^2+b^2+\varepsilon^2c^2}\sqrt{a'^1+b'^2+\varepsilon^2c'^2}}$.

Proof. (*i*), (*ii*), (*iii*) Are immediate.

(*iv*) The relation $\delta = q\Delta$ between the angle of two right lines and the distance implies the property.

(*v*) The formula of angle between two right lines in plane (see Mihăileanu [148], p.37) implies the desired relation. □

Example 8.5 In the Euclidean plane case ($\varepsilon \to 0$), we have

$$d_{ORTH}(u,v) = \frac{|aa' + bb'|}{\sqrt{a^2 + b^2}\sqrt{a'^2 + b'^2}},$$

if $u : ax + by + c = 0$ and $v : a'x + b'y + c' = 0$ or

$$d_{ORTH}(u,v) = \frac{|mm' + 1|}{\sqrt{m^2 + 1}\sqrt{m'^2 + 1}},$$

if $u : y = mx + n$ and $v : y = m'x + n'$. Therefore $d_{ORTH}(u,v) = 0$ if and only if $mm' = -1$, that is exactly the condition of orthogonality of two right lines in Euclidean plane. Also, the maximum value of defect of orthogonality is obtained if and only if $m = m'$, that is the parallelness is a peak of non-orthogonality.

In what follows, we consider $u : ax+by+cz = 0$ and $v : a'x+b'y+c'z = 0$ two right lines in a non-Euclidean plane. By definition, u and v are parallel if their angle is null (see Mihăileanu [148], p. 39). This suggests the following

Definition 8.8 The quantity

$$d_{PAR}(u,v) = |\sin \Delta|$$

is called defect of parallelness of the right lines u and v.

Theorem 8.4 (*i*) $0 \leq d_{PAR}(u,v) \leq 1$.

(*ii*) $d_{PAR}(u,v) = d_{PAR}(v,u)$.

(*iii*) $d_{PAR}(u,u) = 0$.

(*iv*) $d_{PAR}(u,v) = \left|\sin \frac{\delta}{q}\right| = \sqrt{\frac{(ab'-a'b)^2+\varepsilon^2(ac'-a'c)^2+\varepsilon^2(bc'-b'c)^2}{(a^2+b^2+\varepsilon^2c^2)(a'^2+b'^2+\varepsilon^2c'^2)}}$.

Proof. (*i*), (*ii*), (*iii*) Are immediate.

(iv) Because $\delta = q\Delta$ the first part is obvious. The expression of sinus for the angle of two right lines (see Mihăileanu [148], p. 38) implies the second relation. □

Remark. If $\varepsilon \to 0$ (the Euclidean case) then

$$d_{PAR}(u,v) = \frac{|ab' - a'b|}{\sqrt{a^2+b^2}\sqrt{a'^2+b'^2}},$$

where $u : ax + by + c = 0$ and $v : a'x + b'y + c' = 0$. If u and v are given by $y = mx + n$ and $y = m'x + n'$, then

$$d_{PAR}(u,v) = \frac{|m - m'|}{\sqrt{1+m^2}\sqrt{1+m'^2}}.$$

Example 8.6 If $\varepsilon^2 = -1$ and $u : 2x + y + z = 0, v : x + 2y + z = 0$ then $d_{ORTH}(u,v) = \frac{3}{4}$ and $d_{PAR}(u,v) = \frac{\sqrt{7}}{4}$.

Finally, we consider the defect of orthogonality and the defect of parallelness in an abstract frame.

Firstly, we recall the concept of abstract angle in real Banach spaces introduced by Singer [197], as a generalization of the concept of angle in Hilbert spaces.

Let $(E, ||\cdot||)$ be a real Banach space such that $\dim E > 1$ and $x, y \in E$. One define the trigonometric functions in E by

$$\begin{aligned}
\sin_E \frac{\widehat{x,y}}{2} &= \frac{1}{2}\left\|\frac{x}{||x||} - \frac{y}{||y||}\right\| \\
\cos_E \frac{\widehat{x,y}}{2} &= \rho\frac{1}{2}\left\|\frac{x}{||x||} + \frac{y}{||y||}\right\| \\
tg_E \frac{\widehat{x,y}}{2} &= \frac{\sin_E \frac{\widehat{x,y}}{2}}{\cos_E \frac{\widehat{x,y}}{2}},
\end{aligned}$$

where $\widehat{x,y}$ denotes the abstract angle between x and y, and $\frac{\widehat{x,y}}{2}$ is another abstract quantity, named "the half" of $\widehat{x,y}$. Here $\rho = +1$ if in the bidimensional plane determined by x and y, the Euclidean angle $\angle\{x,y\} \in [0,\pi]$ and $\rho = -1$ if the same angle is contained in $(\pi, 2\pi]$. One also defines the equality between abstract angles by

$$\frac{\widehat{x,y}}{2} = \frac{\widehat{x',y'}}{2} \text{ if and only if } \sin_E \frac{\widehat{x,y}}{2} = \sin_E \frac{\widehat{x',y'}}{2}, \cos_E \frac{\widehat{x,y}}{2} = \cos_E \frac{\widehat{x',y'}}{2}$$

and

$$\widehat{x,y} = \widehat{x',y'} \text{ if and only if } \frac{\widehat{x,y}}{2} = \frac{\widehat{x',y'}}{2}.$$

Based on the definitions of trigonometric functions, to each abstract angle in E we can associate a pair of real numbers, the first number being non-negative. The definition of the equality of abstract angles guarantees the unique characterization of these by a pair of real numbers.

Remark. If E is a Hilbertian space, then the definition of abstract angles becomes the usual definition of angles in Hilbert spaces. In fact, the converse statement is also true (see Singer [197]).

By using the result proved in Singer [197], Corollary 2: $x, y \in E$ are parallel if and only if $tg_E \frac{\widehat{x,y}}{2} = 0$, we can introduce the following

Definition 8.9 The defect of parallelness of vectors $x, y \in E \setminus \{0_E\}$ is given by

$$d_{PAR}(x,y) = \left| tg_E \frac{\widehat{x,y}}{2} \right| = \frac{\left\| \frac{x}{\|x\|} - \frac{y}{\|y\|} \right\|}{\left\| \frac{x}{\|x\|} + \frac{y}{\|y\|} \right\|}.$$

The following properties hold:

Theorem 8.5 *(i)*

$$0 \leq \frac{\sqrt{\frac{1}{C_0} - \cos_E^2 \frac{\widehat{x,y}}{2}}}{\left|\cos_E \frac{\widehat{x,y}}{2}\right|} \leq d_{PAR}(x,y) \leq \frac{\sqrt{C_0 - \cos_E^2 \frac{\widehat{x,y}}{2}}}{\left|\cos_E \frac{\widehat{x,y}}{2}\right|} \leq +\infty,$$

for every $x, y \in E$, where C_0 is the Neumann-Jordan constant (see Jordan-Neumann [110]).

(ii) $d_{PAR}(x,y) = 0$ if and only if x and y are parallel and $d_{PAR}(x,y) = +\infty$ if and only if x and $-y$ are parallel.

(iii) $d_{PAR}(x,y) = d_{PAR}(y,x), \forall x, y \in E$.

(iv) $d_{PAR}(ax,by) = d_{PAR}(x,y), \forall x, y \in E \setminus \{0_E\}, \forall a, b \in \mathbf{R} \setminus \{0\}$.

Proof. (i) The inequalities $\frac{\sqrt{\frac{1}{C_0}-\cos_E^2\frac{\widehat{x,y}}{2}}}{\left|\cos_E\frac{\widehat{x,y}}{2}\right|} \leq \left|tg_E\frac{\widehat{x,y}}{2}\right| \leq \frac{\sqrt{C_0-\cos_E^2\frac{\widehat{x,y}}{2}}}{\left|\cos_E\frac{\widehat{x,y}}{2}\right|}$, $\forall x, y \in E$ (see Singer [197]) implies the lower and upper estimations for $d_{PAR}(x,y)$.

(ii) The first part is implied by the above mentioned result. The second part is immediate, because $tg_E\frac{\widehat{x,y}}{2} = +\infty$ if and only if x and $-y$ are parallel (see Singer [197]).

(iii) The formulas (see Singer [197])

$$\begin{aligned}\sin_E\frac{\widehat{y,x}}{2} &= \sin_E\frac{\widehat{x,y}}{2}\\ \cos_E\frac{\widehat{y,x}}{2} &= -\cos_E\frac{\widehat{x,y}}{2}\end{aligned}$$

imply the desired equality.

(iv) Because

$$\begin{aligned}\left\|\frac{ax}{\|ax\|}-\frac{by}{\|by\|}\right\| &= \left\|\frac{x}{\|x\|}-\frac{y}{\|y\|}\right\|,\\ \left\|\frac{ax}{\|ax\|}+\frac{by}{\|by\|}\right\| &= \left\|\frac{x}{\|x\|}+\frac{y}{\|y\|}\right\|, \forall x,y\in E\setminus\{0_E\}, \forall a,b\in\mathbf{R}\setminus\{0\},\end{aligned}$$

we obtain the property. □

Starting from the definition of orthogonality in Banach spaces, we introduce the defect of orthogonality.

Definition 8.10 (Singer [197]) The vectors $x, y \in E \setminus \{0_E\}$ are called orthogonal (we denote this by $x \perp y$) if

$$\left|tg_E\frac{\widehat{x,y}}{2}\right| = 1.$$

Definition 8.11 The defect of orthogonality of vectors $x, y \in E \setminus \{0_E\}$ is given by

$$d_{ORTH}(x,y) = \left|1-\left|tg_E\frac{\widehat{x,y}}{2}\right|\right| = \left|1-\frac{\left\|\frac{x}{\|x\|}-\frac{y}{\|y\|}\right\|}{\left\|\frac{x}{\|x\|}+\frac{y}{\|y\|}\right\|}\right|.$$

The following properties hold:

Theorem 8.6 *(i)*

$$0 \leq d_{ORTH}(x,y) \leq \max\left\{\left|1 - \frac{\sqrt{C_0 - \cos_E^2 \frac{\widehat{x,y}}{2}}}{\left|\cos_E \frac{\widehat{x,y}}{2}\right|}\right|, \left|1 - \frac{\sqrt{\frac{1}{C_0} - \cos_E^2 \frac{\widehat{x,y}}{2}}}{\left|\cos_E \frac{\widehat{x,y}}{2}\right|}\right|\right\}$$

for every $x, y \in E \setminus \{0_E\}$.

(ii) $d_{ORTH}(x,y) = 0$ *if and only if* $x \perp y$.

(iii) $d_{ORTH}(x,y) = |1 - d_{PAR}(x,y)|, \forall x, y \in E \setminus \{0_E\}$.

(iv) $d_{ORTH}(ax, by) = d_{ORTH}(x,y), \forall x, y \in E \setminus \{0_E\}, \forall a, b \in \mathbf{R} \setminus \{0\}$.

(v) If $x, y \in E \setminus \{0_E\}$ *are linear independent vectors, then* $\exists a \in \mathbf{R}$ *such that* $d_{ORTH}(x, ax + y) = 0$.

(vi) $d_I^{\perp}\left(x, y; \frac{\|x\|}{\|y\|}\right) = 0$ *implies* $d_{ORTH}(x,y) = 0$, *where* $d_I^{\perp}(x,y;a)$ *is the defect of orthogonality of* x *with respect to* y, *of a-isosceles type (see Definition 6.2, (i)).*

Proof. *(i)* The positivity of defect is obvious. By using the same inequalities as in the proof of Theorem 8.5, we obtain the right-hand side inequality.

(ii), *(iii)* They are immediate.

(iv) See the proof of Theorem 8.5, *(iv)*.

(v) It is immediate by Singer [197], Proposition 5.

(vi) $d_I^{\perp}\left(x, y; \frac{\|x\|}{\|y\|}\right) = 0$ if and only if $\left\|x - \frac{\|x\|}{\|y\|}y\right\| = \left\|x + \frac{\|x\|}{\|y\|}y\right\|$, which is equivalent to $\left\|\frac{x}{\|x\|} - \frac{y}{\|y\|}\right\| = \left\|\frac{x}{\|x\|} + \frac{y}{\|y\|}\right\|$, that is $d_{ORTH}(x,y) = 0$. □

8.3 Defect of Property in Number Theory

This section discusses various defects of properties in Number Theory: defect of integer number, defect of divisibility, defect of prime number, and so on.

In Number Theory appears as natural the following question: how can be evaluated the deviation of a real number x from rational numbers or from the integers. An answer could be given by

$$d(x, \mathbf{Q}) = \inf\{|x - r|; r \in \mathbf{Q}\}$$

and

$$d(x, \mathbf{Z}) = \inf\{|x - k|; k \in \mathbf{Z}\}.$$

Obviously, the first quantity is always 0. Nevertheless, this idea leads to the following definition of defect.

Definition 8.12 Let $x \in \mathbf{R}$. The quantity

$$d_{\mathbf{RZ}}(x) = \min(x - [x], [x] - x + 1)$$

is called defect of integer number of x, where $[x]$ is the integer part of x.

The main properties of this defect are the following:

Theorem 8.7 *(i)* $d_{\mathbf{RZ}}(x) = \min(\{x\}, 1 - \{x\})$, *where* $\{x\}$ *denotes the fractional part of* x.
(ii) $d_{\mathbf{RZ}}(x) = d(x, \mathbf{Z})$ *and* $d_{\mathbf{RZ}}(x) = 0$ *if and only if* $x \in \mathbf{Z}$.
(iii) $0 \leq d_{\mathbf{RZ}}(x) \leq \frac{1}{2}, \forall x \in \mathbf{R}$.
(iv) $d_{\mathbf{RZ}}(x) = d_{\mathbf{RZ}}(-x), \forall x \in \mathbf{R}$.

Proof. *(i)* It is obvious.
(ii) The previous property implies $d_{\mathbf{RZ}}(x) = 0$ if and only if $\{x\} = 0$, if and only if $x \in \mathbf{Z}$. Also, $[x]$ and $[x] + 1$ are the closest integers to x, which proves the second part.
(iii) Because $0 \leq \{x\} < 1, \forall x \in \mathbf{R}$, we obtain the inequalities.
(iv) If $x \in \mathbf{Z}$ then $-x \in \mathbf{Z}$ and $d_{\mathbf{RZ}}(x) = d_{\mathbf{RZ}}(-x) = 0$. If $x \in \mathbf{R} \setminus \mathbf{Z}$, then the equality $[x] + [-x] = -1$ is well-known. We get

$$\begin{aligned} d_{\mathbf{RZ}}(-x) &= \min(-x - [-x], [-x] + x + 1) \\ &= \min(-x + [x] + 1, -1 - [x] + x + 1) \\ &= \min(x - [x], [x] - x + 1) = d_{\mathbf{RZ}}(x). \end{aligned}$$

□

We recall that (see e.g Klir [120], or Ban-Fechete [20]) a measure of fuzziness is a function $d_c : FS(\Omega) \to \mathbf{R}$ which satisfies the requirements:
(i) $d_c(A) = 0$ if and only if A is crisp, that is $A(x) \in \{0, 1\}, \forall x \in \Omega$.
(ii) If $A(x) \leq B(x)$ for $B(x) \leq \frac{1}{2}$ and $A(x) \geq B(x)$ for $B(x) \geq \frac{1}{2}, \forall x \in \Omega$ (that is $A \prec B$), then $d_c(A) \leq d_c(B)$.
(iii) $d_c(A)$ assumes the maximum value if and only if $A(x) = \frac{1}{2}, \forall x \in \Omega$.

On the other hand, we can consider that a real number is "more integer" if its fractional part is near to 0 or 1 and is "more fractional" if its fractional part is near to $\frac{1}{2}$. These suggest to use the measures of fuzziness to introduce an indicator of integer for a set of real numbers. Let us denote by $X = \{x_1, ..., x_n\}$ a finite set of real numbers and let us consider the fuzzy set

$A_X : \mathbf{R} \to [0,1]$ corresponding to X and defined by $A_X(x) = 0$ if $x \in \mathbf{R} \setminus X$ and $A_X(x) = \{x\}$ if $x \in X$, where $\{x\}$ is the fractional part of x. If d_c is a normalized measure of fuzziness (that is has values in $[0,1]$), then we give the following

Definition 8.13 The quantity

$$d_I(X) = d_c(A_X)$$

is called defect of integer of the whole set X.

Remark. In general, the null values of a fuzzy set do not influence the value of its measure of fuzziness, such that the above fuzzy set A_X can be considered as defined on X.

The defect of integer has the following properties.

Theorem 8.8 *(i)* $d_I(X) = 0$ *(has the minimum value) if and only if* $x_k \in \mathbf{Z}, \forall k \in \{1, ..., n\}$.

(ii) $d_I(X) = 1$ *(has the maximum value) if and only if* $d_{\mathbf{RZ}}(x_k)$ *is maximum,* $\forall k \in \{1, ..., n\}$.

(iii) Let $Y = \{y_1, ..., y_n\}$ *be a finite set of real numbers. If* $\{x_k\} \leq \{y_k\}$ *for* $\{y_k\} \leq \frac{1}{2}$ *and* $\{x_k\} \geq \{y_k\}$ *for* $\{y_k\} \geq \frac{1}{2}, \forall k \in \{1, ..., n\}$ *then* $d_I(X) \leq d_I(Y)$.

Proof. *(i)* $d_I(X) = 0$ if and only if $d_c(A_X) = 0$, that is $A_X(\{x_k\}) \in \{0,1\}, \forall k \in \{1, ..., n\}$, which is equivalent to $\{x_k\} = 0, \forall k \in \{1, ..., n\}$, that is $x_k \in \mathbf{Z}, \forall k \in \{1, ..., n\}$.

(ii) $d_I(X) = 1$ if and only if $A_X(\{x_k\}) = \frac{1}{2}, \forall k \in \{1, ..., n\}$, which is equivalent to $\{x_k\} = \frac{1}{2}, \forall k \in \{1, ..., n\}$, that is $\{x_k\} = 1 - \{x_k\} = \frac{1}{2}, \forall k \in \{1, ..., n\}$, which means $d_{\mathbf{RZ}}(x_k) = \frac{1}{2}, \forall k \in \{1, ..., n\}$.

(iii) The hypothesis imply $A_X \prec B_X$ and by using the properties of measures of fuzziness, we obtain $d_I(X) = d_c(A_X) \leq d_c(A_Y) = d_I(Y)$. □

Example 8.7 A simple normalized measure of fuzziness is that proposed by Kaufmann [114] (see also Klir [120]): $d_c(A) = \frac{1}{n} \sum_{k=1}^{n} |A(x_k) - C(x_k)|$, where $C(x_k) = 0$, if $A(x_k) \leq \frac{1}{2}$ and $C(x_k) = 1$, if $A(x_k) > \frac{1}{2}$ and $X = \{x_1, ..., x_n\}$.If $X = \left\{\frac{1}{2}, \frac{2}{3}, \frac{3}{4}, \frac{4}{5}, \frac{5}{6}, \frac{6}{7}\right\}$ and $Y = \left\{\frac{3}{2}, \frac{4}{3}, \frac{5}{4}, \frac{6}{5}\right\}$, then $d_I(X) = \frac{29}{120}$ and $d_I(Y) = \frac{77}{240}$.

Concerning the property of divisibility of natural numbers, we can introduce the following

Definition 8.14 Let p, q be two natural numbers. The quantity

$$D_{div}(p,q) = \min\{r, q-1-r\},$$

where r is the remainder of division of p by q, is called defect of divisibility of p with respect to q.

Below we give some properties.

Theorem 8.9 *Let p, p_1, p_2, q be natural numbers.*

(i) $D_{div}(p,q) = 0$ if and only if q/p (i.e. q is divisor of p).

(ii) $0 \leq D_{div}(p,q) \leq \left[\frac{q}{2}\right] + 1$, where $[x]$ is the integer part of x.

(iii) If $p_1 \equiv p_2 \,(\text{mod}q)$ then $D_{div}(p_1,q) = D_{div}(p_2,q)$ and $D_{div}(p_1 - p_2, q) = 0$.

(iv) $D_{div}(p_1 + p_2, q) = D_{div}(r_1 + r_2, q)$ and $D_{div}(p_1 p_2, q) = D_{div}(r_1 r_2, q)$, where r_1 and r_2 are the remainders of divisions of p_1 and p_2 by q.

Proof. (i) Because $r \leq q-1$, $D_{div}(p,q) = 0$ if and only if $r = 0$, that is q/p.

(ii) We assume $D_{div}(p,q) > \left[\frac{q}{2}\right] + 1$, which means $r > \left[\frac{q}{2}\right] + 1$ and $q - 1 - r > \left[\frac{q}{2}\right] + 1$. We get

$$q - 1 = q - 1 - r + r > 2\left[\frac{q}{2}\right] + 2 > 2\left(\frac{q}{2} - 1\right) + 2 = q,$$

which is a contradiction.

(iii) If $p_1 \equiv p_2 \,(\text{mod}q)$ then $p_1 = c_1 q + r$ and $p_2 = c_2 q + r$, therefore $D_{div}(p_1,q) = D_{div}(p_2,q) = \min\{r, q-1-r\}$ and $p_1 - p_2 = (c_1 - c_2)\, q$, which implies the second relation.

(iv) Let us assume $p_1 = c_1 q + r_1$ and $p_2 = c_2 q + r_2$. If $r_1 + r_2 < q$, then

$$D_{div}(p_1 + p_2, q) = \min\{r_1 + r_2, q - 1 - r_1 - r_2\} = D_{div}(r_1 + r_2, q).$$

If $r_1 + r_2 \geq q$, then $p_1 + p_2 = (c_1 + c_2 + 1)\, q + r_1 + r_2 - q$ and $r_1 + r_2 = q + r_1 + r_2 - q$ with $r_1 + r_2 - q < q$, therefore

$$D_{div}(p_1 + p_2, q) = \min\{r_1 + r_2 - q, q - 1 - r_1 - r_2 - q\} = D_{div}(r_1 + r_2, q).$$

The proof of the second relation is similar. □

It is obvious that for a given property, we can introduce more defects. For example, starting from well-known results, in what follows we introduce more variants for the defect of prime number.

Theorem 8.10 *(see e.g. Radovici-Mărculescu [167], p.91) Let n be a natural number. If $\varphi(n), \sigma(n), d(n)$ denote the number of numbers from $\{1, ..., n-1\}$ which are prime with n (that is the Euler's totient function), the sum of divisors of n and the number of distinct divisors of n, respectively, then we have:*

(i) n is a prime number if and only if $\sigma(n) = n+1$.
(ii) n is a prime number if and only if $d(n) = 2$.
(iii) n is a prime number if and only if $\varphi(n) = n-1$.
(iv) n is a prime number if and only if $\varphi(n)/n-1$ and $n+1/\sigma(n)$.

Definition 8.15 Let n be a natural number. The quantities

$$\begin{aligned} D^{\sigma}_{pr}(n) &= \sigma(n) - n - 1, \\ D^{d}_{pr}(n) &= d(n) - 2, \\ D^{\varphi}_{pr}(n) &= n - 1 - \varphi(n), \end{aligned}$$

and

$$D^{\varphi\sigma}_{pr}(n) = D_{div}(n-1, \varphi(n)) + D_{div}(\sigma(n), n+1)$$

are called σ-defect, d-defect, φ-defect and $\varphi\sigma$-defect of prime number of n, respectively.

Remark. Of course that the equality to 0 of the above defects is verified if and only if the number n is prime.

Other properties of the functions σ, d, φ imply various properties of these defects. For example, if $n = p_1^{\alpha_1}...p_k^{\alpha_k}$ is the prime factorization of n, then (see *e.g.* Vinogradov [216], p.28)

$$\varphi(n) = n\left(1 - \frac{1}{p_1}\right)...\left(1 - \frac{1}{p_k}\right)$$

and (see *e.g.* Vinogradov [216], p.26)

$$d(n) = (\alpha_1 + 1)...(\alpha_k + 1)$$

therefore

$$D^{\varphi}_{pr}(n) = n - 1 - n\left(1 - \frac{1}{p_1}\right)...\left(1 - \frac{1}{p_k}\right)$$

and

$$D_{pr}^{d}(n) = (\alpha_1 + 1) \ldots (\alpha_k + 1) - 2.$$

Also, the elementary inequalities (see *e.g.* Mitrinovič-Sandor-Crstici [152], p.9) $\varphi(n) \geq \sqrt{n}$ if $n \neq 2, n \neq 6$ and $\varphi(n) \leq n - \sqrt{n}$ if n is composite, imply

$$[\sqrt{n} - 1] \leq \sqrt{n} - 1 \leq D_{pr}^{\varphi}(n) \leq n - 1 - \sqrt{n} \leq [n - 1 - \sqrt{n}] + 1$$

if n is composite and $n \neq 2, n \neq 6$. Because $d(n) \leq 2\sqrt{n}, \forall n \in \mathbf{N}^*$ (see *e.g.* Mitrinovič-Sandor-Crstici [152], p.39), we get

$$D_{pr}^{d}(n) \leq 2(\sqrt{n} - 1), \forall n \in \mathbf{N}^*.$$

Also, $n + \sqrt{n} < \sigma(n) < n\sqrt{n}$ for any $n > 2$ (see *e.g.* Mitrinovič-Sandor-Crstici [152], p. 77), imply

$$\sqrt{n} - 1 < D_{pr}^{\sigma}(n) < n\sqrt{n} - n - 1, \forall n > 2.$$

Other important properties in Number Theory are those of perfect number and of amicable number.

Definition 8.16 (see *e.g.* Mitrinovič-Sandor-Crstici [152], p. 112) Let m and n be positive integers.

(i) n is called perfect if the sum of its divisors is equal to n.

(ii) m and n are called amicable, if the sum of divisors of m is equal to n and the sum of divisors of n is equal to m.

For these concepts, the defects could be introduced as follows.

Definition 8.17 Let m and n be positive integers. The quantities

$$D_{\text{perfect}}(n) = \left| n - \sum_{k/n, k \neq n} k \right|$$

and

$$D_{\text{amicable}}(m, n) = \left| \sum_{i/m, i \neq m} i - \sum_{j/n, j \neq n} j \right|$$

are called defect of perfect number of n and defect of amicable numbers of m, n, respectively.

Obviously, the natural properties

$$D_{\text{perfect}}(n) = 0 \text{ if and only if } n \text{ is perfect}$$

and

$$D_{\text{amicable}}(m, n) = 0 \text{ if and only if } m \text{ and } n \text{ are amicable}$$

are verified.

Remark. We note that for other properties too can be introduced the concept of defect. For example, the defect of multiply perfect, triperfect, quasiperfect, almost perfect, superperfect number (a positive integer is called a multiply perfect, triperfect, quasiperfect, almost perfect, superperfect number if $n/\sigma(n), \sigma(n) = 3n, \sigma(n) = 2n+1, \sigma(n) = 2n-1, \sigma(\sigma(n)) = 2n$, respectively (see *e.g.* Mitrinovič-Sandor-Crstici [152], p.105-110)) could be defined as

$$\begin{aligned}
D_{\text{multiply perfect}}(n) &= D_{div}(\sigma(n), n), \\
D_{\text{triperfect}}(n) &= |\sigma(n) - 3n|, \\
D_{\text{quasiperfect}}(n) &= |\sigma(n) - 2n - 1|, \\
D_{\text{almost perfect}}(n) &= |\sigma(n) - 2n + 1|, \\
D_{\text{superperfect}}(n) &= |\sigma(\sigma(n)) - 2n|,
\end{aligned}$$

where σ is the sum of divisors function.

Estimations for the above introduced defects can be given by using the functions φ, σ and d.

Theorem 8.11 (*i*) $D_{pr}^{\sigma}(n) \leq \inf\{|\sigma(n) - \sigma(p)| + |n - p|;$ *p prime number*$\}$;

(*ii*) $D_{pr}^{d}(n) = \inf\{|d(n) - d(p)|;$ *p prime number*$\}$;

(*iii*) $D_{pr}^{\varphi}(n) \leq \inf\{|\varphi(n) - \varphi(p)| + |n - p|;$ *p prime number*$\}$;

(*iv*) $D_{perfect}(n) \leq \inf\left\{|n - p| + \left|\sum_{j/p, j\neq p} i - \sum_{k/n, k\neq n} j\right|;\right.$ *p perfect number*$\}$;

(*v*) $D_{triperfect}(n) \leq \inf\{|\sigma(n) - \sigma(p)| + 3|n - p|;$ *p triperfect number*$\}$

(*vi*) $D_{quasiperfect}(n) \leq \inf\{|\sigma(n) - \sigma(p)| + 2|n - p|;$ *p quasiperfect number*$\}$;

(*vii*) $D_{almost\ perfect}(n) \leq \inf\{|\sigma(n) - \sigma(p)| + 2|n - p|;$ *p almost perfect number*$\}$;

(*viii*) $D_{superperfect}(n) \leq \inf\{|\sigma(\sigma(n)) - \sigma(\sigma(p))| + 2|n-p|;$ *p superperfect number*$\}$.

Proof. (*i*) – (*iii*) They are consequences of the Definition 8.15 and of the relations

$$\begin{aligned}
\sigma(n) - (n+1) &= |\sigma(n) - (n+1)| \\
&= |\sigma(n) - \sigma(p) + (p+1) - (n+1)| \\
&\leq |\sigma(n) - \sigma(p)| + |p-n|,
\end{aligned}$$

$$d(n) - 2 = |d(n) - 2| = |d(n) - d(p) + 2 - 2| = |d(n) - d(p)|$$

and

$$\begin{aligned}
(n-1) - \varphi(n) &= |(n-1) - \varphi(n)| \\
&= |(n-1) - \varphi(n) + \varphi(p) - (p-1)| \\
&\leq |\varphi(n) - \varphi(p)| + |p-n|,
\end{aligned}$$

respectively, for every prime number p.

(*iv*) – (*viii*) As above, by using the previous remark. □

8.4 Defect of Property in Fuzzy Logic

A fuzzy logic (see *e.g.* Butnariu-Klement-Zafrany [51]) can be described as a $[0,1]$-valued logic, that is one real number $t(p) \in [0,1]$ is assigned to each proposition p. It is called "truth degree" of the proposition p. If T is a triangular norm and S is the triangular conorm associated to T (see Definition 4.13) then the connectives $\vee_S$ (disjunction), $\wedge_T$ (conjunction) and $\neg$ (negation) can be interpreted by extending the evaluation function (or truth assignment function) t, as follows (see *e.g.* Butnariu-Klement-Zafrany [51]):

$$\begin{aligned}
t(p \wedge_T q) &= T(t(p), t(q)), \\
t(p \vee_S q) &= S(t(p), t(q)), \\
t(\neg p) &= 1 - t(p).
\end{aligned}$$

Similarly with the classical logic, each proposition is a propositional form and if A, B are propositional forms then $A \vee_S B, A \wedge_T B, \neg A$ are propositional forms for every triangular norm T and triangular conorm S. Let us abbreviate the propositional form $\neg A \vee_S B$ by $A \rightarrow_T B$.

Three outstanding examples of propositional fuzzy logics are obtained by considering the triangular norms $T(x,y) = T_L(x,y) = \max(x+y-1,0)$ (Lukasiewicz logic, see Lukasiewicz [140], Hájek-Gödel [98]), $T(x,y) = T_M(x,y) = \min(x,y)$ (Gödel logic or min–max logic, see Gödel [90], Hájek-Gödel [98]) and $T(x,y) = T_P(x,y) = xy$ (Product logic, see *e.g.* Hájek-Gödel [98]).

Definition 8.18 (see Hájek-Gödel [98]) The propositional form A is a 1-tautology (or standard tautology) if $t(A) = 1$, for each evaluation.

Definition 8.19 Let A be a propositional form which is represented with propositional forms $A_1, ..., A_n$ and connectives above introduced. The quantity

$$d^1_{TAUT}(A) = 1 - \inf\{t(A_1, ..., A_n, \vee_S, \wedge_T, \neg);$$

$$A_1, ..., A_n \text{ propositional forms}\}$$

is called defect of 1-tautology of the propositional form A.

Remark. It is obvious that $d^1_{TAUT}(A) = 0$ if and only if the propositional form A is a 1-tautology.

Example 8.8 If $T \leq T_L$ (that is $T(x,y) \leq T_L(x,y), \forall x, y \in [0,1]$) then the propositional form $(A_1 \wedge_T A_2) \rightarrow_T A_1$ has the defect of 1-tautology equal to 0. Indeed,

$$t((A_1 \wedge_T A_2) \rightarrow_T A_1) = S(1 - T(t(A_1), t(A_2)), t(A_1))$$

$$\geq S(1 - t(A_1), t(A_1)) \geq S_L(1 - t(A_1), t(A_1)) = 1,$$

which implies $\inf\{t((A_1 \wedge_T A_2) \rightarrow_T A_1); A_1, A_2 \text{ propositional forms}\} = 1$, that is $d^1_{TAUT}((A_1 \wedge_T A_2) \rightarrow_T A_1) = 0$.

Example 8.9 Let us consider the propositional form

$$(A \vee_S A) \rightarrow A,$$

where T is a triangular norm and S is the triangular conorm associated to T. We have

$$t\left(\left(A \vee_S A\right) \rightarrow_T A\right) = t\left(\neg\left(A \vee_S A\right) \rightarrow_T A\right)$$

$$= S\left(t\left(\neg\left(A \vee_S A\right)\right), t\left(A\right)\right) = S\left(1 - S\left(t\left(A\right), t\left(A\right)\right), t\left(A\right)\right).$$

If $T = T_M$ then

$$t\left(\left(A \vee_S A\right) \rightarrow_T A\right) = \max\left(1 - t\left(A\right), t\left(A\right)\right)$$

therefore

$$d^1_{TAUT}\left(\left(A \vee_{S_M} A\right) \rightarrow_{T_M} A\right) = 1 - \frac{1}{2} = \frac{1}{2}.$$

If $T = T_L$ then

$$\begin{aligned} t\left(\left(A \vee_S A\right) \rightarrow_T A\right) &= \min\left(1 - \min\left(2t\left(A\right), 1\right) + t\left(A\right), 1\right) \\ &= \begin{cases} t\left(A\right), & \text{if } t\left(A\right) \geq \frac{1}{2} \\ 1 - t\left(A\right), & \text{if } t\left(A\right) \leq \frac{1}{2}, \end{cases} \end{aligned}$$

therefore

$$d^1_{TAUT}\left(\left(A \vee_{S_L} A\right) \rightarrow_{T_L} A\right) = 1 - \frac{1}{2} = \frac{1}{2}.$$

Another important case is $T = T_P$. We obtain

$$t\left(\left(A \vee_S A\right) \rightarrow_T A\right) = -t^3\left(A\right) + 3t^2\left(A\right) - 2t\left(A\right) + 1.$$

Because the derivative of the function $f : [0,1] \rightarrow \mathbf{R}$ defined by $f\left(x\right) = -x^3 + 3x^2 - 2x + 1$ is positive on $[x_0, 1]$ and negative on $[0, x_0]$, where $x_0 = \frac{3-\sqrt{3}}{3}$, we get

$$\inf\left\{f\left(x\right); x \in [0,1]\right\} = f\left(\frac{3-\sqrt{3}}{3}\right) = \frac{27 - 6\sqrt{3}}{27}.$$

As a conclusion,

$$d^1_{TAUT}\left(\left(A \vee_{S_P} A\right) \rightarrow_{T_P} A\right) = \frac{6\sqrt{3}}{27}.$$

Example 8.10 If we consider the same propositional form as in the previous example, but S and T are not associate, then the minimum value 0 and the maximum value 1 of the defect of 1-tautology can be attained. Indeed, taking $S = S_M$ and $T = T_W$, we get

$$t\left(\left(A \vee_{S_M} A\right) \rightarrow_{T_W} A\right) = S_W\left(1 - S_M\left(t\left(A\right), t\left(A\right)\right), t\left(A\right)\right) = 1,$$

for every $t(A) \in [0, 1]$, therefore

$$d^1_{TAUT}\left(\left(A \vee_{S_M} A\right) \rightarrow_{T_W} A\right) = 0,$$

and taking $S = S_W$ and $T = T_M$ we obtain

$$t\left(\left(A \vee_{S_W} A\right) \rightarrow_{T_M} A\right) = S_M\left(1 - S_W\left(t\left(A\right), t\left(A\right)\right), t\left(A\right)\right)$$

$$= \begin{cases} 1, & \text{if } t\left(A\right) = 0 \\ t\left(A\right), & \text{if } t\left(A\right) > 0, \end{cases}$$

which implies $\inf\{t\left(\left(A \vee_{S_W} A\right) \rightarrow_{T_M} A\right); A \text{ propositional form}\} = 0$, that is

$$d^1_{TAUT}\left(\left(A \vee_{S_W} A\right) \rightarrow_{T_M} A\right) = 1.$$

Among all the fuzzy logics, min – max logic is the most used in practice. In the context of min – max fuzzy logic, Butnariu-Klement-Zafrany [51], Theorem 5.1, proved the following characterization of tautology.

Theorem 8.12 *A propositional form A in the* min – max *fuzzy logic is a tautology, if and only if $t(A) \geq 0.5$, for every evaluation.*

A natural definition of the deviation from tautology of a propositional form A which depends on propositional forms $A_1, ..., A_n$ and connectives $\vee_{S_M}, \wedge_{T_M}, \rightarrow_{T_M}$ (denoted in the following by $\vee, \wedge, \rightarrow$) and $\neg$, is the following.

Definition 8.20 The quantity

$$d_{TAUT}\left(A\right) = \max\{0, 0.5 - \inf\{t\left(A_1, ..., A_n, \vee, \wedge, \rightarrow, \neg\right);$$

$$A_1, ..., A_n \text{ are propositional forms}\}\}$$

is called defect of tautology in the min – max fuzzy logic (or T_M-tautology) of propositional form A.

Remarks. 1) It is immediate that $d_{TAUT}(A) = 0$ if and only if A is a tautology in the min – max fuzzy logic.

2) Let T be a triangular norm which has no zero divisors, that is $T(x,y) = 0$ implies $x = 0$ or $y = 0$. Because a propositional form A is a tautology in a fuzzy logic based on T if and only if, for some $a > 0, t(A) \geq a$ (see Butnariu-Klement-Zafrany [51], Theorem 4.4), we can introduce the concept of defect of T-tautology as above.

Example 8.11 The propositional forms

$$A_1 \to (A_2 \to A_1)$$

$$(\neg A_1 \to \neg A_2) \to (A_2 \to A_1)$$

$$(A_1 \to (A_2 \to A_3)) \to ((A_1 \to A_2) \to (A_1 \to A_3))$$

are tautologies in the min – max fuzzy logic (see Butnariu-Klement-Zafrany [51], the proof of Theorem 5.1), therefore their defect of T_M-tautology is equal to 0.

Example 8.12 The propositional form

$$A_1 \to (A_1 \wedge A_2)$$

has the defect of T_M-tautology equal to 0.5. Indeed,

$$\inf\{t(A_1 \to (A_1 \wedge A_2))\,; A_1, A_2 \text{ propositional forms}\}$$

$$= \inf\{\max(1 - t(A_1), \min(t(A_1), t(A_2)))\,; A_1, A_2 \text{ propositional forms}\}$$

$$\leq \inf\{\max(1 - t(A_1), t(A_2))\,; A_1, A_2 \text{ propositional forms}\} = 0,$$

obtained for $t(A_1) = 1, t(A_2) = 0$.

The intuitionistic fuzzy logic is introduced by Atanassov [6] and then developed by Atanassov [11], [12], [13] and Atanassov-Ban [15]. In the intuitionistic fuzzy logic, two real non-negative numbers, $\mu(p)$ and $\nu(p)$, are assigned to each proposition p, with the following constraint:

$$\mu(p) + \nu(p) \leq 1.$$

They are called "truth degree" and "falsity degree" of the proposition p.

Let this assignment be provided by an evaluation function V defined by

$$V(p) = \langle \mu(p), \nu(p) \rangle .$$

The evaluation of the negation $\neg p$ of proposition p, is defined (see Atanassov [6]-[12]) as

$$V(\neg p) = \langle \nu(p), \mu(p) \rangle .$$

Starting with a triangular norm T and its associated triangular conorm S (see Definition 4.13), when the values $V(p)$ and $V(q)$ of the propositions p and q are known, the evaluation function V can also be extended for the operations $\wedge_T$ and $\vee_T$ by

$$\begin{aligned} V(p) \wedge_T V(q) &= V(p \wedge_T q) = \langle T(\mu(p), \mu(q)), S(\nu(p), \nu(q)) \rangle \\ V(p) \vee_T V(q) &= V(p \vee_T q) = \langle S(\mu(p), \mu(q)), T(\nu(p), \nu(q)) \rangle . \end{aligned}$$

A possibility of modelling the implication in intuitionistic fuzzy logic based on the triangular norm T, is to define

$$p \to_T q = \neg p \vee_T q,$$

therefore

$$V(p \to_T q) = \langle S(\nu(p), \mu(q)), T(\mu(p), \nu(q)) \rangle ,$$

and to introduce

$$V(p) \to_T V(q) = V(p \to_T q) .$$

Similarly with the classical logic, each proposition is a propositional form and if A, B are propositional forms then $A \vee_T B, A \wedge_T B, A \to_T B$ are propositional forms for every triangular norm T.

Definition 8.21 (see *e.g.* Atanassov [12]) The propositional form A is an intuitionistic fuzzy tautology if and only if

$$\mu(A) \geq \nu(A) .$$

In order to measure the deviation from tautology of a propositional form in intuitionistic fuzzy logic, we introduce the following concept.

Definition 8.22 Let A be a propositional form which can be represented with arbitrary propositional forms $A_1, ..., A_n$ and connectives introduced as above. The quantity

$$\begin{aligned} d_{I-TAUT}(A) &= \max\{0, \sup\{\nu(A_1,...,A_n) - \mu(A_1,...,A_n); \\ &\qquad A_1,...,A_n \text{ propositional forms}\}\} \end{aligned}$$

is called defect of intuitionistic fuzzy tautology of the propositional form A.

It is immediate the following.

Theorem 8.13 *(i)* $0 \leq d_{I-TAUT}(A) \leq 1$.
(ii) $d_{I-TAUT}(A) = 0$ *if and only if* A *is an intuitionistic fuzzy tautology.*

Example 8.13 The propositional form

$$((A_1 \rightarrow_T A_2) \rightarrow_T A_1) \rightarrow_T A_1$$

is an intuitionistic fuzzy tautology for $T(x,y) = T_M(x,y) = \min(x,y)$, $\forall x, y \in [0,1]$ (see [12]), that is

$$d_{I-TAUT}(((A_1 \rightarrow_{T_M} A_2) \rightarrow_{T_M} A_1) \rightarrow_{T_M} A_1) = 0.$$

If $T = T_W$, that is

$$T_W(x,y) = \begin{cases} \min(x,y), & \text{if } \max(x,y) = 1 \\ 0, & \text{otherwise,} \end{cases}$$

then the above propositional form is not an intuitionistic fuzzy tautology (see Atanassov-Ban [15]) and its defect of intuitionistic fuzzy tautology is

$$d_{I-TAUT}(((A_1 \rightarrow_{T_W} A_2) \rightarrow_{T_W} A_1) \rightarrow_{T_W} A_1) = 1.$$

Indeed, taking $V(A_1) = \langle a, b\rangle$, $a, b \in (0,1)$, $a < b$, $a+b \leq 1$, $V(A_2) = \langle 0, 1\rangle$, then

$$\nu(((A_1 \rightarrow_{T_W} A_2) \rightarrow_{T_W} A_1) \rightarrow_{T_W} A_1) = b$$

and

$$\mu(((A_1 \rightarrow_{T_W} A_2) \rightarrow_{T_W} A_1) \rightarrow_{T_W} A_1) = a,$$

therefore

$$\begin{aligned} & d_{I-TAUT}(((A_1 \rightarrow_{T_W} A_2) \rightarrow_{T_W} A_1) \rightarrow_{T_W} A_1) \\ = \; & \sup\{b - a; a, b \in (0,1), a < b, a + b \leq 1\} = 1. \end{aligned}$$

Example 8.14 If T, T_1, T_2 are triangular norms and $T_2(x,y) \leq T_1(x,y)$, $\forall x, y \in [0,1]$ then

$$d_{I-TAUT}((A_1 \vee_{T_1} A_2) \rightarrow_T (A_1 \vee_{T_2} A_2)) = 0,$$

because $(A_1 \vee_{T_1} A_2) \rightarrow_T (A_1 \vee_{T_2} A_2)$ is an intuitionistic fuzzy tautology (see Atanassov-Ban [15]). If $T_1(x,y) = T_L(x,y) = \max(x+y-1, 0)$, $T_2(x,y) = T_P(x,y) = xy$ (that is $T_2 \not\leq T_1$) and $T(x,y) = \min(x,y)$, then

$$(A_1 \vee_{T_1} A_2) \rightarrow_T (A_1 \vee_{T_2} A_2)$$

is not intuitionistic fuzzy tautology, because taking $V(A_1) = V(A_2) = \langle \frac{1}{8}, \frac{1}{2} \rangle$, we have

$$\nu((A_1 \vee_{T_1} A_2) \rightarrow_T (A_1 \vee_{T_2} A_2)) = T\left(S_1\left(\frac{1}{8}, \frac{1}{8}\right), T_2\left(\frac{1}{2}, \frac{1}{2}\right)\right) = \frac{1}{4}$$

$$> \frac{15}{64} = S\left(T_1\left(\frac{1}{2}, \frac{1}{2}\right), S_2\left(\frac{1}{8}, \frac{1}{8}\right)\right) = \mu((A_1 \vee_{T_1} A_2) \rightarrow_T (A_1 \vee_{T_2} A_2)).$$

An upper estimation of the defect is

$$d_{I-TAUT}((A_1 \vee_{T_1} A_2) \rightarrow_T (A_1 \vee_{T_2} A_2)) \leq \frac{1}{4}.$$

Indeed,

$$T(S_1(a,c), T_2(b,d)) - S(T_1(b,d), S_2(a,c))$$

$$= \min(\min(a+c, 1), bd) - \max(\max(b+d-1, 0), a+c-ac)$$

$$\leq \min(a+c, bd) - (a+c-ac) \leq ac \leq \frac{1}{4},$$

for every $a, b, c, d \in [0,1], a+b \leq 1, c+d \leq 1$, where $\langle a, b \rangle = V(A_1)$ and $\langle c, d \rangle = V(A_2)$.

8.5 Bibliographical Remarks and Open Problems

All the results in this chapter, excepting those where the authors are mentioned, are completely new.

Open problem 8.1 For a given propositional form depending on the connectives S and T, it would be interesting to find out for which S and T the defect of tautology is the smallest possible.

Bibliography

[1] Alonso, J. and Benitez, C. (1989) Orthogonality in normed linear spaces: a survey. Part II: relations between main orthogonalities, *Extracta Math.*, **4**, 121-131.

[2] Alonso, J. and Benitez, C. (1997) Area orthogonality in normed linear spaces, *Arch. Math. (Basel)*, **68**, 70-76.

[3] Alonso, J. and Soriano, M.L. (1999) On height orthogonality in normed linear spaces, *Rocky Mountain J. Math.*, **29**, 1167-1183.

[4] Amoretti, N. and Cano, J. (1997) A measure of nonconvexity and fixed point theorems, *Bull. Math. Soc. Sci. Math. Roumanie*, **40**(88), No. 3-4, 118-122.

[5] Atanassov, K. (1986) Intuitionistic fuzzy sets, *Fuzzy Sets and Systems*, **20**, 87-96.

[6] Atanassov, K.T. (1988) Two variants of intuitionistic fuzzy propositional calculus, *Preprint IM-MFAIS*-5-88, Sofia.

[7] Atanassov, K. (1989) More on intuitionistic fuzzy sets, *Fuzzy Sets and Systems*, **33**, 37-45.

[8] Atanassov, K.T. (1991) Temporal intuitionistic fuzzy sets, *C. R. Acad. Bulgare Sci.*, **44**(7), 5-8.

[9] Atanassov, K.T. (1993) New operators over the intuitionistic fuzzy sets, *C. R. Acad. Bulgare Sci.*, **46**(11), 5-7.

[10] Atanassov, K.T. (1994) New operations defined over the intuitionistic fuzzy sets, *Fuzzy Sets and Systems*, **61**, 137-142.

[11] Atanassov, K.T. (1994) Some operators of a modal type in intuitionistic fuzzy modal logic, *C. R. Acad. Bulgare Sci.*, **47**, 5-8.

[12] Atanassov, K.T. (1995) Remark on intuitionistic fuzzy logic and intuitionistic logic, *Mathware & Soft Computing*, **2**, 151-156.

[13] Atanassov, K.T. and Gargov, G. (1998) Elements of intuitionistic fuzzy logic. Part I, *Fuzzy Sets and Systems*, **95**, 39-52.

[14] Atanassov, K.T. and Ban, A.I. (1999) On an operator over intuitionistic fuzzy sets, *C. R. Acad. Bulgare Sci.*, **52** (5), .39-42.

[15] Atanassov, K.T. and Ban, A.I. (2001) Triangular norm-based intuitionistic fuzzy propositional calculus, *Notes on Intuitionistic Fuzzy Sets*, **7**, (to appear).

[16] Ban, A.I. (1997-1998) Applications of conormed-seminormed fuzzy integrals, *Anal. Univ. Oradea,* fasc. math., **6**, 100-109.

[17] Ban, A.I. (1998) The convergence of a sequence of intuitionistic fuzzy sets and intuitionistic (fuzzy) measure, *Notes on Intuitionistic Fuzzy Sets*, **4**, 41-47.

[18] Ban, A.I. (1998) Intuitionistic fuzzy measures and intuitionistic entropies, *Notes on Intuitionistic Fuzzy Sets*, **4**, 48-58.

[19] Ban, A. I. (1998) Fuzzy variants of the Poincaré recurrence theorem, *J. Fuzzy Math.,* **7**, 513-519.

[20] Ban, A. I. and Fechete, I. (1998) The application of t-measures of fuzziness, *Korean J. Comp. Appl. Math.*, **5**, 163-169.

[21] Ban, A.I. (1999-2000) On t-measures of fuzziness and applications to description of systems performance, *Anal. Univ. Oradea,* fasc. math., **7**, 101-110.

[22] Ban, A.I. and Gal, S.G. (2000) Measures of noncompactness for fuzzy sets in fuzzy topological spaces, *Fuzzy Sets and Systems,* **109**, 205-216.

[23] Ban, A.I. and Gal, S.G. (2001) Decomposable measures and information measures for intuitionistic fuzzy sets, *Fuzzy Sets and Systems,* **123**, 103-117.

[24] Ban, A.I. and Gal, S.G. On the defect of additivity of fuzzy measures, *Fuzzy Sets and Systems,* (under press).

[25] Ban, A.I.and Gal, S.G. On the defect of complementarity of fuzzy measures, *Fuzzy Sets and Systems,* (to appear).

[26] Ban, A.I.and Gal, S.G. (2001) Defect of monotonicity of fuzzy measures, *Anal. Univ. Oradea,* fasc. math., **8**, 33-45.

[27] Ban, A.I. and Gal, S.G. On the minimal displacement of points under mappings, *Arch. Math. (Brno),* (to appear)

[28] Ban, A.I. and Gal, S.G. On the defect of property of binary operations on metric spaces, *Mathematical Reports (formerly Studii şi Cercetări Matematice)*, (to appear)

[29] Ban, A.I. and Gal, S.G. On the defect of orthogonality in real normed linear spaces, *Bull. Math. Soc. Sci. Math. Roumanie*, (to appear)

[30] Ban, A.I. and Gal, S.G. (2001) On the defect of equality for inequalities, *Octogon Math. Mag.,* **8**, 713-719.

[31] Banaś, J. (1980) On measure of noncompactness in Banach spaces, *Comment.Math. Univ. Carolinae,* **21**, 131-143.

[32] Banas, J. and Goebel, K. (1986) *Measures of Noncompactness in Banach Spaces*, Lecture Notes in Pure and Applied Mathematics, vol. **60**, Marcel Dekker, New York and Basel.

[33] Barbu, V. (1989) *Mathematical Methods in Optimization of Differential Systems* (in Romanian), Ed. Academiei, Bucharest.

[34] Bassanezi, R.C. and Roman-Flores, H.E. (1995) On the continuity of fuzzy

entropies, *Kybernetes,* **24**, 111-120.

[35] Batle, N. and Trillas, E. (1979) Entropy and fuzzy integral, *J. Math. Anal.Appl.*, **69**, 469-474.

[36] Beer, G. (1987) On a measure of noncompactness, *Rend. Mat. Appl. (Roma)*, ser VII, vol. **7**, fasc. 2, 193-205.

[37] Benvenuti, P. and Mesiar, R. (2000) Integrals with respect to a general fuzzy measure, in: M. Grabisch, T. Murofushi and M. Sugeno, (Eds.), *Fuzzy Measures and Integrals: Theory and Applications*, Springer-Verlag, 205-232.

[38] Bertoluzza, C. and Cariolaro, D. (1997) On the measure of a fuzzy set based on continuous t-conorms, *Fuzzy Sets and Systems,* **88**, 355-362.

[39] Birkhoff, G. (1935) Orthogonality in linear metric spaces. *Duke Math. J.,* **1**, 169-172.

[40] Blanchard, N. (1982) Cardinal and ordinal theories about fuzzy sets, in: M.M. Gupta, E. Sanchez, (Eds.), *Fuzzy Information and Decision Processes,* North Holland, Amsterdam, 149-157.

[41] Bocşan, G. and Constantin, G. (1973) The Kuratowski function and some applications to the probabilistic metric spaces, *Atti Accad. Naz. Lincei,* **LV**, fasc. 3-4, 236-240.

[42] Bohm, D. (1951) *Quantum Theory,* Pretince-Hall, Englewood Cliffs, New Jersey.

[43] Borwein, J. and Keener, L. (1980) The Hausdorff metric and Chebyshev centers, *J. Approx. Theory,* **28,** 366-376.

[44] Burgin, M. and Ŝostak, A. (1992) Continuity defect for mappings of topological spaces, *VIII Summer Conference on General Topology,* New York.

[45] Burgin, M. and Ŝostak, A. (1994) Fuzzification of the theory of continuous functions, *Fuzzy Sets and Systems*, **62**, 71-81.

[46] Burillo, P. and Bustince, H. (1995) Two operators on interval-valued intuitionistic fuzzy sets: part II, *C. R. Acad. Bulgare Sci.*, **48**(1), 17-20.

[47] Burillo, P. and Bustince, H. (1996) Entropy on intuitionistic fuzzy sets and on interval-valued fuzzy sets, *Fuzzy Sets and Systems*, **78**, 305-316.

[48] Bustince, H. and Burillo, P. (1995) Correlation of interval intuitionistic fuzzy sets, *Fuzzy Sets and Systems*, **74**, 237-244.

[49] Butnariu, D. (1983) Additive fuzzy measures and integrals, *J. Math. Anal. Appl.,* **93**, 436-452.

[50] Butnariu, D. and Klement, E. P. (1993) *Triangular Norm-Based Measures and Games with Fuzzy Coalitions*, Kluwer Academic Publishers, Dordrecht.

[51] Butnariu, D., Klement, E.P.and Zafrany, S. (1995) On triangular norm-based propositional fuzzy logics, *Fuzzy Sets and Systems,* **69**, 241-255.

[52] Butnariu D. and Klement, E.P. (2000) Measures on triangular norm-based tribes: properties and integral representations, in: M. Grabisch, T. Murofushi and M. Sugeno, (Eds.), *Fuzzy Measures and Integrals: Theory and Applications,* Springer-Verlag, 233-246.

[53] Cano, J. (1990) A measure of non-convexity and another extension of Schauder's theorems, *Bull. Math. Soc. Sci. Math. Roumanie*, **34**(82), No.

1, 3-6.

[54] Chang, C.L. (1968) Fuzzy topological spaces, *J. Math. Anal. Appl.*, **24**, 182-190.

[55] Choquet, G. (1955) Theory of capacities, *Ann. Inst. Fourier,* Tome **V**, 131-295.

[56] Cîrtoaje, V. (1990) On some inequalities with restrictions (in Romanian), *Mathematical Revue of Timişoara*, **1**, 3-7.

[57] Congxin, Wu and Minghu, Ha (1993) Completion of a fuzzy measure, *J. Fuzzy Math.*, **1**, 295-302.

[58] Constantin, G. and Istrătescu, I. (1981) *Elements of Probabilistic Analysis and Applications* (in Romanian), Academic Press, Bucharest.

[59] Crouzeix, J.P. and Lindberg, P.O. (1986) Additively decomposed quasiconvex functions, *Math. Programming*, **27**, 42-57.

[60] Darbo, G. (1955) Punti uniti in transformazioni a codominio non compacto, *Rend. Sem. Math. Univ. Padova,* **24**, 84-92.

[61] Dennenberg, D. (1994) *Non-additive measure and integral,* Kluwer Academic Publishers, Dordrecht, Boston.

[62] Denneberg, D. (2000) Non-additive measure and integral, basic concepts and their role for applications, in: M. Grabisch, T. Murofushi and M. Sugeno, Eds., *Fuzzy Measures and Integrals: Theory and Applications*, Springer Verlag, 42-69.

[63] Diminnie, C.R., Freese, R.W. and Andalafte, E.Z. (1983) An extension of Pythagorean and Isosceles orthogonality and a characterization of inner product spaces, *J. Approx. Theory,* **39** 295-298.

[64] Dubois, D. and Prade, H. (1980) *Fuzzy Sets and Systems: Theory and Applications,* Academic Press, New York.

[65] Dubois D. and Prade, H. (1982) A class of fuzzy measures based on triangular norms, *Internat. J. Gen. Systems,* **8**, 43-61.

[66] Dubois D. and Prade, H. (1985) Fuzzy cardinality and the modeling of imprecise quantification, *Fuzzy Sets and Systems,* **16**, 199-230.

[67] Dubois, D. and Prade, H. (1988) *Possibility Theory,* Plenum Press, New York.

[68] Dumitrescu, D. (1977) A definition of an informational energy in fuzzy sets theory, *Studia Univ. Babeş-Bolyai (ser. math.),* **2**, 57-59.

[69] Egbert, R.J. (1960) Products and quotients of probabilistic metric spaces, *Pacific J. Math.*, **24**, 437-455.

[70] Eisenfeld, J. and Lakshmikantham, V. (1975) On a measure of nonconvexity and applications, The University of Texas at Arlington, *Tehnical Report,* No.**26**.

[71] Ekeland, I. (1974) On the variational principle, *J. Math. Anal. Appl.*, **47**, 324-353.

[72] Ekeland, I. (1979) Nonconvex minimization problems, *Bull. Amer. Math. Soc.*, **1**, 443-474.

[73] Engl, H.W. (1977) Weak convergence of asymptotically regular sequences for

nonexpansive mappings and connections with certain Chebyshef-centers, *Nonlinear Anal.*, **1(5)**, 495-501.

[74] Ewert, J. (1998) The measure of nonconvexity in Fréchet spaces and a fixed point theorem for multivalued maps, *Bull. Math. Soc. Sci. Math. Roumanie*, **41**(89), No. 1, 15-22.

[75] Fan, K. (1969) Extensions of two fixed point theorems of F.E. Browder, *Math. Z.*, **112**, 234-240.

[76] Félix, L. (1973) *Modern Exposure of Elementary Mathematics* (in Romanian), Ed. Ştiinţifică, Bucharest.

[77] Franchetti, C. (1986) Lipschitz maps and the geometry of the unit ball in normed spaces, *Arch. Math. (Basel)*, **46**, 76-84.

[78] Frank, M.J. (1979) On the simultaneous associativity of $F(x, y)$ and $x + y - F(x, y)$, *Aequationes Math.*, **19**, 194-226.

[79] Fujimoto, K. and Murofushi, T. (1997) Some characterizations of the systems represented by Choquet and multi-linear functionals through the use of Möbius inversion, *Internat. J. Uncertain., Fuzziness and Knowledge-Based Systems*, **5**, 547-561.

[80] Fujimoto, K. and Murofushi, T. (2000) Hierarchical decomposition of the Choquet integral, in: M. Grabisch, T. Murofushi and M. Sugeno, (Eds.), *Fuzzy Measures and Integrals: Theory and Applications*, Springer Verlag, 94-103.

[81] Furi, M. and Martelli, M. (1974) On the minimal displacement of points under alpha-Lipschitz maps in normed spaces, *Boll. Un. Mat. Ital.*, **9**, 791-799.

[82] Gal, S.G. (1993) A fuzzy variant of the Weierstrass approximation theorem, *J. Fuzzy Math.*, **1**, 865-872.

[83] Gal, S.G. (1994) Hausdorff distances between fuzzy sets, *J. Fuzzy Math.*, **2**, 623-634.

[84] Gal, S.G. (1995) On some metric concepts for fuzzy sets, *J. Fuzzy Math.*, **3**, 645-657.

[85] Gal, S.G. (1995) Approximate selections for fuzzy-set-valued mappings and applications, *J. Fuzzy Math.*, **3**, 941-947.

[86] Gal, S.G. (1997) Measures of noncompactness for fuzzy sets, *J. Fuzzy Math.*, **5**, 309-320.

[87] Gastinel, N. and Joly, J.L. (1970) Condition number and general projection method, *Linear Algebra Appl.*, **3**, 185-224.

[88] Gelbaum, B.R. and Olmsted, J.M.H. (1973) *Counterexamples in Analysis* (in Romanian), Ed. Ştiinţifică, Bucharest.

[89] Ghermănescu, M. (1960) *Functional Equations* (in Romanian), Edit. Academiei Române, Bucharest.

[90] Gödel, K. (1932) Zum intuitionistichen Aussagenkalkül, Anzeiger Akademie der Wissenschaften Wien, *Math.-naturwissensch. Klasse*, **69**, 65-66.

[91] Goebel, K. (1973) On the minimal displacement of points under lipschitzian mappings, *Pacific J. Math.*, **48**, 151-163.

[92] Goebel, K. and Kirk, W.A. (1990) *Topics in Metric Fixed Point Theory*,

Cambridge University Press, Cambridge.

[93] Grabisch, M. and Nicolas, J.M. (1994) Classification by fuzzy integral-performance and tests, *Fuzzy Sets and Systems,* **65**, 255-271.

[94] Grabisch, M. (1997) Fuzzy measures and integrals for decision making and pattern recognition, *Tatra Mt. Math. Publ.,* **13**, 7-34.

[95] Grabisch, M. and Roubens, M. (2000) Application of the Choquet integral in multicriteria decision making, in: M. Grabisch, T. Murofushi and M. Sugeno, (Eds.), *Fuzzy Measures and Integrals: Theory and Applications*, Springer Verlag, 348-374.

[96] Grabisch, M. (2000) The interaction and Möbius representations of fuzzy measures on finite spaces, k-additive measures: a survey, in: M. Grabisch, T. Murofushi and M. Sugeno, (Eds.), *Fuzzy Measures and Integrals: Theory and Applications*, Springer Verlag, 70-93.

[97] Guay, M.D. and Singh, K.L. (1983) Fixed points of asymptotically regular mappings, *Math. Vesnik,* **35**, 101-106.

[98] Hájek, P. and Godo, L. (1997) Deductive systems of fuzzy logic, *Tatra Mt. Math. Publ.*, **13**, 35-66.

[99] Halmos, P.R. (1956) *Measure Theory,* Van Nostrand, New York.

[100] Hanson, M.A. (1981) On sufficiency of the Kuhn-Tucker conditions, *J. Math. Anal. Appl.*, **80**, 545-550.

[101] Herman, G.T. (1990) On topology as applied to image analysis, *Computer vision, graphics, and image processing,* **52**, 409-415.

[102] Howe, R. (1980) On the role of the Heisenberg group in harmonic analysis, *Bull. Amer. Math. Soc. (N.S.),* **3 (2)**, 821-843.

[103] Ionescu-Tulcea, C.T. (1956) *Hilbert Spaces* (in Romanian), Academic Press, Bucharest.

[104] Ionescu, Gh. D. (1984) *Differential Theory of Curves and Surfaces with Technical Applications* (in Romanian), Ed. Dacia, Cluj-Napoca.

[105] Istrătescu, V.I. (1974) *Introduction to the Theory of Probabilistic Metric Spaces with Applications* (in Romanian), Ed. Tehnică, Bucharest.

[106] Istrătescu, V. (1981) *Fixed Point Theory,* Reidel, Dordrecht.

[107] James, R.C., (1945) Orthogonality in normed linear spaces, *Duke Math. J.,* **12,** 291-301.

[108] Joly, J.L. (1969) Caractérisations d'espaces hilbertiens au moyen de la constante rectangle, *J. Approx. Theory,* **2**, 301-311.

[109] Joly, J.L. (1969) La constante rectangle d'un espace vectorial normé, *C. R. Acad. Sci. Paris, Sér. A-B,* **268**, A36-A38.

[110] Jordan, P. and Neumann, J. (1935) On inner products in linear metric spaces, *Ann. of Math.*, **36**, 719.

[111] Kaleva, O. (1986) Fuzzy performance of a coherent system, *J. Math. Anal. Appl.,* **117**, 234-246.

[112] Kaleva, O. (1988) Fuzzy methods in systems performance, in *Encyclopedia of Systems and Control* (Singh M.G. Ed.).

[113] Kapoor, O.P. and Prasad, J. (1978) Orthogonality and characterizations of

inner product spaces *Bull. Austral. Math. Soc.,* **19**, 403-416.

[114] Kaufmann, A. (1975) *Introduction to the Theory of Fuzzy Subsets,* Academic Press, New York.

[115] Kaufmann, A., Grouchko, D. and Cruon, R. (1977) *Mathematics in Science and Engineering,* Vol. 124, Academic Press, New York.

[116] Kelley, J.L. (1964) *General Topology,* Van Nostrand, New York.

[117] Kimberling, C. (1973) On a class of associative functions, *Publ. Math. Debrecen,* **20**, 21-39.

[118] Klement, E.P. (1980) *Characterization of fuzzy measures constructed by means of triangular norms,* Institutsbericht Nr. 180, Institut für Mathematik, Johaness Kepler Universität Linz.

[119] Klement, P.E. and Mesiar, R. (1997) Triangular norms, *Tatra Mt. Math. Publ.,* **13**, 169-193.

[120] Klir, G. J. (1987) Where do we stand on measures of uncertainty, ambiguity, fuzziness, and the like?, *Fuzzy Sets and Systems,* **24**, 141-160.

[121] Klir, G.J. and Folger, T.A. (1988) *Fuzzy Sets, Uncertainty, and Information,* Pretince-Hall, Englewood Cliffs, New York.

[122] Knophmacher, J. (1975) On measures of fuzziness, *J. Math. Anal. Appl.,* **49**, 529-534.

[123] Kong, T.Y. and Rosenfeld, A. (1989) Digital topology: Introduction and Survey, *Computer vision, graphics, and image processing,* **48**, 357-393.

[124] Kong, T.Y., Kopperman, R. and Meyer, P.R. (1991) A topological approach to digital topology, *Amer. Math. Monthly,* **98**, 901-917.

[125] Kong, T.Y., Roscoe, A.W. and Rosenfeld, A. (1992) Concepts of digital topology, *Topology Appl.,* **46**, 219-262.

[126] Kruse, R. (1982) A note on λ-additive fuzzy measures, *Fuzzy Sets and Systems,* **8**, 219-222.

[127] Kupka, I. (1988) A Banach fixed point theorem for topological spaces, *Rev. Colombiana Mat.,* **26**, 95-99.

[128] Kupka, I. and Toma, V. (1995) Measure of noncompactness in topological spaces and upper semicontinuity of multifunctions, *Rev. Roum. Math. Pures Appl.,* Tome **XL**, No. 5-6, 455-461.

[128] Kuratowski, K. (1930) Sur les espaces complete, *Fund. Math.,* **15**, 301-309.

[130] Kuratowski, K. (1966) *Topology,* Academic Press, New York.

[131] Kwon, S.H. and Sugeno, M. (2000) A hierarchical subjective evaluation model using non-monotonic fuzzy measures and the Choquet integral, in: M. Grabisch, T. Murofushi and M. Sugeno, (Eds.), *Fuzzy Measures and Integrals: Theory and Applications,* Springer Verlag, 375-391.

[132] Lechicki, A. (1985) A generalized measure of noncompactness and Dini's theorem for multifunctions, *Rev. Roumaine Math. Pures Appl.,* **30**, 221-227.

[133] Lee, B.S., Lee, G.M. and Kim, D.S. (1994) Common fixed points of fuzzy mappings in Menger PM-spaces, *J. Fuzzy Math.,* **2**, 859-870.

[134] Lee, K.M and Leekwang, H. (1995) Identification of λ-fuzzy measure by

genetic algorithms, *Fuzzy Sets and Systems,* **75**, 301-309.

[135] Loo, S.G. (1967) Measures of fuzziness, *Cybernetica*, **20**, 201-210.

[136] Lowen, R. (1976) Fuzzy topological spaces and fuzzy compactness, *J. Math. Anal. Appl.*, **56**, 621-633.

[137] Lowen, E. and Lowen, R. (1988) On measures of compactness in fuzzy topological spaces, *J. Math. Anal. Appl.*, **131**, 329-340.

[138] Luca, A.D. and Termini, S. (1972) A definition of a nonprobabilistic entropy in the setting of fuzzy sets theory, *Inform. and Control,* **20**, 301-312.

[139] Luca, A. D. and Termini, S. (1974) Entropy of L-fuzzy sets, *Inform. and Control*, **24**, 55-73.

[140] Lukasiewicz, J. (1970) *Selected Works*, North-Holland Publ. Comp., Amsterdam.

[141] Ma Ming and Hu Haiping (1993) The abstract (S) fuzzy integrals, *J. Fuzzy Math.*, **1**, 89-107.

[142] Mashhour, A.S., Allam, A.A. and El-Saady, K. (1994) Fuzzy uniform structures and order, *J. Fuzzy Math.,* **2**, 57-67.

[143] Mayor, G. and Calvo, T. (1994) On non-associative binary operations on $[0, 1]$, *J. Fuzzy Math.,* **2**, 649-653.

[144] Meghea, C. (1977) *The Bases of Mathematical Analysis* (in Romanian), Ed. Ştiinţ. Enciclop., Bucharest.

[145] Menger, K. (1942) Statistical metrics, *Proc. Nat. Acad. Sci. USA*, **28**, 535-537.

[146] Mesiar, R. (1993) Fundamental triangular norm based tribes and measure, *J. Math. Anal. Appl.*, **177**, 633-640.

[147] Michael, E. (1951) Topologies on spaces of susets, *Trans. Amer. Math. Soc.,* **71**, 152-182.

[148] Mihăileanu, N. (1964) *Non-Euclidean Differential Geometry* (in Romanian), Ed. Acad. R. P. R., Bucharest.

[149] Mihăileanu, N.N. and Neumann, M. (1973) *The Fundaments of Geometry* (in Romanian), Ed. Did. and Ped., Bucharest.

[150] Misra, K.P. and Sharma, A. (1981) Performance index to quantify reliability using fuzzy subsets theory, *Microelectron. Reliab.*, **21**, 543-549.

[151] Mitjagin, B.S. and Semenov, E.M. (1977) Lack of interpolation of linear operators in spaces of smooth functions, *Math. USSR-Izv*, **11**, 1229-1266.

[152] Mitrinovič, D.S., Sandor, J. and Crstici, B. (1996) *Handbook on Number Theory,* Kluwer Academic Publishers, Dordrecht, Boston, London.

[153] Morera, G. (1886) Un teorema fondamentale nella teoria delle funzioni di una variable complessa, *Rend. del R. Istituto Lombardo di Scienze e Lettere,* (2) **19**, 304-307.

[154] Muntean, I. (1973) *Course and Problems in Functional Analysis, vol II* (in Romanian), "Babeş-Bolyai" University Press, Cluj-Napoca.

[155] Muntean, I. (1977) *Course of Functional Analysis, vol II, Spaces of Linear Continuous Mappings* (in Romanian), "Babeş-Bolyai" University Press, Cluj-Napoca.

[156] Murofushi, T. and Sugeno, M. (1989) An interpretation of fuzzy measures and the Choquet integral as an integral with respect to a fuzzy measure, *Fuzzy Sets and Systems,* **29**, 201-227.

[157] Murofushi, T. and Sugeno, M. (1991) A theory of fuzzy measures: Representations, the Choquet integral and null sets, *J. Math. Anal. Appl.,* **159**, 532-549.

[158] Murofushi, T., Sugeno, M. and Machida, M. (1994) Non-monotonic fuzzy measures and the Choquet integral, *Fuzzy Sets and Systems,* **64**, 73-86.

[159] Murofushi, T. and Sugeno, M. (2000) Fuzzy measures and fuzzy integrals, in: M. Grabisch, T. Murofushi and M. Sugeno, (Eds.), *Fuzzy Measures and Integrals: Theory and Applications*, Springer-Verlag, 3-41.

[160] Njastad, O. (1965) On some classes of nearly open sets, *Pacific J. Math.,* **15**, 961-970.

[161] Onisawa, T., Sugeno, M., Nishiwaki, Y., Kawai, H. and Harima, Y. (1986) Fuzzy measure analysis of public attitude towards the use of nuclear energy, *Fuzzy Sets and Systems,* **20**, 259-289.

[162] Pap, E. (1995) Recent results on null-additive set functions, *Tatra Mt. Math. Publ.*, **6**, 141-148.

[163] Pompeiu, D. (1912) Sur une classe de fonctions d'une variable comlexe, *Rend. Circ. Mat. Palermo,* **33**, 108-113.

[164] Pompeiu, D. (1913) Sur une classe de fonctions d'une variable comlexe et sur certaines équations intégrales, *Rend. Circ. Mat. Palermo,* **35**, 277-281.

[165] Popa, E. (1981) *Problems of Functional Analysis* (in Romanian), Ed. Did. Ped., Bucharest.

[166] Prade, H. (1984) Lipski's approach to incomplete information databases restated and generalized in the setting of Zadeh's possibility theory, *Information Systems,* **9**, 27-42.

[167] Radovici-Mărculescu, P. (1986) *Problems in Elementary Number Theory* (in Romanian), Ed. Tehnică, Bucharest.

[168] Ralescu, D. and Adams, G. (1980) The fuzzy integral, *J. Math. Anal. Appl.,* **75**, 562-570.

[169] Ralescu, D. (1995) Cardinality, quantifiers, and the aggregation of fuzzy criteria, *Fuzzy Sets and Systems,* **69**, 355-365.

[170] Rădulescu, S. and Rădulescu, M. (1982) *Theorems and Problems in Analysis* (in Romanian), Ed. Didactică şi Pedagogică, Bucharest.

[171] Reich, S. (1975) Minimal displacement of points under weakly inward pseudo-lipschitzian mappings I, *Atti. Acad. Naz. Linzei Rend. U. Sci. Fis. Mat. Natur.,* **59**, 40-44.

[172] Reich, S. (1976) Minimal displacement of points under weakly inward pseudo-lipschitzian mappings II, *Atti. Acad. Naz. Linzei Rend. U. Sci. Fis. Mat. Natur.,* **60**, 95-96.

[173] Rhoades, B.E., Sessa, S., Khan, M.S. and Swaleh, M. (1987) On fixed points of asymptotically regular mappings, *J. Austral. Math. Soc.* (Series A), **43**, 328-346.

[174] Roberts, A.W. and Varberg, D.E. (1973) *Convex Functions*, Academic Press, New York and London.

[175] Rosenfeld, A. (1976) *Digital Picture Analysis*, Springer, Berlin.

[176] Rosenfeld, A. (1979) Fuzzy digital topology, *Inform. and Control*, **40**, 76-87.

[177] Rosenfeld, A. (1979) Digital topology, *Amer. Math. Monthly*, **86**, 621-630.

[178] Rosenfeld, A. (1983) On connectivity properties of grayscale pictures, *Pattern Recognition Letters*, **16**, 47-50.

[179] Rosenfeld, A. (1984) The fuzzy geometry of image subsets, *Pattern Recognition Letters*, **2**, 311-317.

[180] Roventa E. and Vivona, D. (1996) Representation theorem for transom based measures of fuzziness, *Rendiconti di Matematica (Roma)*, Serie VII, **16**, 289-297.

[181] Rudas, I.J. and Kaynak, M.O. (1998) Minimum and maximum fuzziness generalized operators, *Fuzzy Sets and Systems*, **98**, 83-94.

[182] Rus, I.A. (1979) *Principles and Applications of Fixed Point Theory* (in Romanian), Ed. Dacia, Cluj-Napoca.

[183] Rus, I.A. (1983) On a theorem of Eisenfeld-Lakshmikantham, *Nonlinear Anal.*, **7**, No.3, 279-281.

[184] Rus, A.I. (1984) Measures of non compact-convexity and fixed points, in: *Itinerant Seminar on Functional Equations, Approximation and Convexity*, Cluj-Napoca, 173-180.

[185] Rus, I.A. (1988) Measures of nonconvexity and fixed points, in: *Itinerant Seminar on Functional Analysis, Approximation and Convexity*, Cluj-Napoca, 111-118.

[186] Šabo, M. (1995) On the continuity of t-reverse of t-norms, *Tatra Mt. Math. Publ.*, **6**, 173-178.

[187] Sadowski, B.N. (1967) On a fixed point principle (in Russian), *Funkc. Analiz i ego Priloz.*, **1**, 74-76.

[188] Sadowski, B.N. (1972) Limit-compact and condensing operators (in Russian), *Usp. Mat. Nauk.*, **27**, fasc. 1, 81-146.

[189] Schmeidler, D. (1986) Integral representation without additivity, *Proc. Amer. Math. Soc.*, **97**, 255-261.

[190] Schoenberg, I.J. (1952) A remark on M.M. Day's characterization of inner-product spaces and a conjecture of L.M. Blumenthal, *Proc. Amer. Math. Soc.*, **3**, 961-964.

[191] Schweizer, B. and Sklar, A. (1960) Statistical metric spaces, *Pacific J. Math.*, **10**, 313-334.

[192] Sekita, Y. (1976) A consideration of identifying fuzzy measures, *Osaka University Economy*, **25**, 133-138.

[193] Sendov, B. (1965) An estimate for the approximation of functions by S. N. Berstein polynomials (in Russian), *Mathematica (Cluj)*, **7(30)**, 145-154.

[194] Sendov, Bl. (1969) Several questions on the theory of approximation of functions and sets in Hausdorff metric (in Russian), *Russ. Math. Surv.*, **24**, 141-178.

[195] Shackle, G.L. (1961) *Decision, Order and Time in Human Affairs*, Cambridge Univ. Press, Cambridge.

[196] Shafer, G. (1976) *A Mathematical Theory of Evidence*, Princeton Univ. Press, Princeton, New York.

[197] Singer, I. (1957) Abstract angles and trigonometric functions in Banach spaces (in Romanian), *Bul. Ştiinţ. Acad. R. P. Române, Sect. Şt. Mat. Fiz.*, **9**, 29-42.

[198] Singer, I. (1967) *Best Approximation in Normed Linear Spaces by Elements of Linear Subspaces* (in Romanian), Ed. Ecademiei, Bucharest.

[199] Sireţchi, G. (1985) *Differential and Integral Calculus, vol I* (in Romanian), Ed. Ştiinţ. Enciclop., Bucharest.

[200] Sireţchi, G. (1985) *Differential and Integral Calculus, vol II* (in Romanian), Ed. Ştiinţ. Enciclop., Bucharest.

[201] Soman, K.P. and Krishna, B.M. (1993) Fuzzy fault tree analysis using resolution identity, *J. Fuzzy Math.*, **1**, 193-212.

[202] Suarez Diaz, E. and Suarez Garcia, F. (1999) The fuzzy integral on product spaces for NSA measures, *Fuzzy Sets and Systems,* **103**, 465-472.

[203] Suarez Garcia, F. and Gil Alvarez, P. (1986) Two families of fuzzy integrals, *Fuzzy Sets and Systems*, **18**, 67-81.

[204] Sugeno, M. (1974) *Theory of fuzzy integrals and its applications,* Doctoral Thesis, Tokyo Institute of Technology.

[205] Sugeno, M. and Murofushi, T. (1987) Pseudo-additive measures and integrals, *J. Math. Anal. Appl.,* **122**, 197-222.

[206] Szmidt, E. and Kacprzyk, J. (1997) A concept of entropy for intuitionistic fuzzy sets, *Notes on Intuitionistic Fuzzy Sets*, **3**, 41-52.

[207] Szmidt, E. and Kacprzyk, J. (2001) Entropy for intuitionistic fuzzy sets, *Fuzzy Sets and Systems*, **118**, 467-477.

[208] Szolovitz, P. and Pauker, S.G. (1978) Categorical and probabilistic reasoning in medicine, *Artificial Intelligence,* **11**, 115-144.

[209] Szu-Hoa Min (1944) Non-analytic functions, *Amer. Math. Monthly,* **51**, 510-517.

[210] Tahani, H. and Keller, J.M. (1990) Information fusion on computer vision using the fuzzy integral, *IEEE Trans. Systems Man Cybernet.,* **20**, 733-741.

[211] Tanaka, K. and Sugeno, M. (1991) A study on subjective evaluations of printed color image, *Internat. J. Approx. Reason.,* **5**, 213-222.

[212] Udrişte, C. and Tănăsescu, E. (1980) *Minima and Maxima of Real Functions of Real Variables* (in Romanian), Ed. Tehnică, Bucharest.

[213] Ungar, A.A. (1994) The abstract complex Lorentz transformation group with real metric I, *J. Math. Phys.,* **35 (3)**, 1408-1426.

[214] Vasiu, A. (2000) *The Fundaments of Geometry* (in Romanian), Babeş-Bolyai Univ., Cluj.

[215] Villena, A.R. (1996) Derivations on Jordan-Banach algebras, *Studia Math.,* **118**, 205-229.

[216] Vinogradov, I.M. (1954) *The Basis of Number Theory* (in Romanian), Ed.

Acad. Rom., Bucharest.

[217] Vivona, D. (1996) Mathematical aspects of the theory of measures of fuzziness, *Mathware Soft Comput.*, **3**, 211-224.

[218] Vrănceanu, G. and Teleman, C. (1967) *Euclidean Geometry. Non-Euclidean Geometry. Relativity Theory* (in Romanian), Ed. Tehnică, Bucharest.

[219] Wang, Z. and Klir, G.J. (1992) *Fuzzy Measure Theory*, Plenum Press, New York.

[220] Weiss, M.D. (1975) Fixed points, separation and induced topologies for fuzzy sets, *J. Math. Anal. Appl.*, **50**, 142-150.

[221] Wierzchon, S.T. (1982) On fuzzy measure and fuzzy integral, in: M.M. Gupta and E. Sanchez, (Eds.), *Fuzzy Information and Decision Processes*, North-Holland, 79-86.

[222] Wierzchon, S.T. (1983) An algorithm for identification of fuzzy measure, *Fuzzy Sets and Systems*, **9**, 69-78.

[223] Wu Congxin, Wang Shuli and Ma Ming (1993) Generalized fuzzy integrals: Part I. Fundamental concepts, *Fuzzy Sets and Systems*, **57**, 219-226.

[224] Wygralak, M. (1999) Questions of cardinality of finite fuzzy sets, *Fuzzy Sets and Systems*, **102**, 185-210.

[225] Wygralak, M. (1996) *Vaguely Defined Objects. Representations, Fuzzy Sets, and Nonclassical Cardinality Theory*, Kluwer Academic Publishers, Dordrecht, Boston, London.

[226] Xavier, F.S. (1968) On the product of probabilistic metric spaces, *Portugal Math.*, **27**, 137-147.

[227] Yang, X.Q. and Chen, G.Y. (1992) A class of nonconvex functions and prevariational inequalities, *J. Math. Anal. Appl.*, **169**, 359-373.

[228] Yoneda, M., Fukami, S. and Grabisch, M. (1993) Interactive determination of a utility function represented as a fuzzy integral, *Information Sciences*, **71**, 43-64.

[229] Zadeh, L.A. (1978) Fuzzy sets as a basis for a theory of possibility, *Fuzzy Sets and Systems*, **1**, 3-28.

[230] Zadeh, L.A. (1979) A theory of approximate reasoning, in: J.E. Hayes, D. Michie and L.I. Mikulich, (Eds.), *Machine Intelligence*, Wiley, New York, 149-194.

Index